Lecture Notes in Mathematics 2159

Saint-Flour Probability Summer School

The Saint-Flour volumes are reflections of the courses given at the Saint-Flour Probability Summer School. Founded in 1971, this school is organised every year by the Laboratoire de Mathématiques (CNRS and Université Blaise Pascal, Clermont-Ferrand, France). It is intended for PhD students, teachers and researchers who are interested in probability theory, statistics, and in their applications.

The duration of each school is 13 days (it was 17 days up to 2005), and up to 70 participants can attend it. The aim is to provide, in three high-level courses, a comprehensive study of some fields in probability theory or Statistics. The lecturers are chosen by an international scientific board. The participants themselves also have the opportunity to give short lectures about their research work.

Participants are lodged and work in the same building, a former seminary built in the 18th century in the city of Saint-Flour, at an altitude of 900 m. The pleasant surroundings facilitate scientific discussion and exchange.

The Saint-Flour Probability Summer School is supported by:
- Université Blaise Pascal
- Centre National de la Recherche Scientifique (C.N.R.S.)
- Ministère délégué à l'Enseignement supérieur et à la Recherche

For more information, see

http://recherche.math.univ-bpclermont.fr/stflour/stflour-en.php

Christophe Bahadoran
bahadora@math.univ-bpclermont.fr

Arnaud Guillin
Arnaud.Guillin@math.univ-bpclermont.fr

Laurent Serlet
Laurent.Serlet@math.univ-bpclermont.fr

Université Blaise Pascal – Aubière cedex, France

More information about this series at http://www.springer.com/series/304

Sara van de Geer

Estimation and Testing Under Sparsity

École d'Été de Probabilités de Saint-Flour
XLV – 2015

 Springer

Sara van de Geer
Seminar für Statistik HGG 24.1
ETH Zentrum
Zürich, Switzerland

ISSN 0075-8434   ISSN 1617-9692   (electronic)
Lecture Notes in Mathematics
ISBN 978-3-319-32773-0   ISBN 978-3-319-32774-7   (eBook)
DOI 10.1007/978-3-319-32774-7

Library of Congress Control Number: 2016943427

Mathematics Subject Classification (2010): 62-XX; 60-XX, 68Q87

Printed on acid-free paper

This Springer imprint is published by Springer Nature
The registered company is Springer International Publishing AG Switzerland

To my mother

Preface

These lecture notes were written for the 45th École d'Été de Probabilités de Saint-Flour in 2015. It was a great honour and a pleasure for me to be invited there and enjoy the *Grand Séminaire*, the beautiful surroundings and, most of all, the wonderful participants. The summer was very warm, with temperatures rising up to 40°. However, the old building kept us cool and ready to digest quite a few lectures each day. Indeed, the amount of mathematical activity was impressive, with a scientific programme in both mornings and afternoons and many excellent talks presented by my fellow participants. But there were also numerous other activities: chatting, hiking, dining, ping-pong, and looking after the small kids.

The notes aim to provide an overview of the techniques used to obtain theoretical results for models of high-dimensional data. High-dimensional models have more unknown parameters p than observations n. Sparsity means that the number of relevant—or active—parameters is actually much smaller than n. However, which parameters these are, or how many there are, is not known beforehand. The first goal in these notes is to study methods that perform almost as well as in the hypothetical case with a known active set. The next goal is then to zoom in on certain parameters of interest and test their significance.

An important technique in high-dimensional regression is the Lasso method. It is taken here as a prototype for understanding other methods, such as those inducing structured sparsity or low rank or those based on more general loss functions. The common features are highlighted so that—hopefully—they will serve as a good starting point for the theory of new methods, not treated in this book.

I am very grateful to the Scientific Board for having given me the opportunity to lecture at the Saint-Flour Summer School. I thank all participants and am greatly indebted to Claire Boyer, Yohann De Castro and Joseph Salmon, who spontaneously did a careful reading of the version of the lecture notes available at the time. The mistakes in the current version have entered *after* their proofreading. Special thanks goes to the organisers on the spot, Christophe Bahadoran and Laurent Serlet. I also thank the two other main lecturers, Sourav Chatterjee and Lorenzo Zambotti for our very pleasant time together and for their inspiring courses which opened up new windows with magnificent views.

I thank my colleagues at the Seminar for Statistics in Zürich for their support, for their motivating interest and for providing me with an ideal research environment.

Finally, a very special thanks is extended to my family.

Zürich, Switzerland   Sara van de Geer
March 2016

Contents

Chapter 1
Introduction

1.1 Regularized Empirical Risk Minimization

When there are more measurements per unit of observation than there are observations, data are called "high-dimensional". Today's data *are* often high-dimensional mainly due to the easy way to record or obtain data using the internet, or cameras, or new biomedical technologies, or shopping cards, etc. High-dimensional data can also be "constructed" from only a few variables by considering for example second, third, and higher order interactions.

Models for high-dimensional data contain more parameters p than observations n. In this book, high-dimensional algorithms are studied. Let $R : \mathcal{B} \to \mathbb{R}$ be a "risk function" defined on a space $\mathcal{B} \subset \mathbb{R}^p$. The target is the minimizer $\beta^0 := \arg\min_{\beta \in \mathcal{B}} R(\beta)$. However, the risk function is unknown and thus the target as well. One observes an "empirical risk function" $R_n : \mathcal{B} \to \mathbb{R}$ based on n data points $X_1, \ldots, X_n$ with $n \leq p$. The empirical risk R_n is an estimator of the unknown risk R. One may opt to estimate the target β^0 by the minimizer of the empirical risk R_n but because the parameter space is so large (or "rich") some regularization is in place. This is done by adding a penalty to the empirical risk. The class of penalties studied here are those that induce sparsity: vectors $\beta \in \mathbb{R}^p$ with "many" zero's are favoured. Thus, we consider

$$\hat{\beta} := \arg\min_{\beta \in \mathcal{B}} \left\{ R_n(\beta) + \text{pen}(\beta) \right\},$$

where pen : $\mathcal{B} \to [0, \infty)$ is a given sparsity inducing penalty. The penalty will be taken proportional to a norm Ω on $\mathbb{R}^p$.

First aim of the lecture notes is to provide theory showing that with high probability

$$R(\hat{\beta}) \leq R(\beta^0) + \text{remainder}$$

© Springer International Publishing Switzerland 2016
S. van de Geer, *Estimation and Testing Under Sparsity*,
Lecture Notes in Mathematics 2159, DOI 10.1007/978-3-319-32774-7_1

where the "remainder" is small, depending on how sparse the target β^0 is. This remainder thus depends (among other things) on the unknown target and is therefore unknown. Nevertheless, the theory will show that this remainder can be almost as small as it would be when an oracle told you where the target's non-zero's are.

We will in fact show results of the form: with high probability

$$R(\hat{\beta}) \leq R(\beta) + \text{remainder}(\beta), \forall \beta$$

where for each β the remainder remainder(β) depends (among other things) on the number of non-zero's of β. This is termed a *sharp* oracle inequality. To compare, a non-sharp oracle inequality is of the form

$$R(\hat{\beta}) - R(\beta^0) \leq (1 + \eta)(R(\beta) - R(\beta^0)) + \text{remainder}(\beta), \forall \beta$$

where $\eta > 0$ is some (small) constant. Thus, a sharp oracle inequality makes no reference to the target and is as such in line with a learning point of view. The "remainder" remainder(β) is to be almost as small as when empirical risk minimization was done overall all $\beta' \in \mathscr{B}$ sharing the location of its zero's with those of β. In other words, the remainder is the estimation error and the term $R(\beta) - R(\beta^0)$ is the approximation error.

Second aim of the notes is to establish bounds for the estimation error in Ω-norm (and its dual norm) as well. Final aim is to show how the estimator $\hat{\beta}$ can be invoked as initial estimator for obtaining asymptotically linear estimators of low-dimensional parameters of interest.

The notes are not meant to be an overview of the subject but rather reflect the research interests of the author.

The results are mostly formulated in a non-asymptotic form. However, in the background we do have an asymptotic framework in mind. This is one where the number of observations n and hence also the number of parameters $p = p_n$ tends to infinity. All quantities in the results may depend on n. We sometimes give a "typical" asymptotic setting, where all constants are assumed within fixed bounds except n, p and the "sparsity" (denoted often by s_0).

Computational aspects of the algorithms under investigation are important but not discussed here. Most of the optimization problems studied are convex, so that in principle the computational aspect should not be a major concern. In fact, the Lasso treated in the next chapter was invented as convex relaxation of an original problem requiring searching through exponentially many models. More generally, the algorithms studied are often developed having computational feasibility in mind. As it turns out however, computational feasibility can come with additional theoretical advantages (for example consistency in ℓ_1 of the Lasso).

These lecture notes develop theory for existing methodology. We will not provide computations, simulation studies and real data applications. A good further reading to complement the theory presented here is the recent monograph (Hastie et al. 2015).

1.2 Organization of the Book

Chapter 2 contains the terminology and main line of reasoning for establishing oracle inequalities. Chapter 3 is about oracle inequalities for the square-root Lasso, a method where the tuning parameter does not depend on the (unknown) noise variance. The square-root Lasso is also a great tool for constructing a surrogate inverse of a positive semi-definite singular matrix.

Chapter 4 derives sup-norm bounds for example, and Chap. 5 shows methodology for asymptotic confidence intervals. These two chapters are a next step *after* establishing oracle inequalities.

Chapter 6 looks at more general sparsity inducing norms. This chapter may help to understand the even more general case of Chap. 7. The ingredients of the Lasso-chapters, Chaps. 3, 4 and 5 are to some extent generalized to more general norms in Chap. 6.

So far, it is all about the linear model and least squares. Chapter 7 covers a wide class of loss functions. After this chapter, we deal with one more ingredient: the random part of the problem. With least squares loss, this part boils down to examining the dual norm of a p-dimensional random vector. This is done in Chap. 8 for the norms of Chap. 6. For matrix norms, one needs random matrix theory. We restrict ourselves to the dual norm of the nuclear norm of a matrix (which is the maximal singular value) in Chap. 9, where me mainly cite results from the literature without proof.

For loss functions other than least squares, one needs some further results from empirical process theory. This is elaborated upon in Chap. 10. Instead of providing a full account we concentrate mainly on the case where one can use a dual norm inequality together with symmetrization, contraction and concentration. A description of these latter three beautiful techniques/results from probability theory is postponed to Chap. 16.

Chapter 10 considers the random part of for instance exponential families and generalized linear models. Such models are revisited in Chap. 11 but now with focus on the condition of curvature of the loss function, termed *margin condition*. Both Chaps. 10 and 11 aim at highlighting general structures. Then in Chap. 12 we give more detailed examples and complement previous results with a treatment of the random part.

In Chap. 13 a technique is discussed for models that go beyond the generalized linear model. Main example here is the graphical Lasso and part of this is then applied in Chap. 14 for establishing asymptotic linear estimators of the precision matrix.

In almost all results, so-called *compatibility constants* or more generally the *effective sparsity* plays a prominent role. These are quantities relating one norm with another and often involve norms induced by random quadratic forms. In Chap. 15 random quadratic forms are compared to their expectation leading finally to upper bounds for random effective sparsity numbers.

Chapter 16 is a very brief summary of some symmetrization, contraction and concentration inequalities from the literature, mainly without proofs. When dealing with high-dimensional problems one may have to use further empirical process theory, where typically one needs to apply entropy calculations to handle the richness of parameter space. In these lecture notes, all empirical process theory is hiding in the use of a dual norm inequality together with symmetrization, contraction and concentration. This is of course not always possible. Chapter 17 gives the generic chaining technique as method for obtaining bounds for suprema of empirical processes. But it also serves another purpose. A natural question namely is to ask whether the dual norm bounds are up to constants the same as the ones derived from generic chaining. The answer is, at least in Gaussian case, "yes", as generic chaining is sharp in the Gaussian case. Chapters 17 and 18 are included to discuss this question in some detail. Talagrand's open problem however remains open.

1.3   How to Read This Book

For a study of oracle inequalities, one may read Chap. 2 and then immediately go to Chap. 7 where a broad class of loss functions and norm-penalties is treated. Chapter 6 on structured sparsity may be read before or after Chap. 7. However, since in Chap. 7 both loss function and norms are more general than with the Lasso, one may prefer to do one step at the time and look at Chap. 6 before Chap. 7. Chapters 10 and 11 summarize some approaches for general loss functions, with examples in Chap. 12.

A reader more interested in asymptotic normality and confidence intervals could read Chap. 2 and continue with the subsequent Chaps. 3, 4 and 5. Then a logical next step is Chaps. 13 and 14.

Chapters 8, 9 and 16 contain the needed probabilistic tools. Chapter 15 gives some more insight in constants involved in the results, namely in effective sparsity as a notion replacing minimal eigenvalues.

Chapters 17 and 18 are separate from the main subject and not directly about statistics. They are not needed for the main stream of reasoning, but may be of interest for a reader wondering about the connection with comprehensive empirical process theory. Chapter 17 may also give some insight into the concentration phenomenon cited in Chap. 16.

Chapter 2
The Lasso

Abstract Sharp oracle inequalities for the prediction error and ℓ_1-error of the Lasso are given. We highlight the ingredients for establishing these. The latter is also for later reference where results are extended to other norms and other loss functions.

2.1 The Linear Model with $p < n$

Let X be an $n \times p$ input matrix and $Y \in \mathbb{R}^n$ be an n-vector of responses. The linear model is

$$Y = X\beta^0 + \epsilon,$$

where $\beta^0 \in \mathbb{R}^p$ is an unknown vector of coefficients and $\epsilon \in \mathbb{R}^n$ is a mean-zero noise vector. This is a standard model in regression and $X\beta^0$ is often called the regression of Y on X. The least squares method, usually credited to Gauss, is to estimate the unknown β^0 by minimizing the Euclidean distance between Y and the space spanned by the columns in X:

$$\hat{\beta}_{\mathrm{LS}} := \arg\min_{\beta \in \mathbb{R}^p} \|Y - X\beta\|_2^2.$$

The least squares estimator $\hat{\beta}_{\mathrm{LS}}$ is thus obtained by taking the coefficients of the projection of Y on the column space of X. It satisfies the normal equations

$$X^T Y = X^T X \hat{\beta}_{\mathrm{LS}}.$$

If X has full rank p we can write it as

$$\hat{\beta}_{\mathrm{LS}} = (X^T X)^{-1} X^T Y.$$

The estimated regression is then the projection vector

$$X\hat{\beta}_{\mathrm{LS}} = X(X^T X)^{-1} X^T Y.$$

© Springer International Publishing Switzerland 2016
S. van de Geer, *Estimation and Testing Under Sparsity*,
Lecture Notes in Mathematics 2159, DOI 10.1007/978-3-319-32774-7_2

If the entries $\epsilon_1, \ldots, \epsilon_n$ of the noise vector ϵ are uncorrelated and have common variance σ_0^2 one may verify that

$$\mathbb{E}\|X(\hat{\beta}_{\mathrm{LS}} - \beta^0)\|_2^2 = \sigma_0^2 p$$

(Problem 2.1). We refer to the normalized quantity $\|X(\hat{\beta}_{\mathrm{LS}} - \beta^0)\|_2^2/n$ as the *prediction error*: if we use $X\hat{\beta}_{\mathrm{LS}}$ as prediction of a new (unobserved) response vector Y_{new} when the input is X, then on average the squared error made is

$$\mathbb{E}\|Y_{\mathrm{new}} - (X\hat{\beta}_{\mathrm{LS}})\|_2^2/n = \mathbb{E}\|X(\hat{\beta}_{\mathrm{LS}} - \beta^0)\|_2^2/n + \sigma_0^2.$$

The first term in the above right-hand side is due to the estimation of β^0 whereas the second term σ_0^2 is due to the noise in the new observation. We neglect the unavoidable second term in our terminology. The mean prediction error is then

$$\mathbb{E}\|X(\hat{\beta}_{\mathrm{LS}} - \beta^0)\|_2^2/n = \sigma_0^2 \times \frac{p}{n} = \sigma_0^2 \times \frac{\text{number of parameters}}{\text{number of observations}}.$$

In this monograph we are mainly concerned with models where $p \geq n$ or even $p \gg n$. Clearly, the just described least squares method then breaks down. This chapter studies the so-called Lasso estimator $\hat{\beta}$ when possibly $p > n$. Aim is to show that

$$\|X(\hat{\beta} - \beta^0)\|_2^2/n = \mathscr{O}_{\mathbb{P}}\left(\frac{s_0 \log p}{n}\right) \tag{2.1}$$

where s_0 is the number of non-zero coefficients of β^0 (or the number of in absolute value "large enough" coefficients of β^0). The *active set* $S_0 := \{j : \beta_j^0 \neq 0\}$ is however not assumed to be known, nor its size $s_0 = |S_0|$.

Throughout, we consider estimation methods based on the belief that the *target* is approximately *sparse*. In this chapter, the *target* is the vector β^0, the "true underlying" parameter. More generally, models may be misspecified and the target may be for example the best approximation obtainable by the model. *Sparseness* means that the target can be described with only a few non-zero parameters. In the current situation, this means that β^0 has only a "small" number of non-zero entries or more generally, "most" of the entries are "small". We do not provide a strict definition of sparseness at this point. The concept corresponds to the idea that in nature there are only a few relevant causes of a given phenomenon. We refer to Sect. 2.10 for a notion of *weak sparsity*. *Strong sparsity*, in the context of the linear model, is that the size s_0 of the active set S_0, should be less than the number of observations n or actually in an asymptotic formulation, s_0 should be of smaller order than $n/\log p$. The log-factor in the original number of unknown parameters p is the cost of not knowing the active set S_0 beforehand.

2.2 The Linear Model with $p \geq n$

Let $Y \in \mathbb{R}^n$ be an n-vector of real-valued observations and let X be a given $n \times p$ design matrix. We concentrate from now on mainly on the high-dimensional situation, which is the situation $p \geq n$ or even $p \gg n$.

Write the expectation of the response Y as

$$f^0 := \mathbb{E}Y.$$

The matrix X is fixed in this chapter, i.e., we have fixed design. The entries of the vector f^0 are thus the (conditional) expectation of Y given X. Let $\epsilon := Y - f^0$ be the noise term.

The linear model is

$$f^0 = X\beta^0$$

where β^0 is an unknown vector of coefficients. Thus this model assumes there is a solution β^0 of the equation $f^0 = X\beta^0$. In the high-dimensional situation with rank$(X) = n$ this is always the case: the linear model is never misspecified. When there are several solutions we may take for instance a sparsest solution, that is, a solution with the smallest number of non-zero coefficients. Alternatively one may prefer a basis pursuit solution (Chen et al. 1998)

$$\beta^0 := \arg\min\left\{ \|\beta\|_1 : X\beta = f^0 \right\}$$

where $\|\beta\|_1 := \sum_{j=1}^{p} |\beta_j|$ denotes the ℓ_1-norm of the vector β. We do not express in our notation that basis pursuit may not generate a unique solution.[1]

Aim is to construct an estimator $\hat{\beta}$ of β^0. When $p \geq n$ the least squares estimator $\hat{\beta}_{LS}$ will not work: it will just reproduce the data by returning the estimator $X\hat{\beta}_{LS} = Y$. This is called an instance of *overfitting*. Least squares loss with an ℓ_1-regularization penalty can overcome the overfitting problem. This method is called the Lasso. The Lasso estimator $\hat{\beta}$ is presented in more detail in (2.3) in Sect. 2.4.

---

[1]A suitable notation that expresses the non-uniqueness is $\beta^0 \in \arg\min\{\|\beta\|_1 : X\beta = f^0\}$. In our analysis, non-uniqueness is not a major concern.

2.3 Notation

For a vector $v \in \mathbb{R}^n$ we write $\|v\|_n^2 := v^T v / n = \|v\|_2^2 / n$, where $\|\cdot\|_2$ is the ℓ_2-norm. This abuse of notation has a "historical" background[2] but we believe confusion is not likely. Write the (normalized) Gram matrix as $\hat{\Sigma} := X^T X / n$. Thus $\|X\beta\|_n^2 = \beta^T \hat{\Sigma} \beta$, $\beta \in \mathbb{R}^p$.

For a vector $\beta \in \mathbb{R}^p$ we denote its ℓ_1-norm by $\|\beta\|_1 := \sum_{j=1}^p |\beta_j|$. Its ℓ_∞-norm is denoted by $\|\beta\|_\infty := \max_{1 \le j \le p} |\beta_j|$,

Let $S \subset \{1, \ldots, p\}$ be an index set. The vector $\beta_S \in \mathbb{R}^p$ with the set S as subscript is defined as

$$\beta_{j,S} := \beta_j \mathbb{1}\{j \in S\}, \ j = 1, \ldots, p. \tag{2.2}$$

Thus β_S is a p-vector with entries equal to zero at the indexes $j \notin S$. We will sometimes identify β_S with the vector $\{\beta_j\}_{j \in S} \in \mathbb{R}^{|S|}$. The vector β_{-S} has all entries inside the set S set to zero, i.e. $\beta_{-S} = \beta_{S^c}$ where $S^c = \{j \in \{1, \ldots, p\} : j \notin S\}$ is the complement of the set S. The notation (2.2) allows us to write $\beta = \beta_S + \beta_{-S}$.

The *active set* S_β of a vector $\beta \in \mathbb{R}^p$ is $S_\beta := \{j : \beta_j \neq 0\}$. For a solution β^0 of $X\beta^0 = f^0$, we denote its active set by $S_0 := S_{\beta^0}$ and the cardinality of this active set by $s_0 := |S_0|$.

The jth column of X is denoted by X_j, $j = 1, \ldots, p$ (and if there is little risk of confusion we also write X_i as the ith row of the matrix X, $i = 1, \ldots, n$). For a set $S \subset \{1, \ldots, p\}$ the matrix with only columns in the set S is denoted by $X_S := \{X_j\}_{j \in S}$. To fix the ordering of the columns here, we put them in increasing in j ordering. The "complement" matrix of X_S is denoted by $X_{-S} := \{X_j\}_{j \notin S}$. Moreover, for $j \in \{1, \ldots, p\}$, we let $X_{-j} := \{X_k\}_{k \neq j}$.

2.4 The Lasso, KKT and Two Point Inequality

The Lasso estimator(Tibshirani 1996, Least Absolute Shrinkage and Selection Operator) $\hat{\beta}$ is a solution of the minimization problem

$$\hat{\beta} := \arg \min_{\beta \in \mathbb{R}^p} \left\{ \|Y - X\beta\|_n^2 + 2\lambda \|\beta\|_1 \right\}. \tag{2.3}$$

---

[2] If $X_1, \ldots, X_n$ are n elements of some space $\mathcal{X}$ and $f : \mathcal{X} \to \mathbb{R}$ is some real-valued function on $\mathcal{X}$, one may view $\sum_{i=1}^n f^2(X_i)/n$ as the squared $L_2(P_n)$-norm of f, with $P_n = \sum_{i=1}^n \delta_{X_i}/n$ being the measure that puts equal mass $1/n$ at each X_i ($i = 1, \ldots, n$). Let us denote the $L_2(P_n)$-norm by $\|\cdot\|_{2,P_n}$. We have abbreviated this to $\|\cdot\|_{P_n}$ and then further abbreviated it to $\|\cdot\|_n$. Finally, we identified f with the vector $(f(X_1), \ldots, f(X_n))^T \in \mathbb{R}^n$.

This estimator is the starting point from which we study more general norm-penalized estimators. The Lasso itself will be the object of study in the rest of this chapter and in other chapters as well. Although "Lasso" refers to a method rather than an estimator, we refer to $\hat{\beta}$ as "the Lasso". It is generally not uniquely defined but we do not express this in our notation. This is a justified in the sense that the theoretical results which we will present will hold for any solution of minimization problem (2.3). The parameter $\lambda \geq 0$ is a given tuning parameter: large values will lead to a sparser solution $\hat{\beta}$, that is, a solution with more entries set to zero. Problem 2.2 may help to obtain some insight into this numerical fact. In an asymptotic sense, λ will be "small", it will generally be of order $\sqrt{\log p / n}$.

This Lasso $\hat{\beta}$ satisfies the *Karush-Kuhn-Tucker conditions* or *KKT-conditions* which say that

$$X^T(Y - X\hat{\beta})/n = \lambda \hat{z} \qquad (2.4)$$

where $\hat{z}$ is a p-dimensional vector with $\|\hat{z}\|_\infty \leq 1$ and with $\hat{z}_j = \text{sign}(\hat{\beta}_j)$ if $\hat{\beta}_j \neq 0$. The latter can also be written as

$$\hat{z}^T \hat{\beta} = \|\hat{\beta}\|_1.$$

The KKT-conditions follow from sub-differential calculus which defines the sub-differential of the absolute value function $x \mapsto |x|$ as

$$\partial |x| = \{\text{sign}(x)\}\{x \neq 0\} + [-1, 1]\{x = 0\}.$$

Thus, $\hat{z} \in \partial \|\hat{\beta}\|_1$.

The KKT-conditions may be interpreted as the Lasso version of the *normal equations* which are true for the least squares estimator. The KKT-conditions will play an important role. They imply the *almost orthogonality* of X on the one hand and the residuals $Y - X\hat{\beta}$ on the other, in the sense that

$$\|X^T(Y - X\hat{\beta})\|_\infty / n \leq \lambda.$$

Recall that λ will (generally) be "small". Otherwise put, $X\hat{\beta}$ can be seen as a *surrogate projection* of Y on the column space of X. One easily sees that the KKT-conditions are equivalent to: for any $\beta \in \mathbb{R}^p$

$$(\beta - \hat{\beta})^T X^T(Y - X\hat{\beta})/n \leq \lambda \|\beta\|_1 - \lambda \|\hat{\beta}\|_1.$$

We will often refer to this inequality as the *two point inequality*. As we will see in the proofs this is useful in conjunction with the *two point margin*: for any β and β'

$$2(\beta' - \beta)^T \hat{\Sigma}(\beta' - \beta^0) = \|X(\beta' - \beta^0)\|_n^2 - \|X(\beta - \beta^0)\|_n^2 + \|X(\beta' - \beta)\|_n^2.$$

Thus the two point inequality can be written in the alternative form as

$$\|Y - X\hat{\beta}\|_n^2 - \|Y - X\beta\|_n^2 + \|X(\hat{\beta} - \beta)\|_n^2 \le 2\lambda\|\beta\|_1 - 2\lambda\|\hat{\beta}\|_1, \ \forall \ \beta.$$

The two point inequality was proved more generally by (Güler 1991, Lemma 2.2) and further extended by (Chen and Teboulle 1993, Lemma 3.2), see also Lemma 6.1 in Sect. 6.3 or more generally Lemma 7.1 in Sect. 7.2.

Another important inequality will be the *convex conjugate* inequality: for any $a, b \in \mathbb{R}$

$$2ab \le a^2 + b^2.$$

As a further look-ahead: in the case of loss functions other than least squares, we will be facing convex functions that are not necessarily quadratic and then the convex conjugate inequality is a consequence of Definition 7.2 in Sect. 7.2.

2.5   Dual Norm and Decomposability

As we will see, we will need a bound for the random quantity $\epsilon^T X(\hat{\beta} - \beta^0)/n$ in terms of $\|\hat{\beta} - \beta^0\|_1$, or modifications thereof. Here one may apply the dual norm inequality. The dual norm of $\|\cdot\|_1$ is the ℓ_∞-norm $\|\cdot\|_\infty$. The *dual norm inequality* says that for any two vectors w and β

$$|w^T\beta| \le \|w\|_\infty \|\beta\|_1.$$

Another important ingredient of the arguments to come is the *decomposability* of the ℓ_1-norm:

$$\|\beta\|_1 = \|\beta_S\|_1 + \|\beta_{-S}\|_1 \ \forall \ \beta.$$

The decomposability implies what we call the *triangle property*:

$$\|\beta\|_1 - \|\beta'\|_1 \le \|\beta_S - \beta'_S\|_1 + \|\beta_{-S}\|_1 - \|\beta'_{-S}\|_1,$$

where β and β' are any two vectors and $S \subset \{1, \ldots, p\}$ is any index set. The importance of the triangle property is was highlighted in van de Geer (2001) in the context of adaptive estimation. It has been invoked at first to derive non-sharp oracle inequalities (see Bühlmann and van de Geer 2011 and its references).

2.6 Compatibility

We will need a notion of *compatibility* between the ℓ_1-norm and the Euclidean norm $\| \cdot \|_n$. This allows us to identify β^0 to a certain extent.

Definition 2.1 (van de Geer 2007; Bühlmann and van de Geer 2011) For a constant $L > 1$ and an index set S, the *compatibility constant* is

$$\hat{\phi}^2(L, S) := \min\left\{ |S| \|X\beta_S - X\beta_{-S}\|_n^2 : \|\beta_S\|_1 = 1, \ \|\beta_{-S}\|_1 \le L \right\}.$$

We call L the *stretching factor* (generally $L \ge 1$ in the results to come).

Example 2.1 Let $S = \{j\}$ be the jth variable for some $j \in \{1, \ldots, p\}$. Then

$$\hat{\phi}^2(L, \{j\}) = \min\left\{ \|X_j - X_{-j}\gamma_j\|_n^2 : \gamma_j \in \mathbb{R}^{p-1}, \ \|\gamma_j\|_1 \le L \right\}.$$

Note that the unrestricted minimum $\min\{\|X_j - X_{-j}\gamma_j\|_n : \gamma_j \in \mathbb{R}^{p-1}\}$ is the length of the anti-projection of the first variable X_j on the space spanned by the remaining variables X_{-j}. In the high-dimensional situation this unrestricted minimum will generally be zero. The ℓ_1-restriction $\|\gamma_j\|_1 \le L$ potentially takes care that the ℓ_1-restricted minimum $\hat{\phi}(L, \{j\})$ is strictly positive. The restricted minimization is the dual of the Lagrangian formulation, the latter being used e.g. for the Lasso (2.3).

The compatibility constant $\hat{\phi}^2(L, S)$ measures the distance between the signed convex hull of the variables in X_S and linear combinations of variables in X_{-S} satisfying an ℓ_1-restriction (that is, the latter are restricted to lie within the stretched signed convex hull of $L \times X_{-S}$). Loosely speaking one may think of this as an ℓ_1-variant of "$(1 - \text{canonical correlation})$".

For general S one always has $\hat{\phi}^2(L, \{j\}) \ge \hat{\phi}^2(L, S)/|S|$ for all $j \in S$. The more general case $\underline{S} \subset S$ is treated in the next lemma. It says that the larger the set S the larger the *effective sparsity*[3] $|S|/\hat{\phi}^2(L, S)$.

Lemma 2.1 *For all L and $\underline{S} \subset S$ it holds that*

$$|\underline{S}|/\hat{\phi}^2(L, \underline{S}) \le |S|/\hat{\phi}^2(L, S).$$

Proof of Lemma 2.1 Let

$$\|Xb\|_n^2 := \min\left\{ \|X\beta\|_n^2 : \|\beta_{\underline{S}}\|_1 = 1, \ \|\beta_{-\underline{S}}\|_1 \le L \right\} = \frac{\hat{\phi}^2(L, \underline{S})}{|\underline{S}|}.$$

---

[3]Or non-sparsity actually.

Then $\|b_S\|_1 \geq \|b_{\underline{S}}\|_1 = 1$ and $\|b_{-S}\|_1 \leq \|b_{-\underline{S}}\|_1 \leq L$. Thus, writing $c = b/\|b_S\|_1$, it holds true that $\|c_S\|_1 = 1$ and $\|c_{-S}\|_1 = \|b_{-S}\|_1/\|b_S\|_1 \leq \|b_{-S}\|_1 \leq L$. Therefore

$$\|Xb\|_n^2 = \|b_S\|_1^2 \|Xc\|_n^2$$

$$\geq \|b_S\|_1^2 \min \left\{ \|X\beta\|_n^2 : \|\beta_S\|_1 = 1, \ \|\beta_{-S}\|_1 \leq L \right\}$$

$$= \|b_S\|_1^2 \hat{\phi}^2(L,S)/|S| \geq \hat{\phi}^2(L,S)/|S|.$$

□

2.7 A Sharp Oracle Inequality

Let us summarize what are the main ingredients of the Proof of Theorems 2.1 and 2.2 below:

- the two point margin
- the two point inequality
- the dual norm inequality
- the triangle property, or decomposability
- the convex conjugate inequality
- compatibility

Finally, to control the ℓ_∞-norm of the random vector $X^T \epsilon$ occurring below in Theorem 2.1 (and onwards) we will use

- empirical process theory

but this is postponed to Sect. 17.4 (or see Lemma 8.1 for the case of Gaussian errors ϵ). See also Corollary 12.1 for a complete picture in the Gaussian case.

The paper (Koltchinskii et al. 2011) (see also the excellent monograph Koltchinskii 2011) nicely combines ingredients such as the above to arrive at general sharp oracle inequalities for nuclear-norm penalized estimators for example. Theorem 2.1 below is a special case of their results. The sharpness refers to the constant 1 in front of $\|X(\beta - \beta^0)\|_n^2$ in the right-hand side of the result of the theorem. A sharp oracle inequality says that the estimator behaves up to an additive remainder term as the best in the model class.

Theorem 2.1 (Koltchinskii et al. 2011) *Let λ_ϵ satisfy*

$$\lambda_\epsilon \geq \|X^T \epsilon\|_\infty/n.$$

Define for $\lambda > \lambda_\epsilon$

$$\underline{\lambda} := \lambda - \lambda_\epsilon, \ \bar{\lambda} := \lambda + \lambda_\epsilon$$

and stretching factor

$$L := \bar{\lambda}/\underline{\lambda}.$$

Then

$$\|X(\hat{\beta} - \beta^0)\|_n^2 \leq \min_S \left\{ \min_{\beta \in \mathbb{R}^p,\, S_\beta = S} \|X(\beta - \beta^0)\|_n^2 + \bar{\lambda}^2 |S|/\hat{\phi}^2(L, S) \right\}.$$

Note that the stretching factor L is indeed larger than one and depends on the tuning parameter and the noise level λ_ϵ. If there is no noise, $L = 1$ (as then $\lambda_\epsilon = 0$). However, with noise, it is not always mandatory to take $L > 1$, see Problem 2.3.

The result of the above theorem says that if β^0 can be well-approximated by a sparse vector β, that is, a vector β with S_β small, then the prediction error of the Lasso is also small (with high probability). It is not possible however to verify this "if" without further conditions. In a sense, one must simply "believe" it to be so.

For typical error distributions, a value of order $\sqrt{\log p/n}$ for the tuning parameter λ ensures that the condition $\lambda > \|X^T \epsilon\|_\infty/n$ of the theorem holds with high probability. Thus we see that for a prediction error converging to zero, it must hold that the sparse approximation β has $|S_\beta|$ of small order $n/\log p$. In fact, according to Theorem 2.1, this is when the compatibility constant $\hat{\phi}^2(L, S_\beta)$ stays away from zero (see Chap. 15 for results on this), otherwise the size of the active set S_β should be even smaller. Theorem 2.2 below improves the situation: the prediction error there can be small when compatibility constants are behaving badly but the ℓ_1-norm of the target is not too large.

Theorem 2.1 follows from Theorem 2.2 below by taking there $\delta = 0$. It also follows the general case given in Theorem 7.1. However, a reader preferring to first consult a direct derivation before looking at generalizations may consider the proof given in Sect. 2.11.3.

We call the set of β's over which we minimize, as in Theorem 2.1, "candidate oracles". The minimizer is then called the "oracle".

2.8 Including a Bound for the ℓ_1-Error and Allowing Many Small Values

We will now show that if one increases the stretching factor L in the compatibility constant one can establish a bound for the ℓ_1-estimation error. We moreover will no longer implicitly require that a candidate oracle β is sparse in the strict sense, i.e. in the sense that its active set S_β is small. We allow β to be non-sparse but then its small coefficients should have small ℓ_1-norm. The result is a special case of the results for general loss and penalty given in Theorem 7.1.

Theorem 2.2 *Let λ_ϵ satisfy*

$$\lambda_\epsilon \geq \|X^T \epsilon\|_\infty / n.$$

Let $0 \leq \delta < 1$ be arbitrary and define for $\lambda > \lambda_\epsilon$

$$\underline{\lambda} := \lambda - \lambda_\epsilon, \quad \bar{\lambda} := \lambda + \lambda_\epsilon + \delta\underline{\lambda}$$

and

$$L := \frac{\bar{\lambda}}{(1-\delta)\underline{\lambda}}.$$

Then for all $\beta \in \mathbb{R}^p$ and all sets S

$$2\delta\underline{\lambda}\|\hat{\beta} - \beta\|_1 + \|X(\hat{\beta} - \beta^0)\|_n^2 \leq \|X(\beta - \beta^0)\|_n^2 + \frac{\bar{\lambda}^2|S|}{\hat{\phi}^2(L, S)} + 4\lambda\|\beta_{-S}\|_1. \tag{2.5}$$

The proof of this result invokes the ingredients we have outlined in the previous sections: (two point margin, two point inequality, dual norm, triangle property, convex conjugate, compatibility). Similar ingredients will be used to cook up results with other loss functions and regularization penalties. We remark here that for least squares loss one also may take a different route where the "bias" and "variance" of the Lasso is treated separately. Details are in Problem 2.3.

Proof of Theorem 2.2

- If

$$(\hat{\beta} - \beta)^T \hat{\Sigma}(\hat{\beta} - \beta^0) \leq -\delta\underline{\lambda}\|\hat{\beta} - \beta\|_1 + 2\lambda\|\beta_{-S}\|_1$$

we find from the two point margin

$$2\delta\underline{\lambda}\|\hat{\beta} - \beta\|_1 + \|X(\hat{\beta} - \beta^0)\|_n^2$$
$$= 2\delta\underline{\lambda}\|\hat{\beta} - \beta\|_1 + \|X(\beta - \beta^0)\|_n^2 - \|X(\beta - \hat{\beta})\|_n^2 + 2(\hat{\beta} - \beta)^T\hat{\Sigma}(\hat{\beta} - \beta^0)$$
$$\leq \|X(\beta - \beta^0)\|_n^2 + 4\lambda\|\beta_{-S}\|_1$$

and we are done.
- From now on we may therefore assume that

$$(\hat{\beta} - \beta)^T \hat{\Sigma}(\hat{\beta} - \beta^0) \geq -\delta\underline{\lambda}\|\hat{\beta} - \beta\|_1 + 2\lambda\|\beta_{-S}\|_1.$$

By the two point inequality we have

$$(\hat{\beta} - \beta)^T \hat{\Sigma} (\hat{\beta} - \beta^0) \leq (\hat{\beta} - \beta)^T X^T \epsilon / n + \lambda \|\beta\|_1 - \lambda \|\hat{\beta}\|_1.$$

By the dual norm inequality

$$|(\hat{\beta} - \beta)^T X^T \epsilon| / n \leq \lambda_\epsilon \|\hat{\beta} - \beta\|_1.$$

Thus

$$(\hat{\beta} - \beta)^T \hat{\Sigma} (\hat{\beta} - \beta^0) \leq \lambda_\epsilon \|\hat{\beta} - \beta\|_1 + \lambda \|\beta\|_1 - \lambda \|\hat{\beta}\|_1$$
$$\leq \lambda_\epsilon \|\hat{\beta}_S - \beta_S\|_1 + \lambda_\epsilon \|\hat{\beta}_{-S}\|_1 + \lambda_\epsilon \|\beta_{-S}\|_1 + \lambda \|\beta\|_1 - \lambda \|\hat{\beta}\|_1.$$

By the triangle property and invoking $\underline{\lambda} = \lambda - \lambda_\epsilon$ this implies

$$(\hat{\beta} - \beta)^T \hat{\Sigma} (\hat{\beta} - \beta^0) + \underline{\lambda} \|\hat{\beta}_{-S}\|_1 \leq (\lambda + \lambda_\epsilon) \|\hat{\beta}_S - \beta_S\|_1 + (\lambda + \lambda_\epsilon) \|\beta_{-S}\|_1$$

and so

$$(\hat{\beta} - \beta)^T \hat{\Sigma} (\hat{\beta} - \beta^0) + \underline{\lambda} \|\hat{\beta}_{-S} - \beta_{-S}\|_1 \leq (\lambda + \lambda_\epsilon) \|\hat{\beta}_S - \beta_S\|_1 + 2\lambda \|\beta_{-S}\|_1.$$

Hence, invoking $\bar{\lambda} = \lambda + \lambda_\epsilon + \delta \underline{\lambda}$,

$$(\hat{\beta} - \beta)^T \hat{\Sigma} (\hat{\beta} - \beta^0) + \underline{\lambda} \|\hat{\beta}_{-S} - \beta_{-S}\|_1 + \delta \underline{\lambda} \|\hat{\beta}_S - \beta_S\|_1 \leq \bar{\lambda} \|\hat{\beta}_S - \beta_S\|_1 + 2\lambda \|\beta_{-S}\|_1.$$
$$(2.6)$$

Since $(\hat{\beta} - \beta)^T \hat{\Sigma} (\hat{\beta} - \beta^0) \geq -\delta \underline{\lambda} \|\hat{\beta} - \beta\|_1 + 2\lambda \|\beta_{-S}\|_1$ this gives

$$(1 - \delta) \underline{\lambda} \|\hat{\beta}_{-S} - \beta_{-S}\|_1 \leq \bar{\lambda} \|\hat{\beta}_S - \beta_S\|_1$$

or

$$\|\hat{\beta}_{-S} - \beta_{-S}\|_1 \leq L \|\hat{\beta}_S - \beta_S\|_1.$$

But then by the definition of the compatibility constant

$$\|\hat{\beta}_S - \beta_S\|_1 \leq \sqrt{|S|} \|X(\hat{\beta} - \beta)\|_n / \hat{\phi}(L, S). \qquad (2.7)$$

Continue with inequality (2.6) and apply the convex conjugate inequality:

$$(\hat{\beta} - \beta)^T \hat{\Sigma} (\hat{\beta} - \beta^0) + \underline{\lambda} \|\hat{\beta}_{-S} - \beta_{-S}\|_1 + \delta \underline{\lambda} \|\hat{\beta}_S - \beta_S\|_1$$
$$\leq \bar{\lambda} \sqrt{|S|} \|X(\hat{\beta} - \beta)\|_n / \hat{\phi}(L, S) + 2\lambda \|\beta_{-S}\|_1$$
$$\leq \frac{1}{2} \frac{\bar{\lambda}^2 |S|}{\hat{\phi}^2(L, S)} + \frac{1}{2} \|X(\hat{\beta} - \beta)\|_n^2 + 2\lambda \|\beta_{-S}\|_1.$$

Invoking the two point margin

$$2(\hat{\beta} - \beta)^T \hat{\Sigma} (\hat{\beta} - \beta^0) = \|X(\hat{\beta} - \beta^0)\|_n^2 - \|X(\beta - \beta^0)\|_n^2 + \|X(\hat{\beta} - \beta)\|_n^2,$$

we obtain

$$\|X(\hat{\beta} - \beta^0)\|_n^2 + 2\underline{\lambda} \|\hat{\beta}_{-S} - \beta_{-S}\|_1 + 2\delta \underline{\lambda} \|\hat{\beta}_S - \beta_S\|_1$$
$$\le \|X(\beta - \beta^0)\|_n^2 + \bar{\lambda}^2 |S| / \hat{\phi}^2(L, S) + 4\lambda \|\beta_{-S}\|_1.$$

$\square$

What we see from Theorem 2.2 is firstly that the tuning parameter λ should be sufficiently large to "overrule" the part due to the noise $\|X^T \epsilon\|_\infty / n$. Since $\|X^T \epsilon\|_\infty / n$ is generally random, we need to complete the theorem with a bound for this quantity that holds with large probability. See Corollary 12.1 in Sect. 12.1 for this completion for the case of Gaussian errors. One sees there that one may choose $\lambda \asymp \sqrt{\log p / n}$. Secondly, by taking $\beta = \beta^0$ we deduce from the theorem that the prediction error $\|X(\hat{\beta} - \beta^0)\|_n^2$ is bounded by $\bar{\lambda}^2 |S_0| / \hat{\phi}^2(L, S_0)$ where S_0 is the active set of β^0. In other words, we reached the aim (2.1) of Sect. 2.1, under the conditions that the part due to the noise behaves like $\sqrt{\log p / n}$ and that the compatibility constant $\hat{\phi}^2(L, S_0)$ stays away from zero.

A third insight from Theorem 2.2 is that the Lasso also allows one to bound the estimation error in $\| \cdot \|_1$-norm, provided that the stretching factor L is taken large enough. This makes sense as a compatibility constant that can stand a larger L tells us that we have good identifiability properties. Here is an example statement for the ℓ_1-estimation error.

Corollary 2.1 *As an example, take $\beta = \beta^0$ and take $S = S_0$ as the active set of β^0 with cardinality $s_0 = |S_0|$. Let us furthermore choose $\lambda = 2\lambda_\epsilon$ and $\delta = 1/5$. The following ℓ_0-sparsity based bound holds under the conditions of Theorem 2.2:*

$$\|\hat{\beta} - \beta^0\|_1 \le C_0 \frac{\lambda_\epsilon s_0}{\hat{\phi}^2(4, S_0)},$$

where $C_0 = (16/5)^2 (5/2)$.

Finally, it is important to note that we do not insist that β^0 is sparse. The result of Theorem 2.2 is good if β^0 can be well approximated by a sparse vector β or by a vector β with many smallish coefficients. . The smallish coefficients occur in a term proportional to $\|\beta_{-S}\|_1$. By minimizing the bound over all candidate oracles β and all sets S one obtains the following corollary.

Corollary 2.2 *Under the conditions of Theorem 2.2, and using its notation, we have the following trade-off bound:*

$$2\delta\underline{\lambda}\|\hat{\beta} - \beta^0\|_1 + \|X(\hat{\beta} - \beta^0)\|_n^2$$

$$\leq \min_{\beta\in\mathbb{R}^p} \min_{S\subset\{1,\dots,p\}} \left\{ 2\delta\underline{\lambda}\|\beta - \beta^0\|_1 + \|X(\beta - \beta^0)\|_n^2 + \frac{\bar{\lambda}^2|S|}{\hat{\phi}^2(L,S)} + 4\lambda\|\beta_{-S}\|_1 \right\} \quad (2.8)$$

We will refer to the minimizer (β^*, S_*) in (2.8) as the (or an) *oracle*. Corollary 2.2 says that the Lasso mimics the oracle (β^*, S_*). It trades off approximation error $\|X(\beta^* - \beta^0)\|_n^2$, the sparsity $|S_*|$ and the ℓ_1-norm $\|\beta^*_{-S}\|_1$ of smallish coefficients. In general, we will define oracles in a loose sense, not necessarily the overall minimizer over all candidate oracles and furthermore constants in the various appearances may be (somewhat) different.

Two types of restrictions on the set of candidate oracles will be examined further. The first one, considered in the next section (Sect. 2.9) requires that the pair (β, S) has $S = S_\beta$ so that the term with the smallish coefficients $\|\beta_{-S}\|_1$ vanishes. A second type of restriction is to require $\beta = \beta^0$ but optimize over S, i.e., to consider only candidate oracles (β^0, S). This is done in Sect. 2.10.

2.9 The ℓ_1-Restricted Oracle

Restricting ourselves to candidate oracles (β, S) with $S = S_\beta$ in Corollary 2.2 leads to a trade-off between the the the ℓ_1-error $\|\beta - \beta^0\|_1$, the approximation error $\|X(\beta - \beta^0)\|_n^2$ and the sparsity $|S|$, or rather the *effective sparsity* $|S|/\hat{\phi}^2(L, S)$. To study this let us consider the oracle β^* which trades off approximation error and effective sparsity but is meanwhile restricted to have an ℓ_1-norm at least as large as that of β^0.

Lemma 2.2 *Let for some $\bar{\lambda}$ the vector β^* be defined as*

$$\beta^* := \arg\min\left\{ \|X(\beta - \beta^0)\|_n^2 + \bar{\lambda}^2|S_\beta|/\hat{\phi}^2(L, S_\beta) : \|\beta\|_1 \geq \|\beta^0\|_1 \right\}.$$

Let $S_* := S_{\beta*} = \{j : \beta^*_j \neq 0\}$ be the active set of β^*. Then

$$\bar{\lambda}\|\beta^* - \beta^0\|_1 \leq \|X(\beta^* - \beta^0)\|_n^2 + \frac{\bar{\lambda}^2|S_*|}{\hat{\phi}^2(1, S_*)}.$$

Proof of Lemma 2.2 Since $\|\beta^0\|_1 \leq \|\beta^*\|_1$ we know by the ℓ_1-triangle property

$$\|\beta^0_{-S_*}\|_1 \leq \|\beta^* - \beta^0_{S_*}\|_1.$$

Hence by the definition of the compatibility constant and by the convex conjugate inequality

$$\bar{\lambda}\|\beta^* - \beta^0\|_1 \leq 2\bar{\lambda}\|\beta^* - \beta^0_{S_*}\|_1 \leq \frac{2\bar{\lambda}\|X(\beta^* - \beta^0)\|_n}{\hat{\phi}(1, S_*)} \leq \|X(\beta^* - \beta^0)\|_n^2 + \frac{\bar{\lambda}^2|S_*|}{\hat{\phi}^2(1, S_*)}.$$

□

From Lemma 2.2 we see that an ℓ_1-restricted oracle β^* that trades off approximation error and sparseness is also going to be close in ℓ_1-norm. We have the following corollary for the bound of Theorem 2.2.

Corollary 2.3 *Let*

$$\lambda_\epsilon \geq \|X^T\epsilon\|_\infty/n.$$

Let $0 \leq \delta < 1$ be arbitrary and define for $\lambda > \lambda_\epsilon$

$$\underline{\lambda} := \lambda - \lambda_\epsilon, \ \bar{\lambda} := \lambda + \lambda_\epsilon + \delta\underline{\lambda}$$

and

$$L := \frac{\bar{\lambda}}{(1 - \delta)\underline{\lambda}}.$$

Let the vector β^* with active set S_* be defined as in Lemma 2.2. Then the following holds true:

$$\underline{\lambda}\|\hat{\beta} - \beta^0\|_1 \leq \left(\frac{\bar{\lambda} + 2\delta\underline{\lambda}}{2\delta\bar{\lambda}}\right)\left(\|X(\beta^* - \beta^0)\|_n^2 + \frac{\bar{\lambda}^2|S_*|}{\hat{\phi}^2(L, S_*)}\right).$$

2.10 Weak Sparsity

In the previous section we found a bound for the trade-off in Corollary 2.2 by considering the ℓ_1-restricted oracle. In this section we take an alternative route, where we take in Theorem 2.2 candidate oracles (β, S) with the vector β equal to β^0 as in Corollary 2.1, but now the set S not necessarily equal to the active set $S_0 := \{j : \beta^0_j \neq 0\}$ of β^0. We define

$$\rho_r^r := \sum_{j=1}^p |\beta^0_j|^r, \tag{2.9}$$

where $0 < r < 1$. The constant $\rho_r > 0$ is assumed to be "not too large". This is sometimes called *weak sparsity* as opposed to *strong sparsity* which requires "not too many" non-zero coefficients

$$s_0 := \#\{\beta_j^0 \neq 0\}.$$

Observe that the latter is a limiting case of the former in the sense that

$$\lim_{r \downarrow 0} \rho_r^r = s_0.$$

Lemma 2.3 *Suppose β^0 satisfies the weak sparsity condition (2.9) for some $0 < r < 1$ and $\rho_r > 0$. Then for any $\bar{\lambda}$ and λ*

$$\min_S \left\{ \frac{\bar{\lambda}^2 |S|}{\hat{\phi}^2(L,S)} + 4\lambda \|\beta_{-S}^0\|_1 \right\} \leq \frac{5\bar{\lambda}^{2(1-r)} \lambda^r \rho_r^r}{\hat{\phi}^2(L,S_*)},$$

where $S_* := \{j : |\beta_j^0| > \bar{\lambda}^2/\lambda\}$ and assuming $\hat{\phi}(L,S) \leq 1$ for any L and S (to simplify the expressions).

Proof of Lemma 2.3 Define $\lambda_* := \bar{\lambda}^2/\lambda$. Then $S_* = \{j : |\beta_j^0| > \lambda_*\}$. We get

$$|S_*| \leq \lambda_*^{-r} \rho_r^r = \bar{\lambda}^{2(1-r)} \lambda^r \rho_r^r.$$

Moreover

$$\|\beta_{-S_*}^0\|_1 \leq \lambda_*^{1-r} \rho_r^r = \bar{\lambda}^{2(1-r)} \lambda^{r-1} \rho_r^r \leq \bar{\lambda}^{2(1-r)} \lambda^{r-1} \rho_r^r / \hat{\phi}^2(L,S_*),$$

since by assumption $\hat{\phi}^2(L,S_*) \leq 1$.   □

As a consequence, we obtain bounds for the prediction error and ℓ_1-error of the Lasso under (weak) sparsity. We only present the bound for the ℓ_1-error (because this is what will need in Chap. 5 and beyond).

We make some arbitrary choices for the constants: we set $\lambda = 2\lambda_\epsilon$ and we choose $\delta = 1/5$.

Corollary 2.4 *Assume the ℓ_r-sparsity condition (2.9) for some $0 < r < 1$ and $\rho_r > 0$. Set*

$$S_* := \{j : |\beta_j^0| > 3\lambda_\epsilon\}.$$

Then for $\lambda_\epsilon \geq \|X^T \epsilon\|_\infty / n$ and $\lambda = 2\lambda_\epsilon$, we have the ℓ_r-sparsity based bound

$$\|\hat{\beta} - \beta^0\|_1 \leq C_r \lambda_\epsilon^{1-r} \rho_r^r / \hat{\phi}^2(4, S_*).$$

assuming that $\hat{\phi}(L, S) \leq 1$ for any L and S. The constant $C_r = (16/5)^{2(1-r)}(5^2/2^r)$ depends only on r.

2.11 Complements

2.11.1 A Bound Based on an Alternative Method of Proof

Theorem 7.2 ahead provides an alternative (and "dirty" in the sense that not much care was paid to optimize the constants) way to prove bounds. This alternative route can be useful when considering loss functions more general than least squares. The implied bounds when applied to least squares loss are of the same flavour as in Theorem 2.2 "up to constants". The constants coming from the alternative route are "bad" but on the other hand Theorem 7.2 can deal with some potentially difficult cases. The major difference is however that with this alternative method of proof (it is based on the "argmin" argument[4] rather than the two point inequality) the oracle inequality is not sharp. A further difference is that in the non-sharp oracle inequality the stretching constant is $1/(1-\delta)$ which is never larger than the stretching constant $L = \bar{\lambda}/((1 - \delta)\underline{\lambda})$ in the sharp oracle inequality of Theorem 2.2.

Corollary 2.5 (Corollary of Theorem 7.2.) *Let $\hat{\beta}$ be the Lasso*

$$\hat{\beta} := \arg\min_{\beta \in \mathbb{R}^p} \left\{ \|Y - X\beta\|_n^2 + 2\lambda\|\beta\|_1 \right\}.$$

Take $\lambda_\epsilon \geq \|X^T\epsilon\|_\infty/n$ and $\lambda \geq 8\lambda_\epsilon/\delta$. Then for all $\beta \in \mathbb{R}^p$ and sets S

$$\lambda\delta\|\hat{\beta} - \beta\|_1 \leq M(\beta, S),$$

where

$$M(\beta, S) := \frac{4\lambda^2(1 + \delta)^2|S|}{\hat{\phi}^2(1/(1 - \delta), S)} + 4\|X(\beta - \beta^0)\|_n^2 + 16\lambda\|\beta_{-S}\|_1.$$

Moreover,

$$\|X(\hat{\beta} - \beta^0)\|_n^2 \leq \|X(\beta - \beta^0)\|_n^2 + (\lambda + \lambda_\epsilon)M(\beta, S) + \lambda\|\beta_{-S}\|_1.$$

---

[4]The "argmin" argument takes the inequality: $\|Y - X\hat{\beta}\|_n^2 + 2\lambda\|\hat{\beta}\|_1 \leq \|Y - X\beta\|_n^2 + 2\lambda\|\beta\|_1 \ \forall \ \beta$, as starting point.

2.11.2 When There Are Coefficients Left Unpenalized

In most cases one does not penalize the constant term in the regression. More generally, suppose that the set of coefficients that are not penalized have indices $U \subset \{1, \ldots, p\}$. The Lasso estimator is then

$$\hat{\beta} := \arg\min_{\beta \in \mathbb{R}^p} \left\{ \|Y - X\beta\|_n^2 + 2\lambda \|\beta_{-U}\|_1 \right\}.$$

. The KKT-conditions are now

$$X^T(Y - X\hat{\beta})/n + \lambda \hat{z}_{-U} = 0, \ \|\hat{z}_{-U}\|_\infty \leq 1, \ z_{-U}^T \hat{\beta}_{-U} = \|\hat{\beta}_{-U}\|_1.$$

See also Problem 2.5.

2.11.3 A Direct Proof of Theorem 2.1

Fix some $\beta \in \mathbb{R}^p$. The derivation of Theorem 2.1 is identical to the one of Theorem 2.2 except for the fact that we consider the case $\delta = 0$ and $S = S_\beta$. These restrictions lead to a somewhat more transparent argumentation.

- If

$$(\hat{\beta} - \beta)^T \hat{\Sigma}(\hat{\beta} - \beta^0) \leq 0$$

we find from the two point margin

$$\|X(\hat{\beta} - \beta^0)\|_n^2 = \|X(\beta - \beta^0)\|_n^2 - \|X(\beta - \hat{\beta})\|_n^2 + 2(\hat{\beta} - \beta)^T \hat{\Sigma}(\hat{\beta} - \beta^0)$$

$$\leq \|X(\beta - \beta^0)\|_n^2.$$

Hence then we are done.
- Suppose now that

$$(\hat{\beta} - \beta)^T \hat{\Sigma}(\hat{\beta} - \beta^0) \geq 0.$$

By the two point inequality

$$(\beta - \hat{\beta})^T X^T (Y - X\hat{\beta})/n \leq \lambda \|\beta\|_1 - \lambda \|\hat{\beta}\|_1.$$

As $Y = X\beta^0 + \epsilon$

$$(\hat{\beta} - \beta)^T \hat{\Sigma}(\hat{\beta} - \beta^0) + \lambda \|\hat{\beta}\|_1 \leq (\hat{\beta} - \beta)^T X^T \epsilon/n + \lambda \|\beta\|_1.$$

By the dual norm inequality

$$|(\hat{\beta} - \beta)^T X^T \epsilon|/n \leq (\|X^T \epsilon\|_\infty/n)\|\hat{\beta} - \beta\|_1 \leq \lambda_\epsilon \|\hat{\beta} - \beta\|_1.$$

Thus

$$(\hat{\beta} - \beta)^T \hat{\Sigma}(\hat{\beta} - \beta^0) + \lambda\|\hat{\beta}\|_1 \leq \lambda_\epsilon \|\hat{\beta} - \beta\|_1 + \lambda\|\beta\|_1.$$

By the triangle property this implies

$$(\hat{\beta} - \beta)^T \hat{\Sigma}(\hat{\beta} - \beta^0) + (\lambda - \lambda_\epsilon)\|\hat{\beta}_{-S}\|_1 \leq (\lambda + \lambda_\epsilon)\|\hat{\beta}_S - \beta\|_1.$$

or

$$(\hat{\beta} - \beta)^T \hat{\Sigma}(\hat{\beta} - \beta^0) + \underline{\lambda}\|\hat{\beta}_{-S}\|_1 \leq \bar{\lambda}\|\hat{\beta}_S - \beta\|_1. \tag{2.10}$$

Since $(\hat{\beta} - \beta)^T \hat{\Sigma}(\hat{\beta} - \beta^0) \geq 0$ this gives

$$\|\hat{\beta}_{-S}\|_1 \leq (\bar{\lambda}/\underline{\lambda})\|\hat{\beta}_S - \beta\|_1 = L\|\hat{\beta}_S - \beta\|_1.$$

By the definition of the compatibility constant $\hat{\phi}^2(L, S)$ we then arrive at

$$\|\hat{\beta}_S - \beta\|_1 \leq \sqrt{|S|}\|X(\hat{\beta} - \beta)\|_n/\hat{\phi}(L, S). \tag{2.11}$$

Continue with inequality (2.10) and apply the convex conjugate inequality

$$(\hat{\beta} - \beta)^T \hat{\Sigma}(\hat{\beta} - \beta^0) + \underline{\lambda}\|\hat{\beta}_{-S}\|_1 \leq \bar{\lambda}\sqrt{|S|}\|X(\hat{\beta} - \beta)\|_n/\hat{\phi}(L, S)$$

$$\leq \frac{1}{2}\frac{\bar{\lambda}^2|S|}{\hat{\phi}^2(L, S)} + \frac{1}{2}\|X(\hat{\beta} - \beta)\|_n^2.$$

Since by the two point margin

$$2(\hat{\beta} - \beta^0)^T \hat{\Sigma}(\hat{\beta} - \beta) = \|X(\hat{\beta} - \beta^0)\|_n^2 - \|X(\beta - \beta^0)\|_n^2 + \|X(\hat{\beta} - \beta)\|_n^2,$$

we obtain

$$\|X(\hat{\beta} - \beta^0)\|_n^2 + 2\underline{\lambda}\|\hat{\beta}_{-S}\|_1 \leq \|X(\beta - \beta^0)\|_n^2 + \bar{\lambda}^2|S|/\hat{\phi}^2(L, S).$$

$\square$

Problems

2.1 Let $\hat{\beta}_{LS}$ be the ordinary least squares estimators for the case with $p < n$ and let X have rank p. Check that when the noise is uncorrelated and has common variance σ_0^2, then

$$\mathbb{E}\|X(\hat{\beta} - \beta^0)\|_2^2 = \sigma_0^2 p.$$

2.2 Consider the case of orthogonal design: $\hat{\Sigma} = I$ where I is the $n \times n$ identity matrix (hence $p = n$). The linear model

$$Y = X\beta^0 + \epsilon$$

can then be written as

$$\tilde{Y} = \beta_0 + \tilde{\epsilon},$$

where $\tilde{Y} = X^T Y/n$ and $\tilde{\epsilon} = X^T \epsilon/n$. This is called the *sequence space* model. Check that

$$\|Y - X\beta\|_n^2 = \|Y\|_n^2 - 2\beta^T \tilde{Y} + \|\beta\|_2^2, \ \beta \in \mathbb{R}^p.$$

Show that the Lasso (2.3) is

$$\hat{\beta}_j = \begin{cases} \tilde{Y}_j - \lambda & \tilde{Y}_j > \lambda \\ 0 & |\tilde{Y}_j| \le \lambda, j = 1, \dots, n. \\ \tilde{Y}_j + \lambda & \tilde{Y}_j < -\lambda \end{cases}$$

The operation $\tilde{Y} \mapsto \hat{\beta}$ is called *soft-thresholding*.

2.3 Consider the noiseless problem

$$\beta^* := \arg\min\left\{ \|X(\beta - \beta^0)\|_n^2 + 2\lambda\|\beta\|_1 \right\}.$$

Check that β^* satisfies the KKT-conditions

$$\hat{\Sigma}(\beta^* - \beta^0) = -\lambda z^*, \ z^* \in \partial\|\beta^*\|_1.$$

Let $S_* := \{j : \beta_j^* \ne 0\}$ be the active set of β^*. Verify that the KKT-conditions imply

$$\|\hat{\Sigma}(\beta^* - \beta^0)\|_2^2 \ge \lambda^2 |S_*|.$$

Conclude that $|S_*| \leq \hat{\Lambda}_{\max} \|X(\beta^* - \beta^0)\|_n^2 / \lambda^2$ where $\hat{\Lambda}_{\max}$ is the largest eigenvalue of $\hat{\Sigma}$. Show the corresponding result for the size of the active set of the Lasso in the noisy case (see also Bickel et al. 2009).

2.4 We study the linear model with Gaussian noise $\epsilon \sim \mathcal{N}_n(0, I)$. Let β^* be defined as in Problem 2.3 and let $\hat{\beta}$ be the Lasso (2.3). In this problem, the goal is to prove the following result.

Theorem 2.3 *Suppose that* $\|X_j\|_n = 1$ *for all j. Let* $0 < \alpha < 1$ *and* $0 < \alpha_1 < 1$ *be fixed and* $\lambda_0 := \sqrt{2 \log(2p/\alpha)/n}$. *Let* $0 \leq \eta < 1$ *and* $\eta\lambda > \lambda_0$. *Then with probability at least* $1 - \alpha - \alpha_1$

$$\|X(\hat{\beta} - \beta^*)\|_n \leq \sqrt{\frac{\hat{\Lambda}_{\max}}{n\lambda^2(1-\eta)}} \|X(\beta^* - \beta^0)\|_n + \sqrt{\frac{2\log(1/\alpha_1)}{n}}$$

where $\hat{\Lambda}_{\max}$ *denotes the largest eigenvalue of* $\hat{\Sigma}$.

Thus when $\hat{\Lambda}_{\max} = \mathcal{O}(1)$ and for $\lambda \asymp \sqrt{\log p/n}$ and $p \gg n$,

$$\|X(\hat{\beta} - \beta^0)\|_n = \|X(\beta^* - \beta^0)\|_n (1 + o_{\mathbb{P}}(1)) + \mathcal{O}_{\mathbb{P}}(1/\sqrt{n}).$$

In other words, the "squared bias" of the Lasso $\|X(\beta^* - \beta^0)\|_n^2$ is of larger order than its "variance" $\|X(\hat{\beta} - \beta^*)\|_n^2$. To access the "squared bias" $\|X(\beta^* - \beta^0)\|_n^2$ one may apply Theorem 2.2 with $\lambda_\epsilon = 0$. Advantage of the result is that one sees that one can take any stretching factor $L \geq 1$, i.e., it no longer depends on the noise level and hence not on the choice of the tuning parameter λ. The disadvantage is that the result is essentially only useful when the largest eigenvalue of $\hat{\Sigma}$ is of the same order as its maximal entry.

Here are some hints for the Proof of Theorem 2.3

(i) Show that KKT for $\hat{\beta}$ minus KKT for β^* equals

$$\hat{\Sigma}(\hat{\beta} - \beta^*) + \lambda(\hat{z} - z^*) = X^T \epsilon/n.$$

Multiply by $(\hat{\beta} - \beta^*)^T$ to find

$$\|X(\hat{\beta} - \beta^*)\|_n^2 + \lambda(\hat{\beta} - \beta^*)^T(\hat{z} - z^*) = (\hat{\beta} - \beta^*)^T X^T \epsilon/n.$$

(ii) Define for fixed $0 \leq \eta < 1$

$$\bar{S}_* := \{j : |z_j^*| \geq 1 - \eta\}.$$

Use the arguments of Problem 2.3 to find

$$|\bar{S}_*| \le \frac{\hat{\Lambda}_{max}\|X(\beta^* - \beta^0)\|_n^2}{(1-\eta)\lambda^2}.$$

(iii) Let χ_T^2 be a chi-squared random variable with T degrees of freedom. Apply Lemma 8.6 in Sect. 8.6 to find that all $t > 0$

$$\mathbb{P}(\chi_T \ge \sqrt{T} + \sqrt{2t}) \le \exp[-t].$$

(iv) Apply Corollary 8.1 in Sect. 8.1 to bound $\|X^T \epsilon\|_\infty / n$.

2.5 Consider the case where the coefficients in a set U are not penalized, as in Sect. 2.11.2. Check that the result of Theorem 2.2 is true for all candidate oracles (β, S) with $S \supset U$.

Chapter 3
The Square-Root Lasso

Abstract This chapter presents sharp oracle inequalities for the square-root Lasso, applying essentially the same arguments as in Chap. 2. The main new element is that one needs to make sure that the square-root Lasso does not degenerate. After having dealt with this issue, the chapter continues with a comparison of the square-root Lasso with the scaled Lasso. Furthermore, a multivariate version of the square-root Lasso is introduced. The latter will be invoked in later chapters.

3.1 Introduction

Consider as in the previous chapter the linear model

$$Y = X\beta^0 + \epsilon$$

with $Y \in \mathbb{R}^n$, X a given $n \times p$ matrix, $\beta^0 \in \mathbb{R}^p$ and noise $\epsilon \in \mathbb{R}^n$. In the previous chapter we required that the tuning parameter λ for the Lasso defined in Sect. 2.4 is chosen at least as large as the *noise level* λ_ϵ where λ_ϵ is a bound for $\|\epsilon^T X\|_\infty / n$. Clearly, if for example the entries in ϵ are i.i.d. with variance σ_0^2, the choice of λ will depend on the standard deviation σ_0 which will usually be unknown in practice. To avoid this problem, Belloni et al. (2011) introduced and studied the square-root Lasso

$$\hat{\beta} := \arg \min_{\beta \in \mathbb{R}^p} \left\{ \|Y - X\beta\|_n + \lambda_0 \|\beta\|_1 \right\}.$$

Again, we do not express in our notation that the estimator is in general not uniquely defined by the above equality. The results to come hold for any solution.

The square-root Lasso can be seen as a method that estimates β^0 and the noise variance σ_0^2 simultaneously. Defining the residuals $\hat{\epsilon} := Y - X\hat{\beta}$ and letting $\hat{\sigma}^2 := \|\hat{\epsilon}\|_n^2$ one clearly has

$$(\hat{\beta}, \hat{\sigma}^2) = \arg \min_{\beta \in \mathbb{R}^p, \, \sigma^2 > 0} \left\{ \frac{\|Y - X\beta\|_n^2}{\sigma} + \sigma + 2\lambda_0 \|\beta\|_1 \right\} \tag{3.1}$$

(up to uniqueness) provided the minimum is attained at a non-zero value of σ^2.

© Springer International Publishing Switzerland 2016

S. van de Geer, *Estimation and Testing Under Sparsity*,
Lecture Notes in Mathematics 2159, DOI 10.1007/978-3-319-32774-7_3

We note in passing that the square-root Lasso is *not* a quasi-likelihood estimator as the function $\exp[-z^2/\sigma - \sigma]$, $z \in \mathbb{R}$, is not a density with respect to a dominating measure not depending on $\sigma^2 > 0$. The square-root Lasso is moreover not to be confused with the scaled Lasso. See Sect. 3.7 for our definition of the latter. The scaled Lasso as we define it there *is* a quasi-likelihood estimator. It is studied in e.g. the paper (Sun and Zhang 2010) which comments on Städler et al. (2010). In the rejoinder (Städler et al. 2010) the name scaled Lasso is used. Some confusion arises as for example (Sun and Zhang 2012) call the square-root Lasso the scaled Lasso.

3.2   KKT and Two Point Inequality for the Square-Root Lasso

When $\|Y - X\hat{\beta}\|_n$ ($= \hat{\sigma}$) is non-zero the square-root Lasso $\hat{\beta}$ satisfies the KKT-conditions

$$\frac{X^T(Y - X\hat{\beta})/n}{\|Y - X\hat{\beta}\|_n} = \lambda_0 \hat{z} \tag{3.2}$$

where $\|\hat{z}\|_\infty \leq 1$ and $\hat{z}_j = \text{sign}(\hat{\beta}_j)$ if $\hat{\beta}_j \neq 0$.

These KKT-conditions (3.2) again follow from sub-differential calculus. Indeed, for a fixed $\sigma > 0$ the sub-differential with respect to β of the expression in curly brackets given in (3.1) is equal to

$$-\frac{2X^T(Y - X\beta)/n}{\sigma} + 2\lambda_0 z(\beta)$$

with, for $j = 1, \ldots, p$, $z_j(\beta)$ the sub-differential of $\beta_j \mapsto |\beta_j|$. Setting this to zero at $(\hat{\beta}, \hat{\sigma})$ gives the above KKT-conditions (3.2).

3.3   A Proposition Assuming No Overfitting

If $\|Y - X\hat{\beta}\|_n = 0$ the square-root Lasso returns a degenerate solution which overfits. We assume now at first that $\|Y - X\hat{\beta}\|_n > 0$ and then show in the next section that this is the case under ℓ_1-sparsity conditions. See also Problem 3.2 for the overfitting issue in the sequence space model.

The results to come now depend on a scale free bound $\lambda_{0,\epsilon}$ for $\|X^T\epsilon\|_\infty/(n\|\epsilon\|_n)$ instead of on the noise level $\lambda_\epsilon \geq \|X^T\epsilon\|_n/n$. For the case of normally distributed errors, a probability inequality for $\|X^T\epsilon\|_\infty/(n\|\epsilon\|_n)$ is given in Lemma 8.2. See also Corollary 12.2 for a complete picture for the Gaussian case.

Recall that $\hat{\epsilon} := Y - X\hat{\beta}$, so that $\|\hat{\epsilon}\|_n = \hat{\sigma}$.

Proposition 3.1 *Suppose* $\|\hat{\epsilon}\|_n > 0$. *Let* $\lambda_{0,\epsilon} \geq \|X^T\epsilon\|_\infty/(n\|\epsilon\|_n)$ *for some constant* $\lambda_{0,\epsilon} > 0$. *Let* λ_0 *satisfy*

$$\lambda_0\|\hat{\epsilon}\|_n \geq \lambda_{0,\epsilon}\|\epsilon\|_n.$$

Let $0 \leq \delta < 1$ *be arbitrary and define*

$$\hat{\lambda}_L\|\epsilon\|_n := \lambda_0\|\hat{\epsilon}\|_n - \lambda_{0,\epsilon}\|\epsilon\|_n, \quad \hat{\lambda}_U\|\epsilon\|_n := \lambda_0\|\hat{\epsilon}\|_n + \lambda_{0,\epsilon}\|\epsilon\|_n + \delta\hat{\lambda}_L\|\epsilon\|_n$$

and

$$\hat{L} := \frac{\hat{\lambda}_U}{(1-\delta)\hat{\lambda}_L}.$$

Then

$$
\begin{aligned}
2\delta\hat{\lambda}_L\|\hat{\beta} - \beta^0\|_1\|\epsilon\|_n &+ \|X(\hat{\beta} - \beta^0)\|_n^2 \\
\leq \min_{S\subset\{1,\dots,p\}} \min_{\beta\in\mathbb{R}^p} &\Big\{ 2\delta\hat{\lambda}_L\|\beta - \beta^0\|_1\|\epsilon\|_n + \|X(\beta - \beta^0)\|_n^2 \\
&+ \frac{\hat{\lambda}_U^2\|\epsilon\|_n^2|S|}{\hat{\phi}^2(\hat{L}, S)} + 4\lambda_0\|\hat{\epsilon}\|_n\|\beta_{-S}\|_1 \Big\}.
\end{aligned}
$$

Proof of Proposition 3.1 The estimator $\hat{\beta}$ satisfies the KKT-conditions (3.2) which are exactly the KKT-conditions (2.4) but with λ replaced by $\lambda_0\|\hat{\epsilon}\|_n$. This means we can recycle the Proof of Theorem 2.2 (Problem 3.3). □

3.4 Showing the Square-Root Lasso Does Not Overfit

Proposition 3.1 is not very useful as such as it assumes $\|\hat{\epsilon}\|_n > 0$ and depends also otherwise on the value of $\|\hat{\epsilon}\|_n$. We therefore provide bounds for this quantity.

Lemma 3.1 *Let* λ_0 *be the tuning parameter used for the square-root Lasso. Suppose that for some* $0 < \eta < 1$, *some* $\lambda_{0,\epsilon} > 0$ *and some* $\underline{\sigma} > 0$, *we have*

$$\lambda_0(1 - \eta) \geq \lambda_{0,\epsilon}$$

and

$$\lambda_0\|\beta^0\|_1/\underline{\sigma} \leq 2\left(\sqrt{1 + (\eta/2)^2} - 1\right). \tag{3.3}$$

Then on the set where $\|X^T \epsilon\|_\infty / (n\|\epsilon\|_n) \le \lambda_{0,\epsilon}$ *and* $\|\epsilon\|_n \ge \underline{\sigma}$ *we have*

$$\left| \|\hat{\epsilon}\|_n / \|\epsilon\|_n - 1 \right| \le \eta.$$

The constant $\sqrt{1 + (\eta/2)^2} - 1$ is not essential, one may replace it by a prettier-looking lower bound. Note that it is smaller than $(\eta/2)^2$ but for η small it is approximately equal to $(\eta/2)^2$.

In an asymptotic formulation, say with i.i.d. standard normal noise, the conditions of Lemma 3.1 are met when $\|\beta^0\|_1 = o(\sqrt{n/\log p})$ and $\lambda_0 \asymp \sqrt{\log p / n}$ is suitably chosen.

The proof of the lemma makes use of the convexity of both the least-squares loss function and the penalty.

Proof of Lemma 3.1 Suppose $\|X^T \epsilon\|_\infty / (n\|\epsilon\|_n) \le \lambda_{0,\epsilon}$ and $\|\epsilon\|_n \ge \underline{\sigma}$. First we note that the inequality (3.3) gives

$$\lambda_0 \|\beta^0\|_1 / \|\epsilon\|_n \le 2\left(\sqrt{1 + (\eta/2)^2} - 1 \right).$$

For the upper bound for $\|\hat{\epsilon}\|_n$ we use that

$$\|\hat{\epsilon}\|_n + \lambda_0 \|\hat{\beta}\|_1 \le \|\epsilon\|_n + \lambda_0 \|\beta^0\|_1$$

by the definition of the estimator. Hence

$$\|\hat{\epsilon}\|_n \le \|\epsilon\|_n + \lambda_0 \|\beta^0\|_1 \le \left[1 + 2\left(\sqrt{1 + (\eta/2)^2} - 1 \right) \right] \|\epsilon\|_n \le (1 + \eta)\|\epsilon\|_n.$$

For the lower bound for $\|\hat{\epsilon}\|_n$ we use the convexity of both the loss function and the penalty. Define

$$t := \frac{\eta\|\epsilon\|_n}{\eta\|\epsilon\|_n + \|X(\hat{\beta} - \beta^0)\|_n}.$$

Note that $0 < t \le 1$. Let $\hat{\beta}_t$ be the convex combination $\hat{\beta}_t := t\hat{\beta} + (1-t)\beta^0$. Then

$$\|X(\hat{\beta}_t - \beta^0)\|_n = t\|X(\hat{\beta} - \beta^0)\|_n = \frac{\eta\|\epsilon\|_n \|X(\hat{\beta} - \beta^0)\|_n}{\eta\|\epsilon\|_n + \|X(\hat{\beta} - \beta^0)\|_n} \le \eta\|\epsilon\|_n.$$

Define $\hat{\epsilon}_t := Y - X\hat{\beta}_t$. Then, by convexity of $\|\cdot\|_n$ and $\|\cdot\|_1$,

$$\|\hat{\epsilon}_t\|_n + \lambda_0 \|\hat{\beta}_t\|_1 \le t\|\hat{\epsilon}\|_n + t\lambda_0 \|\hat{\beta}\|_1 + (1-t)\|\epsilon\|_n + (1-t)\lambda_0 \|\beta^0\|_1$$

$$\le \|\epsilon\|_n + \lambda_0 \|\beta^0\|_1$$

where in the last step we again used that $\hat{\beta}$ minimizes $\|Y - X\beta\|_n + \lambda_0\|\beta\|_1$. Taking squares on both sides gives

$$\|\hat{\epsilon}_t\|_n^2 + 2\lambda_0\|\hat{\beta}_t\|_1\|\hat{\epsilon}_t\|_n + \lambda_0^2\|\hat{\beta}_t\|_1^2 \leq \|\epsilon\|_n^2 + 2\lambda_0\|\beta^0\|_1\|\epsilon\|_n + \lambda_0^2\|\beta^0\|_1^2. \quad (3.4)$$

But

$$\|\hat{\epsilon}_t\|_n^2 = \|\epsilon\|_n^2 - 2\epsilon^T X(\hat{\beta}_t - \beta^0)/n + \|X(\hat{\beta}_t - \beta^0)\|_n^2$$

$$\geq \|\epsilon\|_n^2 - 2\lambda_{0,\epsilon}\|\hat{\beta}_t - \beta^0\|_1\|\epsilon\|_n + \|X(\hat{\beta}_t - \beta^0)\|_n^2$$

$$\geq \|\epsilon\|_n^2 - 2\lambda_{0,\epsilon}\|\hat{\beta}_t\|_1\|\epsilon\|_n - 2\lambda_{0,\epsilon}\|\beta^0\|_1\|\epsilon\|_n + \|X(\hat{\beta}_t - \beta^0)\|_n^2.$$

Moreover, by the triangle inequality

$$\|\hat{\epsilon}_t\|_n \geq \|\epsilon\|_n - \|X(\hat{\beta}_t - \beta^0)\|_n \geq (1 - \eta)\|\epsilon\|_n.$$

Inserting these two inequalities into (3.4) gives

$$\|\epsilon\|_n^2 - 2\lambda_{0,\epsilon}\|\hat{\beta}_t\|_1\|\epsilon\|_n - 2\lambda_{0,\epsilon}\|\beta^0\|_1\|\epsilon\|_n$$

$$+ \|X(\hat{\beta}_t - \beta^0)\|_n^2 + 2\lambda_0(1 - \eta)\|\hat{\beta}_t\|_1\|\epsilon\|_n + \lambda_0^2\|\hat{\beta}_t\|_1^2$$

$$\leq \|\epsilon\|_n^2 + 2\lambda_0\|\beta^0\|_1\|\epsilon\|_n + \lambda_0^2\|\beta^0\|_1^2$$

which implies by the assumption $\lambda_0(1 - \eta) \geq \lambda_{0,\epsilon}$

$$\|X(\hat{\beta}_t - \beta^0)\|_n^2 \leq 2(\lambda_0 + \lambda_{0,\epsilon})\|\beta^0\|_1\|\epsilon\|_n + \lambda_0^2\|\beta^0\|_1^2$$

$$\leq 4\lambda_0\|\beta^0\|_1\|\epsilon\|_n + \lambda_0^2\|\beta^0\|_1^2$$

where in the last inequality we used $\lambda_{0,\epsilon} \leq (1 - \eta)\lambda_0 \leq \lambda_0$. But continuing we see that we can write the last expression as

$$4\lambda_0\|\beta^0\|_1\|\epsilon\|_n + \lambda_0^2\|\beta^0\|_1^2 = \left((\lambda_0\|\beta_0\|_1/\|\epsilon_n\|_n + 2)^2 - 4\right)\|\epsilon\|_n^2.$$

Again invoke the ℓ_1-sparsity condition

$$\lambda_0\|\beta^0\|_1/\|\epsilon\|_n < 2\left(\sqrt{1 + (\eta/2)^2} - 1\right)$$

to get

$$\left((\lambda_0\|\beta_0\|_1/\|\epsilon_n\|_n + 2)^2 - 4\right)\|\epsilon\|_n^2 \leq \frac{\eta^2}{4}\|\epsilon\|_n^2.$$

We thus established that

$$\|X(\hat{\beta}_t - \beta^0)\|_n \le \frac{\eta\|\epsilon\|_n}{2}.$$

Rewrite this to

$$\frac{\eta\|\epsilon\|_n\|X(\hat{\beta} - \beta^0)\|_n}{\eta\|\epsilon\|_n + \|X(\hat{\beta} - \beta^0)\|_n} \le \frac{\eta\|\epsilon\|_n}{2},$$

and rewrite this in turn to

$$\eta\|\epsilon\|_n\|X(\hat{\beta} - \beta^0)\|_n \le \frac{\eta^2\|\epsilon\|_n^2}{2} + \frac{\eta\|\epsilon\|_n\|X(\hat{\beta} - \beta^0)\|_n}{2}$$

or

$$\|X(\hat{\beta} - \beta^0)\|_n \le \eta\|\epsilon\|_n.$$

But then, by repeating the argument, also

$$\|\hat{\epsilon}\|_n \ge \|\epsilon\|_n - \|X(\hat{\beta} - \beta^0)\|_n \ge (1 - \eta)\|\epsilon\|_n.$$

$\square$

3.5   A Sharp Oracle Inequality for the Square-Root Lasso

We combine the results of the two previous sections.

Theorem 3.1 *Assume the ℓ_1-sparsity (3.3) for some $0 < \eta < 1$ and $\underline{\sigma} > 0$, i.e.*

$$\lambda_0\|\beta^0\|_1/\underline{\sigma} \le 2\left(\sqrt{1 + (\eta/2)^2} - 1\right).$$

Let λ_0 satisfy for some $\lambda_{0,\epsilon} > 0$

$$\lambda_0(1 - \eta) > \lambda_{0,\epsilon}.$$

Let $0 \le \delta < 1$ be arbitrary and define

$$\underline{\lambda}_0 := \lambda_0(1 - \eta) - \lambda_{0,\epsilon},$$

$$\bar{\lambda}_0 := \lambda_0(1 + \eta) + \lambda_{0,\epsilon} + \delta\underline{\lambda}_0.$$

and

$$L := \frac{\bar{\lambda}_0}{(1-\delta)\underline{\lambda}_0}.$$

Then on the set where $\|X^T\epsilon\|_\infty/(n\|\epsilon\|_n) \leq \lambda_{0,\epsilon}$ and $\|\epsilon\|_n \geq \underline{\sigma}$, we have

$$2\delta\underline{\lambda}_0\|\hat{\beta} - \beta^0\|_1\|\epsilon\|_n + \|X(\hat{\beta} - \beta^0)\|_n^2$$

$$\leq \min_{S\in\{1,\dots,p\}} \min_{\beta\in\mathbb{R}^p} \left\{ 2\delta\underline{\lambda}_0\|\beta - \beta^0\|_1\|\epsilon\|_n + \|X(\beta - \beta^0)\|_n^2 \right.$$

$$\left. + \frac{\bar{\lambda}_0^2|S|\|\epsilon\|_n^2}{\hat{\phi}^2(L,S)} + 4\lambda_0(1+\eta)\|\epsilon\|_n\|\beta_{-S}\|_1 \right\}. \tag{3.5}$$

Proof of Theorem 3.1 This follows from the same arguments as those used for Theorem 2.2, and inserting Lemma 3.1. □

The minimizer (β^*, S_*) in (3.5) is again called the oracle and (3.5) is a sharp oracle inequality. The paper (Sun and Zhang 2013) contains (among other things) similar results as Theorem 3.1, although with different constants and the oracle inequality shown there is not a sharp one.

3.6 A Bound for the Mean ℓ_1-Error

As we shall see in for example Sect. 4.5 (Lemma 4.2), it is of interest to have bounds for the mean ℓ_1-estimation error $\mathbb{E}\|\hat{\beta} - \beta^0\|_1$ (or even for higher moments $\mathbb{E}\|\hat{\beta} - \beta^0\|_1^m$ with $m > 1$). Such bounds are will be important too when aiming at proving so-called strong asymptotic unbiasedness of certain (de-sparsified) estimators, which in turn is invoked for deriving asymptotic lower bounds for the variance of such estimators. We refer to Sect. 4.9 for more details. Actually, Sect. 4.9 is about the case of random design so we will include that case here as well. In the case of random design the compatibility constant $\hat{\phi}^2(L, S)$ is random and we need to lower bound it by a non-random constant.

Lemma 3.2 *Suppose the conditions of Theorem 3.1 are met. Let moreover for some constant* $\underline{\phi}(L, S) > 0$, $\mathcal{T}$ *be the set*

$$\mathcal{T} := \{\|X^T\epsilon\|_\infty/(n\|\epsilon\|_n) \leq \lambda_{0,\epsilon}, \|\epsilon\|_n \geq \bar{\sigma}, \hat{\phi}(L, S) \geq \underline{\phi}(L, S)\}.$$

Let (for the case of random design)

$$\|X\beta\|^2 := \mathbb{E}\|X\beta\|_n^2, \ \beta \in \mathbb{R}^p.$$

Define [as in (3.5)]

$$\eta_n := \min_{S \in \{1,\dots,p\}} \min_{\beta \in \mathbb{R}^p} \left\{ \|\beta - \beta^0\|_1 + \frac{\|X(\beta - \beta^0)\|^2}{2\delta\bar{\sigma}\underline{\lambda}_0} \right.$$

$$\left. + \frac{\bar{\lambda}_0 |S| \sigma_0}{2\delta\underline{\phi}^2(L,S)} + \frac{4\lambda_0(1 + \eta)\|\beta_{-S}\|_1}{2\delta\underline{\lambda}_0} \right\}.$$

Define moreover

$$\zeta_n := \frac{\sigma_0}{\lambda_0}\mathbb{P}^{1/2}(\mathscr{T}^c) + \frac{2\left(\bar{\sigma}\sqrt{1 + (\eta/2)^2} - 1\right) + 1}{\lambda_0}\mathbb{P}(\mathscr{T}^c).$$

Then

$$\mathbb{E}\|\hat{\beta} - \beta^0\|_1 \le \eta_n + \zeta_n.$$

In an asymptotic formulation and with fixed design (where $\hat{\phi}(L, S)$ is fixed), one can choose $\lambda_{0,\epsilon}$ and $\underline{\sigma}$ large such that $\mathbb{P}(\mathscr{T}^c) = \mathcal{O}(p^{-\tau})$ for some $\tau > 0$, but such that the bound η_n for $\|\hat{\beta} - \beta^0\|_1$ is only effected by this in terms of constants. For p large the leading term in the bound $\eta_n + \zeta_n$ for $\mathbb{E}\|\hat{\beta} - \beta^0\|_1$ is then η_n. In other words, the bound in probability for $\|\hat{\beta} - \beta^0\|_1$ is of the same order as the bound in expectation.

To bound $\mathbb{P}(\mathscr{T}^c)$ for the case of fixed design we refer to Lemma 8.2 in Sect. 8.2. For a lower bound on $\hat{\phi}(L, S)$ for the case of random design, we refer to Theorem 15.2 in Sect. 15.3. Then, when for example $s_0 = o(\delta_n\sqrt{n/\log p})$ (say) the overall conclusion is

$$\mathbb{E}\|\hat{\beta} - \beta^0\|_1 = o(\delta_n).$$

Similar conclusions hold under weak sparsity assumptions.

Proof of Lemma 3.2 By Theorem 3.1

$$\mathbb{E}\|\hat{\beta} - \beta^0\|_1 1_{\mathscr{T}} \le \eta_n.$$

Moreover, by the definition of $\hat{\beta}$

$$\|\hat{\beta}\|_1 \le \|\epsilon\|_n/\lambda_0 + \|\beta^0\|_1 \le \|\epsilon\|_n/\lambda_0 + 2\left(\bar{\sigma}\sqrt{1 + (\eta/2)^2} - 1\right)/\lambda_0.$$

It follows that

$$\|\hat{\beta} - \beta^0\|_1 \le \frac{\|\epsilon\|_n}{\lambda_0} + \frac{2\left(\bar{\sigma}\sqrt{1 + (\eta/2)^2} - 1\right) + 1}{\lambda_0}.$$

Therefore

$$\mathbb{E}\|\hat{\beta} - \beta^0\|_1 1_{\mathcal{T}^c} \le \frac{\sigma_0}{\lambda_0} \mathbb{P}^{1/2}(\mathcal{T}^c) + \frac{2\left(\bar{\sigma}\sqrt{1 + (\eta/2)^2} - 1\right) + 1}{\lambda_0} \mathbb{P}(\mathcal{T}^c) = \zeta_n.$$

$\square$

3.7 Comparison with Scaled Lasso

Fix a tuning parameter $\lambda_0 > 0$. Consider the Lasso with scale parameter σ

$$\hat{\beta}(\sigma) := \arg\min_{\beta}\left\{\|Y - X\beta\|_n^2 + 2\lambda_0\sigma\|\beta\|_1\right\}, \tag{3.6}$$

the (scale free) square-root Lasso

$$\hat{\beta}_\sharp := \arg\min_{\beta}\left\{\|Y - X\beta\|_n + \lambda_0\|\beta\|_1\right\}$$

and the *scaled Lasso* (Sun and Zhang 2012)

$$(\hat{\beta}_\flat, \tilde{\sigma}_\flat^2) := \arg\min_{\beta,\sigma}\left\{\frac{\|Y - X\beta\|_n^2}{\sigma^2} + \log\sigma^2 + \frac{2\lambda_0\|\beta\|_1}{\sigma}\right\}.$$

Then one easily verifies that

$$\tilde{\sigma}_\flat^2 = \|Y - X\hat{\beta}_\flat\|_n^2 + \lambda_0\tilde{\sigma}_\flat\|\hat{\beta}_\flat\|_1$$

and that $\hat{\beta}_\flat = \hat{\beta}(\hat{\sigma}_\flat)$. Moreover, if we define

$$\hat{\sigma}_\sharp^2 := \|Y - X\hat{\beta}_\sharp\|_n^2$$

we see that $\hat{\beta}_\sharp = \hat{\beta}(\hat{\sigma}_\sharp)$.

Let us write the residual sum of squares (normalized by n^{-1}) when using σ as scale parameter as

$$\hat{\sigma}^2(\sigma) := \|Y - X\hat{\beta}(\sigma)\|_n^2.$$

Moreover, write the penalized (and normalized) residual sum of squares plus penalty when using σ as scale parameter as

$$\tilde{\sigma}^2(\sigma) := \|Y - X\hat{\beta}(\sigma)\|_n^2 + \lambda_0\sigma\|\hat{\beta}(\sigma)\|_1.$$

Let furthermore

$$\tilde{\sigma}_\sharp^2 := \|Y - X\hat{\beta}_\sharp\|_n^2 + \lambda_0 \hat{\sigma}_\sharp \|\hat{\beta}_\sharp\|_1$$

and

$$\hat{\sigma}_\flat^2 := \|Y - X\hat{\beta}_\flat\|_n^2.$$

The scaled Lasso includes the penalty in its estimator $\tilde{\sigma}_\flat^2$ of the noise variance $\sigma_0^2 := \mathbb{E}\|\epsilon\|_n^2$ (assuming the latter exists). The square-root Lasso does not include the penalty in its estimator $\hat{\sigma}_\sharp^2$ of σ_0^2. It obtains $\hat{\sigma}_\sharp^2$ as a stable point of the equation $\hat{\sigma}_\sharp^2 = \hat{\sigma}^2(\hat{\sigma}_\sharp)$ and the scaled Lasso obtains $\tilde{\sigma}_\flat^2$ as a stable point of the equation $\tilde{\sigma}_\flat^2 = \tilde{\sigma}^2(\tilde{\sigma}_\flat)$. By the mere definition of $\tilde{\sigma}^2(\sigma)$ and $\hat{\sigma}^2(\sigma)$ we also have $\tilde{\sigma}_\sharp^2 = \tilde{\sigma}^2(\hat{\sigma}_\sharp)$ and $\hat{\sigma}_\flat^2 = \hat{\sigma}^2(\tilde{\sigma}_\flat)$.

We end this section with a lemma showing the relation between the penalized residual sum of squares and the inner product between response and residuals.

Lemma 3.3 *It holds that*

$$\tilde{\sigma}^2(\sigma) = Y^T(Y - X\hat{\beta}(\sigma))/n.$$

Proof of Lemma 3.3 We have

$$Y^T(Y - X\hat{\beta}(\sigma))/n = \|Y - X\hat{\beta}(\sigma)\|_n^2 + \hat{\beta}^T(\sigma)X^T(Y - X\hat{\beta}(\sigma))/n$$

and by the KKT-conditions [see (2.4)]

$$\hat{\beta}^T(\sigma)X^T(Y - X\hat{\beta}(\sigma))/n = \lambda_0 \sigma \|\hat{\beta}(\sigma)\|_1.$$

3.8   The Multivariate Square-Root Lasso

For our bounds for the bias of the Lasso in Chap. 4 and also for the construction of confidence sets in Chap. 5 we will consider the regression of X_J on X_{-J} (J being some subset of $\{1, \ldots, p\}$) invoking a multivariate version of the square-root Lasso. Here, we use here a standard notation with X being the input and Y being the response. We will then replace X by X_{-J} and Y by X_J in Sect. 4.5 (and Sect. 5.4).

The matrix X is as before an $n \times p$ input matrix and the response Y is now an $n \times q$ matrix for some $q \geq 1$. For a matrix A we write

$$\|A\|_1 := \sum_{j,k} |A_{j,k}|$$

and we denote its nuclear norm by

$$\|A\|_{\text{nuclear}} := \text{trace}((A^T A)^{1/2}).$$

We define the *multivariate square-root Lasso*

$$\hat{B} := \arg\min_B \left\{ \|Y - XB\|_{\text{nuclear}}/\sqrt{n} + \lambda_0 \|B\|_1 \right\} \tag{3.7}$$

with $\lambda_0 > 0$ again a tuning parameter. The minimization is over all $p \times q$ matrices B. We consider $\hat{\Sigma} := (Y - X\hat{B})^T (Y - X\hat{B})/n^1$ as estimator of the noise co-variance matrix.

The KKT-conditions for the multivariate square-root Lasso will be a major ingredient of the proof of the result in Theorem 5.3. We present these KKT-conditions in the following lemma in Eq. (3.8).

Lemma 3.4 *We have*

$$(\hat{B}, \hat{\Sigma}) = \arg\min_{B,\ \Sigma > 0} \left\{ \text{trace}\left((Y - XB)^T (Y - XB) \Sigma^{-1/2} \right)/n \right.$$

$$\left. + \text{trace}(\Sigma^{1/2}) + 2\lambda_0 \|B\|_1 \right\}$$

where the minimization is over all symmetric positive definite matrix Σ (this being denoted by $\Sigma > 0$) and where it is assumed that the minimum is indeed attained at some $\hat{\Sigma} > 0$. The multivariate Lasso satisfies the KKT-conditions

$$X^T (Y - X\hat{B}) \hat{\Sigma}^{-1/2}/n = \lambda_0 \hat{Z}, \tag{3.8}$$

where $\hat{Z}$ is a $p \times q$ matrix with $\|\hat{Z}\|_\infty \leq 1$ and with $\hat{Z}_{k,j} = \text{sign}(\hat{B}_{k,j})$ if $\hat{B}_{k,j} \neq 0$ $(k = 1, \ldots, p, j = 1, \ldots, q)$.

Proof of Lemma 3.4 Let us write, for a $p \times q$ matrix B, the residuals as $\Sigma_B := (Y - XB)^T (Y - XB)/n$. Let $\Sigma_{\min}(B)$ be the minimizer of

$$\text{trace}(\Sigma_B \Sigma^{-1/2}) + \text{trace}(\Sigma^{1/2}) \tag{3.9}$$

over Σ. Then $\Sigma_{\min}(B)$ equals Σ_B. To see this we invoke the reparametrization $\Omega := \Sigma^{-1/2}$ so that $\Sigma^{1/2} = \Omega^{-1}$. We now minimize

$$\text{trace}(\Sigma_B \Omega) + \text{trace}(\Omega^{-1})$$

---

[1]In this subsection $\hat{\Sigma}$ is not the Gram matrix $X^T X/n$.

over $\Omega > 0$. The matrix derivative with respect to Ω of trace($\Sigma_B\Omega$) is Σ_B. The matrix derivative of trace(Ω^{-1}) with respect to Ω is equal to $-\Omega^{-2}$. Hence the minimizer $\Omega_{\min}(B)$ satisfies the equation

$$\Sigma_B - \Omega_{\min}^{-2}(B) = 0,$$

giving

$$\Omega_{\min}(B) = \Sigma_B^{-1/2}.$$

so that

$$\Sigma_{\min}(B) = \Omega_{\min}^{-2}(B) = \Sigma_B.$$

Inserting this solution back in (3.9) gives 2trace($\Sigma_B^{1/2}$) which is equal to $2\|Y - XB\|_{\text{nuclear}}/\sqrt{n}$. This proves the first part of the lemma.

Let now for each $\Sigma > 0$, B_Σ be the minimizer of

$$\text{trace}(\Sigma_B\Sigma^{-1/2}) + 2\lambda_0\|B\|_1.$$

By sub-differential calculus we have

$$X^T(Y - XB_\Sigma)\Sigma^{-1/2}/n = \lambda_0 Z_\Sigma$$

where $\|Z_\Sigma\|_\infty \leq 1$ and $(Z_\Sigma)_{k,j} = \text{sign}((B_\Sigma)_{k,j})$ if $(B_\Sigma)_{k,j} \neq 0$ ($k = 1,\ldots,p$, $j = 1,\ldots q$). The KKT-conditions (3.8) follow from $\hat{B} = B_{\hat{\Sigma}}$.   □

Problems

3.1 Verify that the sub-differential of the ℓ_1-norm is

$$\partial\|\beta\|_1 = \{z \in \mathbb{R}^p : \|z\|_\infty \leq 1, z_j = \text{sign}(\beta_j) \text{ if } \beta_j \neq 0, j = 1,\ldots,p\}$$
$$= \{z \in \mathbb{R}^p : \|z\|_\infty \leq 1, z^T\beta = \|\beta\|_1\}, \beta \in \mathbb{R}^p.$$

Now comes a preparation for the next problem, something that can also be useful when studying the group Lasso (see Example 6.1 in Sect. 6.9). Let $\|\cdot\|_2$ be the ℓ_2-norm:

$$\|\beta\|_2^2 = \sum_{j=1}^p \beta_j^2, \beta \in \mathbb{R}^p.$$

Verify that its sub-differential is

$$\partial\|\beta\|_2 = \{z \in \mathbb{R}^p : \|z\|_2 \leq 1,\ z = \beta/\|\beta\|_2 \text{ if } \beta \neq 0\}.$$

3.2 Consider the sequence space model

$$\tilde{Y}_j = \beta_j^0 + \tilde{\epsilon}_j,\ j = 1,\ldots,n,$$

as in Problem 2.2. Let $\hat{\beta}$ be the square-root Lasso

$$\hat{\beta} := \arg\min_{\beta\in\mathbb{R}^p}\left\{ \|\tilde{Y} - \beta\|_n + \lambda_0\|\beta\|_1\right\}.$$

Use sub-differential calculus to find that $\hat{\beta}$ is a solution of the KKT-conditions

$$\hat{z}_2 + \lambda_0\hat{z}_1 = 0$$

where $\hat{z}_2 \in \partial\|\tilde{Y} - \hat{\beta}\|_2$ and $\hat{z}_1 \in \partial\|\hat{\beta}\|_1$. Assuming that all $\tilde{Y}_j$ are non-zero ($j = 1,\ldots,n$) show that if $\lambda_0 > 1/\sqrt{n}$ the square-root Lasso does not overfit: $\tilde{Y} - \hat{\beta} \neq 0$.

3.3 Verify Proposition 3.1 by following the steps of Theorem 2.2.

3.4 Use similar arguments as in Lemma 3.2 to bound higher order moments

$$\mathbb{E}\|\hat{\beta} - \beta^0\|_1^m,\ 1 < m < \infty.$$

Chapter 4
The Bias of the Lasso and Worst Possible Sub-directions

Abstract Bounds for the bias of the Lasso are derived. These bounds are based on so-called worst possible sub-directions or surrogate versions thereof. Both random design as well as fixed design is considered. In the fixed design case the bounds for groups of variables may be different than the ones for single variables due to a different choice of the surrogate inverse. An oracle inequality for subsets of the variables is presented, where it is assumed that the ℓ_1-operator norm of the worst possible sub-direction is small. It is shown that the latter corresponds to the irrepresentable condition. It is furthermore examined under what circumstances variables with small coefficients are de-selected by the Lasso. To explain the terminology "worst possible sub-direction", a section on the semi-parametric lower bound is added.

4.1 Introduction

Consider an $n \times p$ design matrix X and an n-vector Y of responses. We assume the linear model $\mathbb{E}(Y|X) = X\beta^0$, and we write the noise as $\epsilon = Y - X\beta^0$.

Recall the Lasso from (2.3)

$$\hat{\beta} := \arg\min\left\{ \|Y - X\beta\|_n^2 + 2\lambda \|\beta\|_1 \right\}.$$

In this chapter we study the bias of the Lasso, we establish oracle inequalities for subsets of variables and we examine the de-selection of variables. The KKT-conditions will play a key role. Recall that they read [see (2.4)]

$$X^T(Y - X\hat{\beta})/n = \lambda\hat{z} \tag{4.1}$$

with $\|\hat{z}\|_\infty \leq 1$ and $\hat{z}^T\hat{\beta} = 1$.

© Springer International Publishing Switzerland 2016
S. van de Geer, *Estimation and Testing Under Sparsity*,
Lecture Notes in Mathematics 2159, DOI 10.1007/978-3-319-32774-7_4

4.2 Matrix Algebra

As we will see later, in linear regression the worst possible sub-direction for estimating a coefficient, say β_j^0, is given by the coefficients of that part of the column space of X_j that is orthogonal to all other X_k, $k \neq j$ (see Definition 4.1 later on), that is, by the coefficients of the "anti-projection" of X_j, which is the residuals after projecting X_j on all the others. In the high-dimensional case, orthogonality is relaxed to "almost orthogonality".

Let us recall here some projection theory. We show the inverse of a symmetric positive definite matrix Σ_0 in terms of projections.

Let $X_0 \in \mathbb{R}^p$ be a random row-vector with distribution P. The inner-product matrix of X_0 is $\Sigma_0 := \mathbb{E}X_0^T X_0$. If $\mathbb{E}X_0 = 0$ the matrix Σ_0 is the co-variance matrix of X_0. (We will often refer to Σ_0 as co-variance matrix even when X_0 is not centered.) We assume that Σ_0 is invertible and write its inverse as $\Theta_0 := \Sigma_0^{-1}$. The matrix Θ_0 is called the *precision matrix* of X_0. Let $\| \cdot \|$ be the $L_2(P)$-norm.

For each $j \in \{1, \ldots, p\}$ we define $X_{-j,0}\gamma_j^0$ as the projection in $L_2(P)$ of $X_{j,0}$ on $X_{-j,0} := \{X_{k,0}\}_{k \neq j}$. Thus

$$\gamma_j^0 = \arg \min_{\gamma \in \mathbb{R}^{p-1}} \|X_{j,0} - X_{-j,0}\gamma\|.$$

This projection will alternatively be denoted as

$$X_{j,0}\mathrm{P}X_{-j,0} := X_{-j,0}\gamma_j^0$$

(with "P" standing for "projection"). Define further for all j

$$C_{k,j}^0 := \begin{cases} 1 & k = j \\ -\gamma_{k,j}^0 & k \neq j \end{cases}$$

and let $C_0 := (C_{k,j}^0)$. The columns of C_0 are written as $C_j^0, j = 1, \ldots, p$. In other words

$$X_{j,0} - X_{-j,0}\gamma_j^0 = X_0 C_j^0.$$

We call $X_0 C_j^0$ the *anti-projection* of $X_{j,0}$, or the vector of *residuals* and write it alternatively as

$$X_{j,0}\mathrm{A}X_{-j,0} := X_0 C_j^0$$

("A" standing for "anti-projection"). Thus

$$X_{j,0} = X_{j,0}\mathrm{P}X_{-j,0} + X_{j,0}\mathrm{A}X_{-j,0},$$

and by Pythagoras' theorem

$$\|X_{j,0}\|^2 = \|X_{j,0}PX_{-j,0}\|^2 + \|X_{j,0}AX_{-j,0}\|^2.$$

The squared length of the anti-projection or *residual variance* is denoted by

$$\tau_{j,0}^2 := \|X_0 C_j^0\|^2 = \|X_{j,0}AX_{-j,0}\|^2.$$

Then the precision matrix is $\Theta_0 := (\Theta_1^0, \ldots, \Theta_p^0)$ where the columns Θ_j^0 are equal to

$$\Theta_j^0 := C_j^0/\tau_{j,0}^2, \, j = 1, \ldots, p.$$

Note thus that

$$\Theta_{jj}^0 = \frac{1}{\tau_{j,0}^2} = \frac{1}{\|X_{j,0}AX_{-j,0}\|^2}, \, j = 1, \ldots, p.$$

Let now J be a group of variables, say $J = \{1, \ldots, |J|\}$. The multivariate version of the above gives an expression for the inverse when Σ_0 is partitioned into blocks, an upper left block of size $|J| \times |J|$, a lower right block of size $(p - |J|) \times (p - |J|)$ and off-diagonal blocks of sizes $|J| \times (p - |J|)$ and $(p - |J|) \times |J|$. We let

$$\Gamma_J^0 := \arg \min_{\Gamma_J} \text{trace}\left(\mathbb{E}(X_{J,0} - X_{-J,0}\Gamma_J)^T(X_{J,0} - X_{-J,0}\Gamma_J) \right)$$

where the minimization is over all $(p - |J|) \times |J|$-matrices. Then $X_{-J,0}\Gamma_J^0$ is the projection of $X_{J,0}$ on the space spanned by $X_{-J,0}$:

$$X_{-J,0}\Gamma_J^0 =: X_{J,0}PX_{-J,0}.$$

Set

$$C_J^0 := \begin{pmatrix} I \\ -\Gamma_J^0 \end{pmatrix},$$

where I is the identity matrix of dimension $|J| \times |J|$. The anti-projection (or residual matrix) is

$$X_0 C_J^0 := X_{J,0}AX_{-J,0}.$$

Let the residual co-variance matrix be

$$T_J^0 := \mathbb{E}(X_0 C_J^0)^T(X_0 C_J^0).$$

Then one may write

$$\Theta_0 = \begin{pmatrix} (T_J^0)^{-1} & -\Gamma_{-J}^0(T_{-J}^0)^{-1} \\ -\Gamma_J^0(T_J^0)^{-1} & (T_{-J}^0)^{-1} \end{pmatrix}.$$

4.3   Notation

Consider a matrix X with n rows and p columns and write $\hat{\Sigma} := X^T X/n$. The columns of X are denoted by $(X_1, \ldots, X_p)$. We let for a vector $v \in \mathbb{R}^n$ the normalized Euclidean norm be $\|v\|_n := \sqrt{v^T v/n}$.

For a matrix A we denote its ℓ_∞-norm by $\|A\|_\infty := \max_{kj} |A_{kj}|$. We define the ℓ_1-operator norm

$$\|\|A\|\|_1 := \max_j \sum_k |A_{kj}|.$$

For matrices A and B a dual norm inequality is

$$\|AB\|_\infty \leq \|A\|_\infty \|\|B\|\|_1.$$

The *nuclear norm* (or *trace norm*) of a matrix A is

$$\|A\|_{\text{nuclear}} := \text{trace}((A^T A)^{1/2}).$$

(The dual norm of $\|\cdot\|_{\text{nuclear}}$ is $\Lambda_{\max}(\cdot)$ where $\Lambda_{\max}^2(A)$ is the largest eigenvalue of the matrix $A^T A$ (Watson 1992).

If $Z \in \mathbb{R}^q$ (for some $q \geq 1$) is a random vector which is standard multivariate normal this is denoted by $Z = \mathcal{N}_q(0, I)$ or $Z \sim \mathcal{N}_q(0, 1)$.

Asymptotics   To clarify the results we present asymptotic statements ($n \to \infty$) at places in this and subsequent chapters. For a sequence $z_n \in \mathbb{R}$ we write $z_n = \mathcal{O}(1)$ if $\limsup_{n\to\infty} |z_n| < \infty$. We write $z_n \asymp 1$ if both $z_n = \mathcal{O}(1)$ and $1/z_n = \mathcal{O}(1)$. We write $z_n = o(1)$ if $\lim_{n\to\infty} z_n = 0$.

4.4   Bias and Sup-Norm Bounds for the Lasso with Random Design

In this section a bound is shown for the bias of the Lasso for the case of random design. The bias for fixed design is similar but the fact that we then need to choose a *surrogate* inverse conceals the main argument: see Sect. 4.5 (and also Sect. 5.1).

We view the matrix X as being random and assume that its rows are i.i.d. copies of a random row vector X_0 with distribution P. We write $\Sigma_0 := \mathbb{E} X_0^T X_0 = \mathbb{E} \hat{\Sigma}$. Let $\Theta_0 := \Sigma_0^{-1} := (\Theta_1^0, \ldots, \Theta_p^0)$. Define $\tau_{j,0}^2 := 1/\Theta_{j,j}^0$.

Definition 4.1 For the case of random design we call $C_j^0 := \Theta_j^0 \tau_{j,0}^2$ the *worst possible sub-direction* for estimating β_j^0 $(j = 1, \ldots, p)$.

See Sect. 4.9 for a motivation of the terminology used in Definition 4.1.

Lemma 4.1 *We have*

$$\underbrace{\left\| \mathbb{E}(\hat{\beta} - \beta^0) \right\|_\infty}_{bias} \leq \|\!|\Theta_0|\!\|_1 \left[\mathbb{E} \left(\|\hat{\Sigma} - \Sigma_0\|_\infty \|\hat{\beta} - \beta^0\|_1 \right) + \lambda \right].$$

Furthermore for $\lambda_\epsilon \geq \|X^T \epsilon\|_\infty / n$,

$$\|\hat{\beta} - \beta^0\|_\infty \leq \|\!|\Theta_0|\!\|_1 \left[\lambda + \lambda_\epsilon + \|\hat{\Sigma} - \Sigma_0\|_\infty \|\hat{\beta} - \beta^0\|_1 \right].$$

In the lemma we applied the bound $\|\Theta_0(\hat{\Sigma} - \Sigma_0)\|_\infty \leq \|\!|\Theta_0|\!\|_1 \|\hat{\Sigma} - \Sigma_0\|_\infty$ although one may prefer to bound $\|\Theta_0(\hat{\Sigma} - \Sigma_0)\|_\infty$ directly as maximum of mean-zero random variables (see also Problem 14.2).

The Proof of Lemma 4.1 makes use of the KKT-conditions which are in essence saying that at the minimum, the derivative is zero (with sub-differential calculus obscuring this somewhat). We then multiply KKT by Θ_j^{0T}. This step corresponds to looking—for $t \in \mathbb{R}$—at the derivative in the sub-direction $\hat{\beta} + t\Theta_j^0$ at $t = 0$ and this explains why we call $C_j^0 = \Theta_j^0 \tau_{j,0}^2$ a "sub-direction". The scaling with $\tau_{j,0}^2$ in this terminology is rather arbitrary.

Proof of Lemma 4.1 By the KKT-conditions [see (4.1)]

$$\hat{\Sigma}(\hat{\beta} - \beta^0) + \lambda \hat{z} = X^T \epsilon / n$$

where $\hat{z}^T \hat{\beta} = \|\hat{\beta}\|_1$ and $\|\hat{z}\|_\infty \leq 1$. It follows that

$$\Theta_j^{0T} \hat{\Sigma}(\hat{\beta} - \beta^0) + \Theta_j^{0T} \lambda \hat{z} = \Theta_j^{0T} X^T \epsilon / n.$$

But

$$\Theta_j^{0T} \Sigma_0 (\hat{\beta} - \beta^0) = \hat{\beta}_j - \beta_j^0.$$

So we find

$$\hat{\beta}_j - \beta_j^0 = -\Theta_j^{0T}(\hat{\Sigma} - \Sigma_0)(\hat{\beta} - \beta^0) - \Theta_j^{0T} \lambda \hat{z} + \Theta_j^{0T} X^T \epsilon / n$$

or

$$\hat{\beta} - \beta^0 = -\Theta_0(\hat{\Sigma} - \Sigma_0)(\hat{\beta} - \beta^0) - \lambda\Theta_0\hat{z} + \Theta_0 X^T \epsilon/n.$$

Since $\|\hat{z}\|_\infty \le 1$ we have $\|\Theta_0\hat{z}\|_\infty \le \|\Theta_0\|_1$. Moreover

$$\|\Theta_0(\hat{\Sigma} - \Sigma_0)(\hat{\beta} - \beta^0)\|_\infty \le \|\Theta_0\|_1 \|(\hat{\Sigma} - \Sigma_0)(\hat{\beta} - \beta^0)\|_\infty$$

$$\le \|\Theta_0\|_1 \|\hat{\Sigma} - \Sigma_0\|_\infty \|\hat{\beta} - \beta^0\|_1.$$

$\square$

Asymptotics Assume that $\lambda \asymp \sqrt{\log p/n}$, that $\mathbb{E}\|\hat{\Sigma} - \Sigma_0\|_\infty^2 = \mathcal{O}(\log p/n)$ and that $\mathbb{E}\|\hat{\beta} - \beta^0\|_1^2$ converges to zero (see also Sect. 3.6). Then

$$\left\|\mathbb{E}(\hat{\beta} - \beta^0)\right\|_\infty \le \lambda\|\Theta_0\|_1(1 + o(1)) \asymp \sqrt{\log p/n}\|\Theta_0\|_1.$$

Hence when $\|\Theta_0\|_1 = \mathcal{O}(1)$ then the bias is $\mathcal{O}(\sqrt{\log p/n})$ uniformly for the components. The condition $\|\Theta_0\|_1 = \mathcal{O}(1)$ may be compared with the irrepresentable condition defined in Sect. 4.7, see Problem 4.1.

Example 4.1 (Equal Correlation) Let $0 \le \rho < 1$ and

$$\Sigma_0 := \begin{pmatrix} 1 & \rho & \cdots & \rho \\ \rho & 1 & \cdots & \rho \\ \vdots & \vdots & \ddots & \vdots \\ \rho & \rho & \cdots & 1 \end{pmatrix} = (1 - \rho)I + \rho\iota\iota^T, \quad \iota := \begin{pmatrix} 1 \\ 1 \\ \vdots \\ 1 \end{pmatrix}.$$

Then

$$\Theta_0 = \frac{1}{1 - \rho}\left\{ I - \frac{\rho\iota\iota^T}{1 - \rho + p\rho} \right\}$$

and

$$\|\Theta_0\|_1 = \frac{1}{1 - \rho}\left\{ \frac{1 + (2p - 3)\rho}{1 + (p - 1)\rho} \right\} \le \frac{2}{1 - \rho}.$$

Hence when ρ stays away from 1 the bias of the Lasso in the above asymptotic setup is $\mathcal{O}(\sqrt{\log p/n})$ and also $\|\hat{\beta} - \beta^0\|_\infty = \mathcal{O}_{\mathbb{P}}(\sqrt{\log p/n})$. Note that we reached this result without the "incoherence conditions" imposed in Lounici (2008), but with the price of assuming consistency in ℓ_1-norm.

4.5 Bias and Sup-Norm Bounds for the Lasso with Fixed Design

In the rest of this chapter, but with the exception of the last section on a semi-parametric lower bound, the matrix X is a fixed $n \times p$ matrix with columns X_j $(j = 1, \ldots, p)$. If $p > n$ then necessarily the Gram matrix $\hat{\Sigma}$ is singular. In fact, then (when all X_j are distinct, $j = 1, \ldots, p$) for any j the projection of $X_j \in \mathbb{R}^n$ on the column space of X_{-j} is X_j itself, as X_j is some linear combination of the vectors in X_{-j}. Therefore, we need to do some surrogate type of projection. Here, the surrogate proposed is the square-root Lasso. The reason for applying the square-root Lasso (and not for example the standard Lasso) as method for obtaining surrogate projections and inverses is because it gives in the next chapter the right the scaling: see Theorem 5.3 in Sect. 5.4. This section can be seen as a preparation for Sect. 5.4.

We consider surrogate projections for groups. Consider some subset $J \subset \{1, \ldots, p\}$. We have in mind the case where J is "small". Let $\hat{\Gamma}_J$ be the multivariate square-root Lasso (see Sect. 3.8)

$$\hat{\Gamma}_J := \arg\min_{\Gamma_J} \left\{ \|X_J - X_{-J}\Gamma_J\|_{\text{nuclear}}/\sqrt{n} + \lambda_{\sharp}\|\Gamma_J\|_1 \right\}$$

where $\lambda_{\sharp}$ is a tuning parameter. The surrogate projection is

$$X_J \hat{P} X_{-J} := X_{-J}\hat{\Gamma}_J$$

and the surrogate anti-projection is

$$X_J \hat{A} X_{-J} := X_J - X_J \hat{P} X_{-J} := X\hat{C}_J.$$

Define

$$\tilde{T}_J := (X_J \hat{A} X_{-J})^T X_J / n$$
$$= (X_J - X_{-J}\hat{\Gamma}_J)^T X_J / n,$$

and

$$\hat{T}_J := (X_J \hat{A} X_{-J})^T (X_J \hat{A} X_{-J}) / n$$
$$= (X_J - X_{-J}\hat{\Gamma}_J)^T (X_J - X_{-J}\hat{\Gamma}_J) / n.$$

We assume that the inverse $\tilde{T}_J^{-1}$ exists. This means indeed that J has to be "small". Write

$$\hat{\Theta}_J := \hat{C}_J \tilde{T}_J^{-1} = \{\hat{\Theta}_j\}_{j \in J}.$$

One may think of $\hat{\Theta}_j$ as the jth column of a surrogate inverse for the matrix $\hat{\Sigma}$ (only defined now for $j \in J$).

Definition 4.2 For the case of fixed design, we call $\hat{C}_J$ the *surrogate worst possible sub-direction*—based on the multivariate square-root Lasso—for estimating β_J^0.

The KKT-conditions for the multivariate square-root Lasso (see Lemma 3.4) give

$$X_{-J}^T(X_J - X_{-J}\hat{\Gamma}_J)\hat{T}_J^{-1/2}/n = \lambda_\sharp \hat{Z}_J,$$

where $\hat{Z}_J$ is a $(p - |J|) \times |J|$ matrix with $\|\hat{Z}_J\|_\infty \leq 1$.

Lemma 4.2 *We have*

$$\| \mathbb{E}\hat{\beta}_J - \beta_J^0\|_\infty \leq \lambda_\sharp \|\!|\hat{T}_J^{1/2}\tilde{T}_J^{-1}\|\!|_1 \mathbb{E}\|\hat{\beta}_{-J} - \beta_{-J}^0\|_1 + \lambda\|\!|\hat{\Theta}_J\|\!|_1.$$

Furthermore, for $\lambda_\epsilon \geq \|X^T\epsilon\|_\infty/n$,

$$\|\hat{\beta}_J - \beta_J^0\|_\infty \leq \lambda_\sharp \|\!|\hat{T}_J^{1/2}\tilde{T}_J^{-1}\|\!|_1 \|\beta_{-J} - \beta_{-J}^0\|_1 + (\lambda + \lambda_\epsilon)\|\!|\hat{\Theta}_J\|\!|_1.$$

Proof of Lemma 4.2 By the KKT-conditions for the Lasso

$$-\hat{\Sigma}(\hat{\beta} - \beta^0) + X^T\epsilon/n = \lambda\hat{z}$$

where $\hat{z}$ is a p-vector satisfying $\|\hat{z}\|_\infty \leq 1$. Multiplying both sides by $\hat{\Theta}_J^T = \tilde{T}_J^{-T}\hat{C}_J^T$ gives

$$-\hat{\Theta}_J^T\hat{\Sigma}(\hat{\beta} - \beta^0) + \hat{\Theta}_J^T X^T\epsilon/n = \lambda\hat{\Theta}_J^T\hat{z}.$$

For the first term on the left, we have

$$\hat{\Theta}_J^T\hat{\Sigma}(\hat{\beta} - \beta^0) = \tilde{T}_J^{-T}\hat{C}_J^T X^T X_J(\hat{\beta}_J - \beta_J^0) + \tilde{T}_J^{-T}\hat{C}_J^T X^T X_{-J}(\hat{\beta}_{-J} - \beta_{-J}^0)$$

$$= \hat{\beta}_J - \beta_J^0 + \lambda_\sharp \tilde{T}_J^{-T}\hat{T}_J^{1/2}\hat{Z}_J^T(\hat{\beta}_{-J} - \beta_{-J}^0)$$

and

$$\|\tilde{T}_J^{-T}\hat{T}_J^{1/2}\hat{Z}_J^T(\hat{\beta}_{-J} - \beta_{-J}^0)\|_\infty \leq \|\!|\hat{T}_J^{1/2}\tilde{T}_J^{-1}\|\!|_1 \|\hat{Z}_J^T(\hat{\beta}_{-J} - \beta_{-J}^0)\|_\infty$$

$$\leq \|\!|\hat{T}_J^{1/2}\tilde{T}_J^{-1}\|\!|_1 \|\hat{\beta}_{-J} - \beta_{-J}^0\|_1$$

since $\|\hat{Z}_J\|_\infty \leq 1$. The second term on the left has mean zero and can moreover be bounded by

$$\|\hat{\Theta}_J^T X^T\epsilon/n\|_\infty \leq \lambda_\epsilon\|\!|\hat{\Theta}_J\|\!|_1.$$

The term on the right an be bounded by

$$\lambda \|\hat{\Theta}_j^T \hat{z}\|_\infty \le \lambda \|\hat{\Theta}_j\|_1$$

since $\|\hat{z}\|_\infty \le 1$.   □

Corollary 4.1 *In the special case of J being a single variable, say $J = \{j\}$, we find*

$$\left| \mathbb{E}\hat{\beta}_j - \beta_j^0 \right| \le \frac{\hat{\tau}_j}{\hat{\tau}_j^2} \|\hat{\beta}_{-j} - \beta_{-j}^0\|_1 + \lambda \|\hat{\Theta}_j\|_1,$$

with $\hat{\tau}_j^2 = \|X_j - \hat{\gamma}_j\|_n^2$ and $\tilde{\tau}_j^2 = \hat{\tau}_j^2 + \lambda_\sharp \hat{\tau}_j \|\hat{\gamma}_j\|_1$ (see also Sect. 5.2).

4.6   Oracle Inequalities for Groups

We now examine the bias for "larger" groups J. Because for larger groups the matrices $\tilde{T}_J$ and $\hat{T}_J$ obtained from the multivariate square-root Lasso may not be invertible, we replace it by the "standard" Lasso. We use the same notation as in the previous section but now for the standard Lasso instead of the multivariate square-root Lasso. Thus we let

$$\hat{\Gamma}_J := \arg \min_{\Gamma_J} \left\{ \text{trace}\left((X_J - X_{-J}\Gamma_J)^T (X_J - X_{-J}\Gamma_J) \right)/n + \lambda_\diamond \|\Gamma_J\|_1 \right\}$$

with $\lambda_\diamond \ge 0$ a tuning parameter. The surrogate projection is

$$X_J \hat{P} X_{-J} := X_{-J}\hat{\Gamma}_J$$

with surrogate anti-projection

$$X_J \hat{A} X_{-J} := X_J - X_J \hat{P} X_{-J} := X\hat{C}_J.$$

We let

$$\hat{T}_J := (X\hat{C}_J)^T (X\hat{C}_J)/n.$$

Note that

$$\beta_J^T \hat{T}_J \beta_J = \|(X_J \hat{A} X_{-J})\beta_J\|_n^2, \ \beta_J \in \mathbb{R}^{|J|}.$$

Definition 4.3 For the case of fixed design, we call $\hat{C}_J$ the *surrogate worst possible sub-direction*—based on the standard Lasso—for estimating β_J^0.

We now define the compatibility constant by only looking at that part of X_J that is "almost orthogonal" to X_{-J}.

Definition 4.4 Define, for $S \subset J$, the *compatibility constant restricted to the set J* as

$$\hat{\phi}^2(L, S, J) := \min\left\{ |S| \|(X_J \hat{A} X_{-J}) \beta_J\|_n^2 : \|\beta_{J\backslash S}\|_1 \le L, \ \|\beta_S\|_1 = 1 \right\}.$$

Note that

$$\hat{\phi}^2(L, S, J) \ge \hat{\Lambda}_{\min}(\hat{T}_J),$$

where $\hat{\Lambda}_{\min}(\hat{T}_J)$ is the smallest eigenvalue of the matrix $\hat{T}_J$. However, for large groups this smallest eigenvalue may be zero. It is therefore of interest to relate $\hat{\phi}^2(L, S, J)$ with other already known quantities. The next lemma relates it with the standard compatibility constant $\hat{\phi}^2(\tilde{L}, S)$ for a suitable value of the stretching factor $\tilde{L}$.

Lemma 4.3 *For all L and all $S \subset J$*

$$\hat{\phi}^2(L, S, J) \ge \hat{\phi}^2\left((L+1)\|\!|\hat{C}_J|\!\|_1 - 1, S \right).$$

Proof of Lemma 4.3 By the definitions, for $\beta_J \in \mathbb{R}^{|J|}$,

$$\|(X_J \hat{A} X_{-J})\beta_J\|_n^2 = \|X \hat{C}_J \beta_J\|_n^2 = \|X_J \beta_J - X_{-J} \hat{\Gamma}_J \beta_J\|_n^2$$
$$= \|X_S \beta_S + X_{J\backslash S} \beta_{J\backslash S} - X_{-J} \hat{\Gamma}_J \beta_J\|_n^2.$$

If $\|\beta_S\|_1 = 1$ and $\|\beta_{J\backslash S}\|_1 \le L$ we have

$$\|\beta_{J\backslash S} - \hat{\Gamma}_J \beta_J\|_1 \le L + \|\!|\hat{\Gamma}_J|\!\|_1(L+1) = (L+1)\|\!|\hat{C}_J|\!\|_1 - 1.$$

$\square$

In the following theorem we derive rates for groups assuming $\|\!|\hat{\Gamma}_J|\!\|_1 < 1$ is small enough. For large groups this corresponds to the so-called irrepresentable condition: see Sect. 4.7 for more details.

We make use of the KKT-conditions for the surrogate projections. For the standard Lasso, they read

$$X_{-J}^T(X_J - X_{-J}\hat{\Gamma}_J)/n = \lambda_\circ \hat{Z}_J$$

where $\hat{Z}_J$ is a $(p - |J|) \times |J|$ matrix with $\|\hat{Z}_J\|_\infty \le 1$ and with $(\hat{Z}_J)_{k,j} = \text{sign}((\hat{\Gamma}_J)_{k,j})$ if $(\hat{\Gamma}_J)_{k,j} \ne 0$.

Theorem 4.1 *Let*

$$\lambda_\epsilon \geq \|X^T\epsilon\|_\infty/n.$$

Define

$$\lambda_J := (\lambda + \lambda_\epsilon + \lambda_\circ\|\hat\beta - \beta^0\|_1)\|\|\hat{C}_J\|\|_1 - \lambda.$$

Suppose that $\lambda > \lambda_J$. *Define for some* $0 \leq \delta < 1$

$$\underline{\lambda} := \lambda - \lambda_J, \ \bar{\lambda} := \lambda + \lambda_J + \delta\underline{\lambda}$$

and

$$L := \frac{\bar{\lambda}}{(1-\delta)\underline{\lambda}}.$$

Then for all β_J *and* $S \subset J$

$$2\delta\underline{\lambda}\|\hat\beta_J - \beta_J\|_1 + \|(X_J\hat{A}X_{-J})(\hat\beta_J - \beta_J^0)\|_n^2$$

$$\leq \|(X_J\hat{A}X_{-J})(\hat\beta_J - \beta_J^0)\|_n^2 + \frac{\bar{\lambda}^2|S|}{\hat\phi^2(L,S,J)} + 4\lambda\|\beta_{J\setminus S}\|_1.$$

On may optimize the result over all candidate oracles (β_J, S) as in Theorem 2.2 (but now with $S \subset J$). If one takes $J = \{1,\ldots,p\}$ and $\lambda_\circ = 0$ in Theorem 4.1, one sees that $\|\|C_J\|\|_1 = 1$ so that $\lambda_J = \lambda_\epsilon$ and $\hat\phi^2(L,S,J) = \hat\phi^2(L,S)$ (see Lemma 4.3). Hence then Theorem 4.1 reduces to Theorem 2.2.

The theorem shows that by taking $S \subset J$ and β_J in a trade-off fashion one gets adaptive rates over groups which may be faster than the global rate.

Proof of Theorem 4.1 We start again with the KKT-conditions for the Lasso:

$$\hat\Sigma(\hat\beta - \beta^0) = X^T\epsilon/n - \lambda\hat{z}$$

where $\hat{z}$ is a p-vector satisfying $\|\hat{z}\|_\infty \leq 1$ and $\hat{z}^T\hat\beta = \|\hat\beta\|_1$. Now multiply both sides with $(\hat\beta_J - \beta_J)^T\hat{C}_J$ to find

$$(\hat\beta_J - \beta_J)^T\hat{T}_J(\hat\beta_J - \beta_J^0) = (\hat\beta_J - \beta_J)^T\hat{C}_J^TX^T\epsilon/n - \lambda_\circ(\hat\beta_J - \beta_J)^T\hat{Z}_J^T(\hat\beta_{-J} - \beta_{-J}^0)$$

$$-\lambda(\hat\beta_J - \beta_J)^T\hat{C}_J^T\hat{z} + (\hat\beta_J - \beta_J)^T\lambda_\circ\hat{Z}_J^T\hat{\Gamma}_J(\hat\beta_J - \beta_J^0).$$

It holds that

$$|(\hat{\beta}_J - \beta_J)^T \hat{C}_J^T X^T \epsilon/n| \leq \|\hat{C}_J(\hat{\beta}_J - \beta_J)\|_1 \|X^T \epsilon\|_\infty/n$$
$$\leq \lambda_\epsilon \|\|\hat{C}_J\|\|_1 \|\hat{\beta}_J - \beta_J\|_1.$$

Moreover

$$\lambda_\diamond (\hat{\beta}_J - \beta_J)^T \hat{Z}_J^T (\hat{\beta}_{-j} - \beta^0_{-j}) \leq \lambda_\diamond \|\hat{\beta}_J - \beta_J\|_1 \|\hat{Z}_J^T (\hat{\beta}_{-J} - \beta^0_{-J})\|_\infty$$
$$\leq \lambda_\diamond \|\hat{\beta}^0_{-J} - \beta^0_{-J}\|_1 \|\hat{\beta}_J - \beta_J\|_1$$
$$\leq \lambda_\diamond \|\hat{\beta} - \beta^0\|_1 \|\hat{\beta}_J - \beta_J\|_1$$

and

$$\lambda_\diamond |(\hat{\beta}_J - \beta_J)^T \hat{Z}_J^T \hat{\Gamma}_J (\hat{\beta}_J - \beta^0_J)| \leq \lambda_\diamond \|\hat{\beta}_J - \beta_J\|_1 \|\|\hat{\Gamma}_J\|\|_1 \|\hat{\beta}_J - \beta^0_J\|_1$$
$$\leq \lambda_\diamond \|\hat{\beta} - \beta^0\|_1 \|\|\hat{\Gamma}_J\|\|_1 \|\hat{\beta}_J - \beta_J\|_1.$$

Finally

$$\lambda(\hat{\beta}_J - \beta_J)^T \hat{C}_J^T \hat{z} \geq \lambda \|\hat{\beta}_J\|_1 - \lambda\|\beta_J\|_1 - \lambda \|\|\hat{\Gamma}_J\|\|_1 \|\hat{\beta}_J - \beta_J\|_1.$$

Hence we have the inequality

$$(\hat{\beta}_J - \beta_J)^T \hat{T}_J (\hat{\beta}_J - \beta^0_J) \leq \lambda_\epsilon \|\|\hat{C}_J\|\|_1 \|\hat{\beta}_J - \beta_J\|_1 + \lambda_\diamond \|\hat{\beta} - \beta^0\|_1 \|\hat{\beta}_J - \beta_J\|_1$$
$$+ \lambda_\diamond \|\hat{\beta} - \beta^0\|_1 \|\|\hat{\Gamma}_J\|\|_1 \|\hat{\beta}_J - \beta_J\|_1 - \lambda\|\hat{\beta}_J\|_1$$
$$+ \lambda\|\beta_J\|_1 + \lambda \|\|\hat{\Gamma}_J\|\|_1 \|\hat{\beta}_J - \beta_J\|_1$$
$$= \lambda_J \|\hat{\beta}_J - \beta_J\|_1 - \lambda\|\hat{\beta}_J\|_1 + \lambda\|\beta_J\|_1$$

or

$$(\hat{\beta}_J - \beta_J)^T \hat{T}_J (\hat{\beta}_J - \beta^0_J) \leq \lambda_J \|\hat{\beta}_J - \beta_J\|_1 - \lambda\|\hat{\beta}_J\|_1 + \lambda\|\beta_J\|_1. \qquad (4.2)$$

The result follows now by the same arguments as used for the Proof of Theorem 2.2.
                                                                                    □

We end this section with the counterpart of Lemma 4.2 for the current case where matrices like $\hat{T}_J$ and $\tilde{T}_J$ may not be invertible. (Problem 4.2 has the situation where they *are* invertible.) In contrast to Theorem 4.1 we do not assume here that $\|\|\hat{\Gamma}_J\|\|_1$ is small.

Lemma 4.4 *We have for $\lambda_\epsilon \geq \|X^T\epsilon\|_\infty/n$ the inequality*

$$\|(X_J\hat{A}X_{-J})(\hat{\beta}_J - \beta_J^0)\|_n^2 \leq \left(\lambda + \lambda_\epsilon + \lambda_\circ\|\hat{\beta} - \beta^0\|_1\right)^2 \frac{|J|\,\|\hat{C}_J\|_1^2}{\hat{\phi}^2(\|\hat{\Gamma}_J\|_1, J)}.$$

Proof of Lemma 4.4 Define

$$\lambda_J := (\lambda + \lambda_\epsilon + \lambda_\circ\|\hat{\beta} - \beta^0\|_1)\|\hat{C}_J\|_1 - \lambda.$$

The result follows from Eq. (4.2) inserting the value $\beta_J = \beta_J^0$, then applying the triangle inequality and then using the definition of the restricted compatibility constant:

$$\|(X_J\hat{A}X_{-J})(\hat{\beta}_J - \beta_J^0)\|_n^2 \leq (\lambda_J + \lambda)\|\hat{\beta}_J - \beta_J^0\|_1$$

$$\leq \frac{(\lambda_J + \lambda)\sqrt{J}}{\hat{\phi}(0, J, J)}\|(X_J\hat{A}X_{-J})(\hat{\beta}_J - \beta_J^0)\|_n.$$

The restricted compatibility constant $\hat{\phi}^2(0, J, J)$ can be bounded by $\hat{\phi}^2(\|\hat{\Gamma}_J\|_1, J)$. $\qquad\square$

4.7 Worst Possible Sub-directions and the Irrepresentable Condition

Consider now the situation where J is a rather large group, say $J = S_*^c$ where X_{S_*} has rank $|S_*|$. Then we can choose $\lambda_\circ = 0$ to find

$$\hat{\Gamma}_{-S_*} = (X_{S_*}^T X_{S_*})^{-1} X_{S_*}^T X_{-S_*},$$

and

$$\|\hat{\Gamma}_{-S_*}\|_1 = \|(X_{S_*}^T X_{S_*})^{-1} X_{S_*}^T X_{-S_*}\|_1$$

$$= \max_{j \notin S_*} \max_{\|z_{S_*}\|_\infty \leq 1} |z_{S_*}^T (X_{S_*}^T X_{S_*})^{-1} X_{S_*}^T X_j|.$$

The condition $\|\hat{\Gamma}_{-S_*}\|_1 < 1$ is called an *irrepresentable condition* on the set S_*, see (Bühlmann and van de Geer 2011, Sect. 7.5.1) and its references. The condition $\lambda > \lambda_J$ of Theorem 4.1 reads

$$\|\hat{\Gamma}_{-S_*}\|_1 < \frac{\lambda - \lambda_\epsilon}{\lambda + \lambda_\epsilon}.$$

We thus arrive at the following lemma which reproves the result in for example Bühlmann and van de Geer (2011, Problem 7.5).

Lemma 4.5 Let $\hat{\Gamma}_{-S_0} := (X_{S_0}^T X_{S_0})^{-1} X_{S_0}^T X_{-S_0}$. Suppose for some $\lambda_\epsilon \geq \|X^T \epsilon\|_\infty / n$ the irrepresentable condition

$$\|\hat{\Gamma}_{-S_0}\|_1 < \frac{\lambda - \lambda_\epsilon}{\lambda + \lambda_\epsilon}.$$

Then we have no false alarms: $\hat{\beta}_{-S_0} = 0$.

Proof of Lemma 4.5 Take $J = S_0^c$ in Theorem 4.1 and $S = \emptyset$, $\beta_J = \beta_{-S_0}^0 = 0$. Then we find

$$2\delta\underline{\lambda}\|\hat{\beta}_{-S_0}\|_1 + \|(X_{-S_0}\hat{A}X_{S_0})\hat{\beta}_{-S_0}\|_n^2 \leq 0$$

and hence $\hat{\beta}_{-S_0} = 0$. □

Remark 4.1 Let $\hat{\Gamma}_{-S_0}$ be defined as in Lemma 4.5. Then for $L \leq 1/\|\hat{\Gamma}_{-S_0}\|_1$

$$\hat{\phi}^2(L, S_0) \geq (1 - L\|\hat{\Gamma}_{-S_0}\|_1)\Lambda_{\min}(X_{S_0}^T X_{S_0})/n$$

where $\Lambda_{\min}(X_{S_0}^T X_{S_0})$ is the smallest eigenvalue of the matrix $X_{S_0}^T X_{S_0}$: see van de Geer and Bühlmann (2009).

4.8   De-selection of Variables in the Approximately Sparse Case

In Lemma 4.5 it was shown that under the irrepresentable condition on S_0 there are no false positives, i.e. the Lasso puts $\hat{\beta}_j$ to zero when j is not in the active set $S_0 = \{j : \beta_j^0 \neq 0\}$. But what if certain β_j^0 are very small? Then it may be desirable that the Lasso estimates these as being zero as well. This issue is the theme of the next lemma. It is an extension of Lemma 4.5: the coefficients β_j^0 are assumed to be small within a certain set S^c and under appropriate conditions the Lasso puts these coefficients to zero. We require that $S \subset \{1,\ldots,p\}$ is a set where X_S has rank S, define $\hat{\Gamma}_{-S} := (X_S^T X_S)^{-1} X_S^T X_{-S}$ and define the projection in $\mathbb{R}^n$

$$X_{-S}\hat{P}X_S := X_S\hat{\Gamma}_{-S} = X_S(X_S^T X_S)^{-1}X_S^T X_{-S}$$

and the anti-projection

$$X_{-S}\hat{A}X_S := X_{-S} - X_{-S}\hat{P}X_S.$$

Let

$$\hat{T}_{-S} := (X_{-S}\hat{A}X_S)^T(X_{-S}\hat{A}X_S)/n.$$

Lemma 4.6 *If*

$$(\lambda + \lambda_\epsilon)\|\|\hat{\Gamma}_{-S}\|\|_1 + \|\hat{T}_{-S}\beta^0_{-S}\|_\infty \leq \lambda - \lambda_\epsilon$$

then there is a solution $\hat{\beta}$ of the KKT-conditions (4.1) such that $\hat{\beta}_{-S} = 0$.

Proof of Lemma 4.6 Let $\tilde{\beta}_S$ be a solution of the KKT-conditions

$$X_S^T(Y - X_S\tilde{\beta}_S)/n = \lambda\tilde{z}_S,$$

where $\|\tilde{z}_S\|_\infty \leq 1$ and $\tilde{z}_S^T\tilde{\beta}_S = \|\tilde{\beta}_S\|_1$.

We now show that $\|X_{-S}^T(Y - X_S\tilde{\beta}_S)/n\|_\infty \leq \lambda$. We have

$$\tilde{\beta}_S = (X_S^TX_S)^{-1}(X_S^TY) - \lambda(X_S^TX_S/n)^{-1}\tilde{z}_S$$

and hence

$$X_{-S}^TX_S\tilde{\beta}_S/n = (X_{-S}\hat{P}X_S)^TY/n - \lambda\hat{\Gamma}_{-S}^T\tilde{z}_S.$$

Then

$$X_{-S}^T(Y - X_S\tilde{\beta}_S)/n = (X_{-S}\hat{A}X_S)^TY/n - \lambda\hat{\Gamma}_{-S}^T\tilde{z}_S$$

$$= (X_{-S}\hat{A}X_S)^T\epsilon/n - \hat{T}_{-S}\beta^0_{-S} - \lambda\hat{\Gamma}_{-S}^T\tilde{z}_S.$$

Now

$$\|(X_{-S}\hat{A}X_S)^T\epsilon\|_\infty/n \leq \|X_{-S}^T\epsilon\|_\infty/n + \|(X_{-S}\hat{P}X_S)^T\epsilon\|_\infty/n$$

$$\leq \|X_{-S}^T\epsilon\|_\infty/n + \|\|\hat{\Gamma}_{-S}\|\|_1\|X_S^T\epsilon\|_\infty/n$$

$$\leq \lambda_\epsilon + \lambda_\epsilon\|\|\hat{\Gamma}_{-S}\|\|_1.$$

Moreover

$$\lambda\|\hat{\Gamma}_{-S}^T\tilde{z}_S\|_\infty \leq \lambda\|\|\hat{\Gamma}_{-S}\|\|_1$$

since $\|\tilde{z}_S\|_\infty \leq 1$. Hence

$$\|X_{-S}^T(Y - X\tilde{\beta}_S)\|_\infty/n \leq \lambda_\epsilon + (\lambda + \lambda_\epsilon)\|\|\hat{\Gamma}_{-S}\|\|_1 + \|\hat{T}_{-S}\beta^0_{-S}\|_\infty \leq \lambda$$

by the assumption of the lemma. So if we define

$$\lambda \tilde{z}_{-S} := X_{-S}^T (Y - X_S \tilde{\beta}_S)/n.$$

we know that $\|\tilde{z}_{-S}\|_\infty \leq 1$. It follows that $\tilde{\beta} := (\tilde{\beta}_S, 0)^T$ is a solution of the KKT-conditions

$$X^T(Y - X\tilde{\beta})/n = \lambda \tilde{z}, \ \tilde{z} = (\tilde{z}_S, \tilde{z}_{-S})^T, \ \|\tilde{z}\|_\infty \leq 1, \ \tilde{z}^T \tilde{\beta} = \|\tilde{\beta}\|_1.$$

□

4.9 A Semi-parametric Lower Bound

The terminology "worst possible sub-direction" is based on the Cramér-Rao lower bound for unbiased estimators in higher dimensions. One can prove that the variance of each entry of the estimator is at least the corresponding diagonal element of the inverse Fisher information, by considering directional derivatives and then taking the direction where the thus obtained lower bound for the variance is maximal.

For estimators that are possibly biased the *asymptotic* Cramér-Rao lower bound has been studied, see Bickel et al. (1993) and van der Vaart (2000). In our context, the results from literature may not be immediately applicable because we need to consider asymptotics with $n \to \infty$ where the models change with n. Therefore, we redo some of this work for our context, but only in rudimentary form and only for the linear model with Gaussian design and noise.

The linear model studied in this section is

$$Y = X\beta^0 + \epsilon,$$

where $\epsilon \sim \mathcal{N}_n(0, I)$ and X is a random Gaussian $n \times p$ matrix independent of ϵ, with i.i.d. rows with mean zero and covariance matrix Σ_0.

The vector $\beta^0 \in \mathbb{R}^p$ is unknown. The Fisher-information matrix for the model is then the matrix Σ_0. We assume the inverse $\Theta_0 := \Sigma_0^{-1}$ exists. Let $g : \mathbb{R}^p \to \mathbb{R}$ be a given function and let T_n be some estimator of $g(\beta^0)$. In this section we give some direct arguments for an asymptotic lower bound for the variance of T_n when T_n is strongly asymptotically unbiased. The latter property is defined as follows.

Definition 4.5 Let $a \in \mathbb{R}^p$ be such that $a^T \Sigma_0 a = 1$ and let $0 < \delta_n \downarrow 0$. We call T_n a *strongly asymptotically unbiased estimator* at β^0 in the direction a with rate δ_n if $\mathrm{var}_{\beta^0}(T_n) = \mathcal{O}(1/n)$ and for $m_n := n/\delta_n$ and for $\beta := \beta^0 + a/\sqrt{m_n}$ as well as for $\beta = \beta^0$ it holds that

$$\sqrt{m_n}(\mathbb{E}_\beta(T_n) - g(\beta)) = o(1).$$

Theorem 4.2 *Let $a \in \mathbb{R}^p$ be such that $a^T \Sigma_0 a = 1$. Suppose that T_n is a strongly asymptotically unbiased estimator at β^0 in the direction a with rate δ_n. Assume moreover that for some $\dot{g}(\beta^0) \in \mathbb{R}^p$ and for $m_n = n/\delta_n$*

$$\sqrt{m_n}\left(g(\beta^0 + a/\sqrt{m_n}) - g(\beta^0)\right) = a^T \dot{g}(\beta^0) + o(1).$$

Then

$$n\mathrm{var}_{\beta^0}(T_n) \geq [a^T \dot{g}(\beta^0)]^2 - o(1).$$

Corollary 4.2 *It is easily seen that the maximum of $[a^T \dot{g}(\beta^0)]^2$ over vectors $a \in \mathbb{R}^p$ satisfying $a^T \Sigma_0 a = 1$ is obtained at the value*

$$a^0 := \Theta_0 \dot{g}(\beta^0) / \sqrt{\dot{g}(\beta^0)^T \Theta_0 \dot{g}(\beta^0)}.$$

Hence if we assume the conditions of Theorem 4.2 with the value a^0 for the direction a we get

$$n\mathrm{var}_{\beta^0}(T) \geq \dot{g}(\beta^0)^T \Theta_0 \dot{g}(\beta^0) - o(1).$$

Definition 4.6 Let g be differentiable at β^0 with derivate $\dot{g}(\beta^0)$. We call

$$c^0 := \frac{\Theta_0 \dot{g}(\beta^0)}{\dot{g}(\beta^0)^T \Theta_0 \dot{g}(\beta^0)}$$

the *worst possible sub-direction* for estimating $g(\beta^0)$.

The motivation for the terminology in Definition 4.6 is given by Corollary 4.2. The normalization here by $\dot{g}(\beta^0)^T \Theta_0 \dot{g}(\beta^0)$ is quite arbitrary yet natural from a projection theory point of view.

As a special case, consider $g(\beta^0) := \beta_j^0$ for some fixed value of j. Then $\dot{g}(\beta^0) = e_j$, the jth unit vector in $\mathbb{R}^p$. Clearly $\Theta_0 e_j = \Theta_j^0$ and $e_j^T \Theta_0 e_j = \Theta_{j,j}^0$, where Θ_j^0 is the jth column of Θ_0 and $\Theta_{j,j}^0$ is its jth diagonal element. It follows that $C_j^0 = \Theta_j^0 / \Theta_{j,j}^0$ is the worst possible sub-direction for estimating β_j^0.

For proving the strong asymptotic unbiasedness of certain de-sparsified estimators, one may combine for example Lemma 4.2 in Sect. 4.5 with Lemma 3.2 in Sect. 3.6. One then sees that strong asymptotic unbiasedness basically follows from suitable sparsity assumptions on β^0 and Θ_0. See van de Geer and Janková (2016) for more details.

To prove Theorem 4.2 the two auxiliary Lemmas 4.7 and 4.8 below are useful.

Lemma 4.7 *Let $Z \sim \mathcal{N}(0, 1)$. Then for all $t \in \mathbb{R}$*

$$\mathbb{E}\left[\exp[tZ - t^2/2] - 1 - tZ\right]^2 = \exp[t^2] - 1 - t^2.$$

Moreover for $2t^2 < 1$ we have

$$\mathbb{E}\exp[t^2 Z^2] = \frac{1}{\sqrt{1 - 2t^2}}.$$

Proof of Lemma 4.7 By direct calculation, for any $t > 0$

$$\mathbb{E}\left[\exp[tZ - t^2/2]\right]^2 = \mathbb{E}\exp[2tZ - t^2] = \exp[t^2], \quad \mathbb{E}\exp[tZ - t^2/2] = 1$$

and

$$\mathbb{E}Z\exp[tZ - t^2/2] = t\mathbb{E}\exp[tZ - t^2/2] = t.$$

The first result now follows immediately. The second result is also easily found by standard calculations:

$$\mathbb{E}\exp[t^2 Z^2] = \int \exp[t^2 z^2]\phi(z)dz = \int \phi\left(z\sqrt{1 - 2t^2}\right) = \frac{1}{\sqrt{1 - 2t^2}}.$$

$\square$

Write the density of Y given X as $\mathbf{p}_{\beta^0}(\cdot|X)$, i.e.

$$\mathbf{p}_{\beta^0}(y|X) := \prod_{i=1}^{n}\phi(y_i - x_i\beta^0), \quad y = (y_1, \ldots, y_n)^T,$$

where ϕ is the standard normal density.

Lemma 4.8 *Suppose $h \in \mathbb{R}^p$ satisfies $2h^T \Sigma_0 h < 1$. Then it holds that*

$$\mathbb{E}_{\beta^0}\left[\frac{\mathbf{p}_{\beta^0}(Y - Xh|X) - \mathbf{p}_{\beta^0}(Y|X)}{\mathbf{p}_{\beta^0}(Y|X)} - \epsilon^T Xh\right]^2 = (1 - 2h^T \Sigma_0 h)^{-n/2} - 1 - nh^T \Sigma_0 h.$$

Proof of Lemma 4.8 We have that given X the random variable $\epsilon^T Xh$ has the $\mathcal{N}(0, nh^T \hat{\Sigma} h)$-distribution. It follows therefore from the first result of Lemma 4.7 that

$$\mathbb{E}_{\beta^0}\left[\frac{\mathbf{p}_{\beta^0}(Y - Xh|X) - \mathbf{p}_{\beta^0}(Y|X)}{\mathbf{p}_{\beta^0}(Y|X)} - \epsilon^T Xh\right]^2 = \mathbb{E}\exp[nh^T \hat{\Sigma} h] - 1 - nh^T \Sigma_0 h.$$

Since $X_i h \sim \mathcal{N}(0, h^T \Sigma_0 h)$ $(i = 1, \dots, n)$ we have by the second result of Lemma 4.7

$$\mathbb{E} \exp[(X_i h)^2] = \frac{1}{\sqrt{1 - 2h^T \Sigma_0 h}}.$$

Whence the result.   □

Proof of Theorem 4.2 We apply Lemma 4.8 with $h = a/\sqrt{m_n}$. One then sees that

$$\mathbb{E}_{\beta^0} \left[\frac{\sqrt{m_n}(\mathbf{p}_{\beta^0}(Y - Xa/\sqrt{m_n}|X) - \mathbf{p}_{\beta^0}(Y|X))}{\mathbf{p}_{\beta^0}(Y|X)} - \epsilon^T Xa \right]^2 = \mathcal{O}(n\delta_n) = o(n).$$

Using the strong asymptotic unbiasedness and standard arguments we find that

$$\sqrt{m_n}\left(g(\beta^0 + a/\sqrt{m_n}) - g(\beta^0) \right) = \mathbb{E}_{\beta^0 + a/\sqrt{m_n}} T_n - \mathbb{E}_{\beta^0}(T_n) + o(1)$$

$$= \mathrm{cov}_{\beta^0}(T_n, \epsilon^T Xa) + o(1).$$

Hence

$$a^T \dot{g}(\beta^0) = \mathrm{cov}_{\beta^0}(T_n, \epsilon^T Xa) + o(1).$$

But

$$\mathrm{cov}^2_{\beta^0}(T_n, \epsilon^T Xa) \leq \mathrm{var}_{\beta^0}(T_n)\mathrm{var}(\epsilon^T Xa) = n\mathrm{var}_{\beta^0}(T_n).$$

□

Problems

4.1 We use the notation of Sect. 4.2. and suppose the smallest eigenvalue of Σ_0 stays away from zero. First, fix some $j \in \{1, \dots, p\}$. Lemma 4.1 shows that under some distributional assumptions and when $\|\Theta_j^0\|_1 = \mathcal{O}(1)$, then the bias of the Lasso estimator $\hat{\beta}_j$ of β_j^0 is $\mathcal{O}(\sqrt{\log p/n})$ (see the asymptotics following the lemma). Clearly since $\|\Theta_j^0\|_1 = (1 + \|\gamma_j^0\|_1)/\tau_{j,0}^2$ we have

$$\|\gamma_j^0\|_1 = \mathcal{O}(1) \implies \|\Theta_j^0\|_1 = \mathcal{O}(1).$$

Let us see how this works out for the matrix version. Let Γ_J^0 be defined as in Sect. 4.2, so that the projection of $X_{J,0}$ on the space spanned by $X_{-J,0}$ is

$$X_{J,0}PX_{-J,0} = X_{-J,0}\Gamma_J^0.$$

Let the anti-projection be

$$X_{J,0}AX_{-J,0} = X_{J,0} - X_{J,0}PX_{-J,0}$$

and let

$$T_J^0 := \mathbb{E}(X_{J,0}AX_{-J,0})^T(X_{J,0}AX_{-J,0}),$$

and write

$$(T_J^0)^{-1} := (T_J^{0,j,k}), \quad D_J^0 := \sum_j \max_k |T_J^{0,j,k}|.$$

Show that

$$\||\Theta_j^0\||_1 \le D_J^0 \||\Gamma_J^0\||_1, \quad j \in J.$$

So when J is a small set, for example $|J| = \mathcal{O}(1)$, we still have

$$\||\Gamma_J^0\||_1 = \mathcal{O}(1) \quad\Rightarrow\quad \||\Theta_J^0\||_1 = \mathcal{O}(1)$$

where $\Theta_J^0 = \{\Theta_j^0\}_{j \in J}$.

Assume now that J is indeed "small" in some sense. The irrepresentable condition on the set J is (see Sect. 4.7)

$$\||\Gamma_{-J}^0\||_1 < 1.$$

This is a condition on the coefficients of the projection of the *large* set J^c on the on the small set J.

4.2 Give a formulation of Lemma 4.2 when instead of the (multivariate) square-root Lasso, the "standard" Lasso

$$\hat{\Gamma}_J := \arg\min_{\Gamma_J}\left\{ \text{trace}\left((X_J - X_{-J}\Gamma_J)^T(X_J - X_{-J}\Gamma_J) \right)/n + 2\lambda\|\Gamma_J\|_1 \right\}$$

was used. Assume here that $\hat{T}_J$ and $\tilde{T}_J$ are invertible.

Chapter 5
Confidence Intervals Using the Lasso

Abstract Asymptotic linearity of a de-sparsified Lasso is established. This implies asymptotic normality under certain conditions and therefore can be used to construct confidence intervals for parameters of interest. Asymptotic linearity of groups of parameters, leading to confidence sets for groups, is also presented. Here, a the multivariate version of the square-root Lasso is invoked. The case of a linearized loss—applicable when the covariance matrix of the design is known—is briefly addressed as well. Throughout the chapter except for the last section, the design is considered as fixed.

5.1 A Surrogate Inverse for the Gram Matrix

The Lasso depends on a tuning parameter λ which in turn depends on (an estimator of) the standard deviation of the noise which we call here the "scale". We will now consider node-wise regression using the Lasso. There are p nodes, and each has its own scale. Hence the dependence on the scale needs to be carefully addressed. Therefore we employ here a notation where this dependence is explicit.

Consider a $n \times p$ input matrix X with columns $\{X_j\}_{j=1}^p$. We define $X_{-j} := \{X_k\}_{k \neq j}$, $j = 1, \ldots, p$. Let $\hat{\Sigma} = X^T X / n$ be the (normalized) Gram matrix. We consider for each j

$$\hat{\gamma}_j(\tau_j) := \arg\min_{\gamma_j} \left\{ \|X_j - X_{-j}\gamma_j\|_n^2 + 2\lambda_\sharp \tau_j \|\gamma_j\|_1 \right\}$$

the Lasso for node j on the remaining nodes X_{-j} with tuning parameter $\lambda_\sharp$ and scale parameter τ_j (Meinshausen and Bühlmann 2006). The aim to use a single tuning parameter $\lambda_\sharp$ for all p node-wise Lasso's. In the next section we will employ the square-root node-wise Lasso which corresponds to a particular choice of the scales. As we will see this approach leads to a final scale free result.

Denote the normalized residual sum of squares as $\hat{\tau}_j^2(\tau_j) := \|X_j - X_{-j}\hat{\gamma}_j(\tau_j)\|_n^2$. We assume that $\hat{\tau}_j(\tau_j) \neq 0$ for all j. For the square-root node-wise Lasso the equality $\hat{\tau}_j^2(\tau_j) = \tau_j^2$ holds (see Sect. 3.7).

© Springer International Publishing Switzerland 2016
S. van de Geer, *Estimation and Testing Under Sparsity*,
Lecture Notes in Mathematics 2159, DOI 10.1007/978-3-319-32774-7_5

Writing $\tau := \mathrm{diag}(\tau_1, \ldots, \tau_p)$ define the matrix $\hat{\Theta}(\tau)$ as

$$\hat{\Theta}_{j,j}(\tau_j) := 1/\tilde{\tau}_j^2(\tau_j),\ j \in \{1, \ldots, p\},$$

$$\hat{\Theta}_{k,j}(\tau_j) := -\hat{\gamma}_{k,j}(\tau_j)/\tilde{\tau}_j^2(\tau_j),\ k \neq j \in \{1, \ldots, p\},$$

with

$$\tilde{\tau}_j^2 := \|X_j - X_{-j}\hat{\gamma}_j(\tau_j)\|_n^2 + \lambda_\sharp \tau_j \|\hat{\gamma}(\tau_j)\|_1,\ j \in \{1, \ldots, p\}.$$

Let e_j be the jth unit vector and let $\hat{\Theta}_j(\tau_j)$ be the jth column of $\hat{\Theta}(\tau)$. The following lemma states that $\hat{\Theta}(\tau)$ can be viewed as *surrogate* inverse for the matrix $\hat{\Sigma}$. The lemma is mainly a reformulation of the *almost orthogonality* resulting from the KKT-conditions (as explained in Sect. 2.4).

Lemma 5.1 *Fix some $j \in \{1, \ldots, p\}$. We have*

$$\|e_j - \hat{\Sigma}\hat{\Theta}_j(\tau_j)\|_\infty \leq \lambda_\sharp \tau_j / \tilde{\tau}_j^2(\tau_j)$$

and in fact, for the jth entry,

$$e_{j,j} - \left(\hat{\Sigma}\hat{\Theta}_j(\tau_j)\right)_j = 0.$$

Proof From Lemma 3.3

$$X_j^T(X_j - X_{-j}\hat{\gamma}_j(\tau_j))/n = \tilde{\tau}_j^2(\tau_j)$$

so that

$$X_j^T X \hat{\Theta}_j(\tau_j)/n = 1.$$

Moreover from the KKT-conditions [see (2.4)]

$$X_{-j}^T(X_j - X_{-j}\hat{\gamma}_j(\tau_j))/n = \lambda_\sharp \tau_j \hat{z}_j(\tau_j),$$

where $\|\hat{z}_j(\tau_j)\|_\infty \leq 1$. One can rewrite this as

$$X_{-j}^T X \hat{\Theta}_j(\tau_j)/n = \lambda_\sharp \tau_j \hat{z}_j / \tilde{\tau}_j^2(\tau_j)$$

yielding $\|X_{-j}^T X \hat{\Theta}_j(\tau_j)\|_\infty / n \leq \lambda_\sharp \tau_j / \tilde{\tau}_j^2(\tau_j)$. $\square$

Lemma 5.2 *For all j it holds that*

$$\left(\hat{\Theta}(\tau)^T \hat{\Sigma}\hat{\Theta}(\tau)\right)_{j,j} = \frac{\hat{\tau}_j^2(\tau_j)}{\tilde{\tau}_j^4(\tau_j)}.$$

Proof This is simply rewriting the expressions. We have

$$\left(\hat{\Theta}(\tau)^T \hat{\Sigma} \hat{\Theta}(\tau)\right)_{j,j} = \hat{\Theta}_j^T(\tau) \hat{\Sigma} \hat{\Theta}_j(\tau) = \|X\hat{\Theta}_j(\tau_j)\|_n^2 = \|X\hat{C}_j(\tau_j)\|_n^2 / \tilde{\tau}_j^4(\tau_j),$$

where $\hat{C}_j(\tau_j) := \hat{\Theta}_j(\tau_j)\tilde{\tau}_j^2(\tau_j)$. But

$$\|X\hat{C}_j(\tau_j)\|_n^2 = \|X_j - X_{-j}\hat{\gamma}_j(\tau_j)\|_n^2 = \hat{\tau}_j^2(\tau_j).$$

$\square$

5.2 Asymptotic Linearity of the De-sparsified Lasso

Let X be a fixed $n \times p$ input matrix and Y an n-vector of outputs. We let $f^0 := \mathbb{E}(Y)$ be the expectation of Y (given X) and write the noise as $\epsilon = Y - f^0$. We assume X has rank n and let β^0 be any solution of the equation $f^0 = X\beta^0$.

Consider the Lasso with scale parameter σ

$$\hat{\beta}(\sigma) := \arg\min_{\beta}\left\{ \|Y - X\beta\|_n^2 + 2\lambda_0\sigma\|\beta\|_1 \right\}. \tag{5.1}$$

We define as in Zhang and Zhang (2014) or van de Geer et al. (2014) (see also Javanmard and Montanari 2014 for a similar approach) the *de-sparsified Lasso*.

Definition 5.1 The *de-sparsified* estimator of β^0—or *de-sparsified Lasso*—is

$$\hat{b}(\sigma, \tau) = \hat{\beta}(\sigma) + \hat{\Theta}^T(\tau)X^T(Y - X\hat{\beta}(\sigma))/n.$$

For the matrix $\hat{\Theta}(\tau)$ we choose the square-root node-wise Lasso $\hat{\Theta}_\sharp = \hat{\Theta}(\hat{\tau}_\sharp)$ which for all j has $\hat{\tau}_{j,\sharp}^2$ as stable point of the equation

$$\hat{\tau}_{j,\sharp}^2 = \hat{\tau}_j^2(\hat{\tau}_{j,\sharp}) = \|X_j - X_{-j}\hat{\gamma}_j(\hat{\tau}_{j,\sharp})\|_n^2$$

which is assumed to be non-zero for all j. Denote the corresponding de-sparsified Lasso as

$$\hat{b}_\sharp(\sigma) = \hat{\beta}(\sigma) + \hat{\Theta}_\sharp^T X^T(Y - X\hat{\beta}(\sigma))/n.$$

The reason for this choice (and not for instance for the scaled node-wise Lasso or a node-wise Lasso with cross-validation) is that the problem becomes scale free (see also Problem 5.1). There remains however the choice of the tuning parameter $\lambda_\sharp$. Simulations leads to recommending a value which is smaller than the common choice for the tuning parameter λ_0 for the (standard with $\sigma = \sigma_0$, square-root or scaled) Lasso for estimating β^0 (see (3.6) at the beginning of Sect. 3.7 for the general definition of $\hat{\beta}(\sigma)$ with tuning parameter λ_0).

We now show that up to a remainder term the estimator $\hat{b}_\sharp(\sigma)$ is linear.

Theorem 5.1 *For all j and for certain known non-random vectors $v_j \in \mathbb{R}^n$ (depending on $\hat{\Sigma}$ and $\lambda_\sharp$ only) with $\|v_j\|_n = 1$ we have*

$$\frac{\tilde{\tau}_{j,\sharp}^2}{\hat{\tau}_{j,\sharp}}\left(\hat{b}_{j,\sharp}(\sigma) - \beta_j^0\right) = \underbrace{v_j^T\epsilon/n}_{\text{linear term}} + \underbrace{\text{rem}_j(\sigma)}_{\text{remainder}}$$

where the remainder satisfies

$$|\text{rem}_j(\sigma)| \le \lambda_\sharp\|\hat{\beta}_{-j}(\sigma) - \beta_{-j}^0\|_1.$$

Proof We have

$$\hat{b}_\sharp(\sigma) = \hat{\Theta}_\sharp^T X^T \epsilon/n + \hat{\beta}(\sigma) - \hat{\Theta}_\sharp^T \hat{\Sigma}(\hat{\beta}(\sigma) - \beta^0)$$

so for all j

$$\hat{b}_{j,\sharp}(\sigma) = \hat{\Theta}_{j,\sharp}^T X^T \epsilon/n + \hat{\beta}_j(\sigma) - \hat{\Theta}_{j,\sharp}^T \hat{\Sigma}(\hat{\beta}(\sigma) - \beta^0)$$

$$= \beta_j^0 + \hat{\Theta}_{j,\sharp}^T X^T \epsilon/n + (e_j^T - \hat{\Theta}_{j,\sharp}^T \hat{\Sigma})(\hat{\beta}(\sigma) - \beta^0).$$

We thus find

$$\frac{\tilde{\tau}_{j,\sharp}^2}{\hat{\tau}_{j,\sharp}}\left(\hat{b}_{j,\sharp}(\sigma) - \beta_j^0\right) = \underbrace{\frac{\tilde{\tau}_{j,\sharp}^2}{\hat{\tau}_{j,\sharp}}\hat{\Theta}_{j,\sharp}^T X^T \epsilon/n}_{:=v_j^T\epsilon/n} + \underbrace{\frac{\tilde{\tau}_{j,\sharp}^2}{\hat{\tau}_{j,\sharp}}(e_j^T - \hat{\Theta}_{j,\sharp}^T \hat{\Sigma})(\hat{\beta}(\sigma) - \beta^0)}_{:=\text{rem}_j(\sigma)}$$

where

$$v_j := \frac{\tilde{\tau}_{j,\sharp}^2}{\hat{\tau}_{j,\sharp}}X\hat{\Theta}_{j,\sharp}, \ \text{rem}_j(\sigma) := \frac{\tilde{\tau}_{j,\sharp}^2}{\hat{\tau}_{j,\sharp}}(e_j^T - \hat{\Theta}_{j,\sharp}^T \hat{\Sigma})(\hat{\beta}(\sigma) - \beta^0).$$

Invoking Lemma 5.2 we see that

$$\hat{\Theta}_{j,\sharp}^T \hat{\Sigma}\hat{\Theta}_{j,\sharp} = (\hat{\Theta}_\sharp \hat{\Sigma}\hat{\Theta}_\sharp)_{jj} = \frac{\tilde{\tau}_{j,\sharp}^2}{\tilde{\tau}_{j,\sharp}^4}.$$

Therefore $\|v_j\|_n = 1$. Moreover by Lemma 5.1, $\|\text{rem}(\sigma)\|_\infty \le \lambda_\sharp\|\hat{\beta}_{-j}(\sigma) - \beta_{-j}^0\|_1$.
$\qquad\square$

If the remainder term in Theorem 5.1 can be neglected (which means it should be of order $o_{\mathbb{P}}(\sigma_0/\sqrt{n})$), one has asymptotic linearity of the "studentized" estimator, invoking some consistent estimator of σ_0 (possibly by applying the square-root Lasso for estimating $\hat{\beta}$, see also Remark 5.3). Once asymptotic linearity is established, the next step is to show asymptotic normality. Because we are dealing with triangular arrays, one then needs to verify the Lindeberg conditions. To avoid these technicalities, we only present the case where the noise is Gaussian.

Theorem 5.2 *Suppose that* $\epsilon \sim \mathcal{N}_n(0, \sigma_0^2 I)$. *Then for all* j

$$\frac{\sqrt{n}(\hat{b}_{j,\sharp}(\sigma) - \beta_j^0)}{\sigma_0 \hat{\tau}_{j,\sharp}/\tilde{\tau}_{j,\sharp}^2} = \mathcal{N}(0,1) + \Delta_j$$

where $|\Delta_j| \le \sqrt{n}\lambda_{\sharp}\|\hat{\beta}_{-j}(\sigma) - \beta_{-j}^0\|_1/\sigma_0$.

Proof This follows immediately from Theorem 5.1. □

Asymptotics Suppose that $\epsilon \sim \mathcal{N}_n(0, \sigma_0^2 I)$, that $\lambda_{\sharp} \asymp \sqrt{\log p/n}$ and that $\|\hat{\beta}(\sigma) - \beta^0\|_1/\sigma_0 = o_{\mathbb{P}}(1/\sqrt{\log p})$. Then

$$\frac{\sqrt{n}(\hat{b}_{j,\sharp}(\sigma) - \beta_j^0)}{\sigma_0 \hat{\tau}_{j,\sharp}/\tilde{\tau}_{j,\sharp}^2} = \mathcal{N}(0,1) + o_{\mathbb{P}}(1).$$

Remark 5.1 Observe that the remainder term only depends on the design X via the ℓ_1-estimation error $\|\hat{\beta}_{-j} - \beta_{-j}^0\|_1$. Thus, in order to show that this term is negligible one needs conditions that ensure that the ℓ_1-error of the Lasso for estimating β^0 is sufficiently small, but no additional assumptions on the design for the de-sparsifying step. In particular, one need not assume that the inverse of some invertible approximation of the Gram matrix has sparsity properties. The "scale-freeness" of the remainder term is a particular feature of the square-root Lasso being used in the surrogate inverse.

Remark 5.2 In Sect. 4.9 a lower bound for the asymptotic variance of a "strongly asymptotically unbiased" estimator of a coefficient β_j^0 was presented. Some tools for proving strong asymptotic unbiasedness are Lemma 3.2 in Sect. 3.6 and for example Lemma 4.2 in Sect. 4.5. As a conclusion, for random Gaussian design for example, with co-variance matrix Σ_0, one sees that the asymptotic variance of $\hat{b}_{j,\sharp}(\sigma)$ reaches the lower bound when $\Theta_0 = \Sigma_0^{-1}$ has sufficiently sparse jth column. In other words, although Theorem 5.2 does not require sparseness conditions on the design, such conditions do appear when showing validity of the lower bound for the asymptotic variance.

Remark 5.3 Assuming that the remainder terms Δ_j in Theorem 5.2 are negligible one can apply its result for the construction of confidence intervals. One then needs a consistent estimator of σ_0. For example a preliminary estimator $\hat{\sigma}_{\text{pre}}$ may be invoked for scaling the tuning parameter for the estimation of β^0 and then the normalized residual sum of squares $\|Y - X\hat{\beta}(\hat{\sigma}_{\text{pre}})\|_n^2$ can be applied as variance estimator for the studentizing step. Alternatively one may choose the tuning parameter by cross-validation resulting in an estimator $\hat{\beta}_{\text{cross}}$ and then studentize using $\|Y - X\hat{\beta}_{\text{cross}}\|_n$ as estimator of scale. Another approach would be to apply the square-root Lasso or the scaled Lasso for the estimation of β^0 and σ_0 simultaneously.

Remark 5.4 The parameter β^0 is generally not identified as we may take it to be any solution of the equation $f^0 = X\beta^0$ (in $\mathbb{R}^n$). However, we can formulate conditions (see also the next remark) depending on the particular solution β^0 such that $\|\hat{\beta}(\sigma) - \beta^0\|_1/\sigma_0$ converges to zero. Such β^0 are thus *nearly identifiable* and Theorem 5.2 can be used to construct confidence intervals for nearly identifiable β^0 which have $\|\hat{\beta}(\sigma) - \beta^0\|_1/\sigma_0$ converging to zero fast enough (see for example Corollary 2.4 in Sect. 2.10).

Remark 5.5 Recall the notation $\hat{\beta}(\sigma)$ given in (3.6) at the beginning of Sect. 3.7. We note that $\|\hat{\beta}(\sigma) - \beta^0\|_1/\sigma_0$ can be viewed as a properly scaled ℓ_1-estimation error. One may invoke Theorem 2.2 to bound it. According to this theorem if $\lambda_{0,\epsilon}$ is some universal (i.e., not depending on σ_0) high probability upper bound for the scaled noise $\|\epsilon^T X\|_\infty/(n\sigma_0)$ and moreover $\sigma = \sigma_0$ and $\lambda_0 > \lambda_{0,\epsilon}$, one gets

$$\frac{\|\hat{\beta}(\sigma_0) - \beta^0\|_1}{\sigma_0} \leq \min_S \min_\beta \left\{ \frac{\|\beta - \beta^0\|_1}{\sigma_0} + \frac{\|X(\beta - \beta^0)\|_n^2}{2\delta(\lambda_0 - \lambda_{0,\epsilon})\sigma_0^2} \right.$$
$$\left. + \frac{2\delta L^2(\lambda_0 - \lambda_{0,\epsilon})}{(1 - \delta)^2} \frac{|S|}{\hat{\phi}^2(L, S)} + \frac{4\lambda_0\|\beta_{-S}\|_1}{\sigma_0} \right\},$$

with $L := [\lambda_0 + \lambda_{0,\epsilon} + \delta(\lambda_0 - \lambda_{0,\epsilon})]/[(1 - \delta)(\lambda_0 - \lambda_u)]$ (which does not depend on σ_0). In particular we have for all S

$$\frac{\|\hat{\beta}(\sigma_0) - \beta^0\|_1}{\sigma_0} \leq \frac{2\delta L^2(\lambda_0 - \lambda_{0,\epsilon})}{(1 - \delta)^2} \frac{|S_0|}{\hat{\phi}^2(L, S_0)} + \frac{4\lambda_0\|\beta_{-S_0}^0\|_1}{\sigma_0}.$$

With the choice $S = S_0$ the dependence on σ_0 in the above bound disappears altogether. But see also Corollary 2.4 where a weak sparsity condition on β^0 is imposed.

Asymptotics If we take take $\lambda_{\sharp} \asymp \lambda_0 \asymp \lambda_\epsilon \asymp \sqrt{\log p/n}$ and assume $1/\hat{\phi}(L, S_0) = \mathcal{O}(1)$ and $\sigma_0 \asymp 1$, then (non-)sparseness $s_0 := |S_0|$ of small order $\sqrt{n}/\log p$ ensures that $\|\Delta\|_\infty = o_{\mathbb{P}}(1)$. In other words the remainder term in the linear approximation is negligible if β^0 is sufficiently ℓ_0-sparse. But also more generally, if β^0 is not ℓ_0-sparse one can still have a small enough remainder term by the above trade-off.

5.3 Confidence Intervals for Linear Contrasts

For a given $a \in \mathbb{R}^p$ we aim at constructing a confidence interval for the linear contrast $a^T \beta^0$. For instance, one may be interested in predicting the response y_{new} at a value x_{new} of the input (a row vector). Then $a^T = x_{\text{new}}$.

The surrogate inverse of $\hat{\Sigma}$ is in this section again based on the square root nodewise Lasso $\hat{\Theta}_\sharp := \hat{\Theta}(\hat{\tau}_\sharp)$ with tuning parameter $\lambda_\sharp$. For ease of notation the subscript $\sharp$ is omitted: $\hat{\Theta} := \hat{\Theta}_\sharp$, $\hat{\gamma}_j := \hat{\gamma}_j(\hat{\tau}_{j,\sharp})$ $\hat{\tau}_j := \hat{\tau}_{j,\sharp}$ and $\tilde{\tau}_j := \tilde{\tau}_{j,\sharp}$ $(j = 1, \ldots, p)$. We let $\hat{\tau} := \text{diag}(\hat{\tau}_1, \ldots, \hat{\tau}_p)$ and $\tilde{\tau} := \text{diag}(\tilde{\tau}_1, \ldots, \tilde{\tau}_p)$.

Define as in Sect. 5.2 the de-sparsified Lasso

$$\hat{b} := \hat{\beta} + \hat{\Theta}^T X^T (Y - X\hat{\beta})/n.$$

Lemma 5.3 *Suppose that $\epsilon \sim \mathcal{N}_n(0, \sigma_0^2 I)$. Then*

$$\frac{\sqrt{n} a^T (\hat{b} - \beta^0)}{\sigma_0 \|\bar{\tau}^{-1} a\|_1} = \mathcal{N}(0, \iota) + \Delta$$

where $\bar{\tau} := \tilde{\tau}^2 \hat{\tau}^{-1}$, $0 < \iota \leq 1$ and $|\Delta| \leq \sqrt{n} \lambda_\sharp \|\hat{\beta} - \beta^0\|_1 / \sigma_0$.

This lemma states that a "studentized" version is asymptotically normally distributed under conditions on the remainder term which are essentially the same as in Theorem 5.2. However, the studentization is possibly an "over-studentization" as the asymptotic variance ι may be strictly less than one.

Proof of Lemma 5.3 By the KKT-conditions

$$\hat{\Sigma}\hat{\Theta} - I = \lambda_\sharp \hat{Z}\bar{\tau}^{-1}$$

where $\hat{Z} = (\hat{Z}_1, \ldots, \hat{Z}_p)$, $\hat{z}_{j,j} = 0$ and $\hat{z}_{j,-j}^T \hat{\gamma}_j = \|\hat{\gamma}_j\|_1$. Thus

$$\|(\hat{\Sigma}\hat{\Theta} - I)a\|_\infty \leq \lambda_\sharp \|\bar{\tau}^{-1} a\|_1.$$

Moreover

$$\text{var}(a^T \hat{\Theta}^T X^T \epsilon)/n = a^T \hat{\Theta}^T \hat{\Sigma}\hat{\Theta} a \sigma_0^2 = a^T \bar{\tau}^{-1} \hat{R} \bar{\tau}^{-1} a \sigma_0^2,$$

where

$$\hat{R} = \hat{\tau}^{-1} \hat{C}^T \hat{\Sigma} \hat{C} \hat{\tau}^{-1}.$$

Here $\hat{C} = (\hat{C}_1, \dots \hat{C}_p)$ and $\hat{C}_j$ is defined by $X_j - X_{-j}\hat{\gamma}_j = X\hat{C}_j$ ($j = 1, \dots, p$). Note that $\|\hat{R}\|_\infty \le 1$ (Problem 5.2). So

$$a^T \bar{\tau}^{-1} \hat{R} \bar{\tau}^{-1} a \le \|\bar{\tau}^{-1} a\|_1 \|\hat{R} \bar{\tau}^{-1} a\|_\infty \le \|\bar{\tau}^{-1} a\|_1^2.$$

It follows that

$$a^T(\hat{b} - \beta^0) = \|\bar{\tau}^{-1} a\|_1 \{\mathcal{N}(0, \iota)\sigma_0 / \sqrt{n} + \mathrm{rem}\}$$

for some $\iota < 1$ and where

$$|\mathrm{rem}| \le \lambda_\sharp \|\hat{\beta} - \beta^0\|_1.$$

$\square$

5.4 Confidence Sets for Groups

Let $J \subset \{1, \dots, p\}$. This section is about building a "chi-squared"-confidence set for $\beta_J^0 := \{\beta_j^0 : j \in J\}$. To this end, compute as in Sect. 4.5 the multivariate ($|J|$-dimensional) square-root Lasso

$$\hat{\Gamma}_J := \arg\min_{\Gamma_J} \left\{ \|X_J - X_{-J}\Gamma_J\|_{\mathrm{nuclear}} / \sqrt{n} + \lambda_\sharp \|\Gamma_J\|_1 \right\} \tag{5.2}$$

where $\lambda_\sharp > 0$ is a tuning parameter. The minimization is over all $(p - |J|) \times |J|$ matrices Γ_J. Let as before the surrogate projection be

$$X_J \hat{P} X_{-J} := X_{-J} \hat{\Gamma}_J$$

and surrogate anti-projection

$$X_J \hat{A} X_{-J} := X_J - X_J \hat{P} X_{-J}.$$

Let

$$\tilde{T}_J := (X_J \hat{A} X_{-J})^T X_J / n \tag{5.3}$$

and

$$\hat{T}_J := (X_J \hat{A} X_{-J})^T (X_J \hat{A} X_{-J}) / n, \tag{5.4}$$

The "hat" matrix $\hat{T}_J$ is throughout this section assumed to be non-singular. The "tilde" matrix $\tilde{T}_J$ only needs to be non-singular in order that the de-sparsified estimator $\hat{b}_J$ given below in Definition 5.2 is well-defined. However, the "studentized" version given there is also defined for singular $\tilde{T}_J$.

We define the normalization matrix

$$M := \hat{T}_J^{-1/2} \tilde{T}_J. \tag{5.5}$$

Let furthermore $\hat{\beta} := \hat{\beta}(\sigma)$ be the Lasso as defined in (5.1) with scale parameter σ.

Definition 5.2 The *de-sparsified* estimator of β_J^0 is

$$\hat{b}_J := \hat{\beta}_J + \tilde{T}_J^{-1}(X_J - X_{-J}\hat{\Gamma}_J)^T(Y - X\hat{\beta})/n,$$

with $\hat{\Gamma}_J$ the multivariate square-root Lasso given in (5.2) and the matrix $\tilde{T}_J$ given in (5.3). The *studentized* de-sparsified estimator is $M\hat{b}_J$ with M the normalization matrix given in (5.5). (The studentization is modulo the standardization with an estimator of the scale σ_0 of the noise.)

The KKT-conditions (3.8) appear in the form

$$X_{-J}^T(X_J - X_{-J}\hat{\Gamma}_J)\hat{T}_J^{-1/2}/n = \lambda_\sharp \hat{Z}_J, \tag{5.6}$$

where $\hat{Z}_J$ is a $(p - |J|) \times |J|$ matrix with $(\hat{Z}_J)_{k,j} = \mathrm{sign}(\hat{\Gamma}_J)_{k,j}$ if $(\hat{\Gamma}_J)_{k,j} \neq 0$ and $\|\hat{Z}_J\|_\infty \leq 1$.

With the help of the KKT-conditions, it is now shown that inserting the multivariate square-root Lasso for de-sparsifying, and then "studentizing", results in a well-scaled "asymptotic pivot". This is up to the estimation of σ_0. The terminology "studentized" should thus be understood up to the estimation of σ_0.

Theorem 5.3 *Consider the model* $Y \sim \mathcal{N}_n(f^0, \sigma_0^2)$ *where* $f^0 = X\beta^0$. *Let* $\hat{b}_J$ *be the de-sparsified estimator given in Definition 5.2 and let* $M\hat{b}_J$ *be its studentized version. Then*

$$M(\hat{b}_J - \beta_J^0) = \sigma_0 \mathcal{N}_{|J|}(0, I)/\sqrt{n} + \mathrm{rem}$$

where $\|\mathrm{rem}\|_\infty \leq \lambda_\sharp \|\hat{\beta}_{-J} - \beta_{-J}^0\|_1$.

Observe that as in Theorem 5.2 in Sect. 5.2, the remainder term only depends on the design X via the ℓ_1-estimation error. No additional assumptions on the design are needed for the for the de-sparsifying step (see also Remark 5.1 following Theorem 5.2).

From Theorem 5.3 one may derive an asymptotic chi-squared confidence interval for the group J:

$$\|M(\hat{b}_J - \beta_J^0)\|_2^2 = \sigma_0^2 \chi_{|J|}^2/n + \|\mathrm{rem}\|_2^2,$$

where $\chi_{|J|}^2$ is a chi-squared random variable with $|J|$ degrees of freedom, and where $\|\mathrm{rem}\|_2^2 \le \lambda_\sharp^2 |J| \|\hat{\beta}_{-J} - \beta_{-J}^0\|_1$.

Proof of Theorem 5.3 The following equalities are true

$$\begin{aligned} M(\hat{b}_J - \hat{\beta}_J) &= \hat{T}_J^{-1/2}(X_J - X_{-J}\hat{\Gamma}_J)^T \epsilon/n - \hat{T}_J^{-1/2}(X_J - X_{-J}\hat{\Gamma}_J)^T X(\hat{\beta} - \beta^0)/n \\ &= \hat{T}_J^{-1/2}(X_J - X_{-J}\hat{\Gamma}_J)^T \epsilon/n - \hat{T}_J^{-1/2}(X_J - X_{-J}\hat{\Gamma}_J)^T X_J(\hat{\beta}_J - \beta_J^0)/n \\ &\quad -\hat{T}_J^{-1/2}(X_J - X_{-J}\hat{\Gamma}_J)^T X_{-J}(\hat{\beta}_{-J} - \beta_{-J}^0)/n \\ &= \hat{T}_J^{-1/2}(X_J - X_{-J}\hat{\Gamma}_J)^T \epsilon/n - M(\hat{\beta}_J - \beta_J^0) - \lambda_\sharp \hat{Z}_J^T(\hat{\beta}_{-J} - \beta_{-J}^0) \end{aligned}$$

where we invoked the KKT-conditions (5.6). One thus arrives at

$$M(\hat{b}_J - \beta_J^0) = \hat{T}_J^{-1/2}(X_J - X_{-J}\hat{\Gamma}_J)^T \epsilon/n + \mathrm{rem}, \qquad (5.7)$$

where

$$\mathrm{rem} = -\lambda_\sharp \hat{Z}_J^T(\hat{\beta}_{-J} - \beta_{-J}^0)/\sigma_0.$$

The co-variance matrix of the first term $\hat{T}_J^{-1/2}(X_J - X_{-J}\hat{\Gamma}_J)^T \epsilon/n$ in (5.7) is equal to

$$\sigma_0^2 \hat{T}_J^{-1/2}(X_J - X_{-J}\hat{\Gamma}_J)^T(X_J - X_{-J}\hat{\Gamma}_J)\hat{T}_J^{-1/2}/n^2 = \sigma_0^2 I/n$$

where I is the identity matrix with dimensions $|J| \times |J|$. It follows that this term is $|J|$-dimensional standard normal scaled with $\sigma_0/\sqrt{n}$. The remainder term can be bounded using the dual norm inequality for each entry:

$$|\mathrm{rem}_j| \le \lambda_\sharp \max_{k \notin J} |(\hat{Z}_J)_{k,j}| \|\hat{\beta}_{-J} - \beta_{-J}^0\|_1 \le \lambda_\sharp \|\hat{\beta}_{-J} - \beta_{-J}^0\|_1$$

since by the KKT-conditions (5.6), it hold that $\|\hat{Z}_J\|_\infty \le 1$. $\square$

5.5   Random Design with Known Co-variance Matrix

Consider linear model

$$Y = X\beta^0 + \epsilon,$$

with X a random $n \times p$ design matrix. Assume that the noise $\epsilon \sim \mathcal{N}_n(0, \sigma_0^2 I)$ is independent of X. Let $\hat{\Sigma} := X^T X/n$ and $\Sigma_0 := \mathbb{E}\hat{\Sigma}$. To simplify the terminology we refer to Σ_0 as the co-variance matrix of the design (although we need not insist that the rows of X are i.i.d. mean zero random variables). Important in this section is that it is assumed here that Σ_0 is known.

We briefly sketch an alternative de-sparsified estimator for this case. Assume that the inverse $\Theta_0 := \Sigma_0^{-1}$ of Σ_0 exists and write its columns as Θ_j^0 ($j = 1, \ldots, p$). Let $\hat{\beta}$ be an initial estimator of β^0 and take as de-sparsified estimator

$$\hat{b} := \Theta_0 X^T Y/n - (\Theta_0 \hat{\Sigma} - I)\hat{\beta}. \tag{5.8}$$

The properties of this alternative estimator are in line with earlier results. We have

$$\hat{b} - \beta^0 = \Theta_0 X^T \epsilon/n + \Theta_0 \hat{\Sigma} \beta^0 - (\Theta_0 \hat{\Sigma} - I)\hat{\beta} - \beta^0$$

$$= \Theta_0 X^T \epsilon/n - \underbrace{(\Theta_0 \hat{\Sigma} - I)(\hat{\beta} - \beta^0)}_{\text{rem}}.$$

The first term $\Theta_0 X^T \epsilon/n$ is conditionally on X the linear term. It holds that

$$\frac{1}{\sqrt{n}}(\Theta_0 X^T \epsilon)_j \Big| X \sim \mathcal{N}(0, (\Theta_0 \hat{\Sigma} \Theta^0)_{j,j}), \forall j.$$

The second term rem is the remainder term. Clearly

$$|\text{rem}_j| \leq \|(\hat{\Sigma} - \Sigma_0)\Theta_j^0\|_\infty \|\hat{\beta} - \beta^0\|_1, \forall j.$$

So again the remainder term will generally be negligible if the ℓ_1-error of $\hat{\beta}$ is small enough (see Problem 14.2 about bounding $\|(\hat{\Sigma} - \Sigma_0)\Theta_j^0\|_\infty$ for all j).

Example 5.1 Section 12.5 examines trace regression with a nuclear norm penalty, in particular the matrix completion problem. This is a case where the co-variance matrix Σ_0 is indeed known, and one may ask the question whether a de-sparsifying step can yield confidence intervals for the matrix completion problem. Further details of the matrix completion problem can be found in Sect. 12.5. The point we want to make here is that in such a particular case, when the number of parameters[1] is less than n, de-sparsified estimator (5.8) is asymptotically equivalent to the standard least squares estimator. More on this in Problem 5.4.

---

[1]The number of parameters is pq in the matrix completion problem of Sect. 12.5.

In this example the design is in fact orthogonal. For simplicity we assume that the residual variance σ_0^2 is equal to 1. We also assume

$$(p \log p)/n = o(1).$$

This example is related to Problem 12.4, where an oracle result is given for the ℓ_1-penalized least absolute deviations estimator.

Let $\{1, \ldots, p\}$ be certain labels and $Z_1, \ldots, Z_n$ be i.i.d. uniformly distributed on $\{1, \ldots, p\}$. Set

$$X_{i,j} = 1\{Z_i = j\}, \ i = 1, \ldots, n, \ j = 1, \ldots, p.$$

The number of observations with label j is denoted by $N_j := \sum_{i=1}^{n} X_{i,j}, j = 1, \ldots, p$.

Lemma 5.4 *Fix some j and let $\hat{\beta}_{j,\mathrm{LS}}$ be the least-squares estimator, $\hat{\beta}_j$ be the Lasso estimator (with non-normalized design) with tuning parameter $\lambda \asymp \sqrt{\log p/np}$, and let $\hat{b}_j$ be the de-sparsified estimator given in (5.8) in this section. Then*

$$\hat{b}_j - \beta_j^0 = (\hat{\beta}_{j,\mathrm{LS}} - \beta_j^0)(1 + o_{\mathbb{P}}(1)).$$

Proof of Lemma 5.4 The standard least squares estimator is

$$\hat{\beta}_{j,\mathrm{LS}} := \frac{1}{N_j} \sum_{i=1}^{n} Y_i 1\{Z_i = j\}.$$

We moreover have

$$\Theta_0 \hat{\Sigma} = \mathrm{diag}(N_1, \ldots, N_p) p/n.$$

- Suppose that $\hat{\beta}_{j,\mathrm{LS}} N_j/n > \lambda$. For non-normalized design the Lasso estimator is then

$$\hat{\beta}_j = \frac{\hat{\beta}_{j,\mathrm{LS}} N_j/n - \lambda}{N_j/n} \tag{5.9}$$

(Problem 5.3). Thus

$$\hat{b}_j = \hat{\beta}_{j,\mathrm{LS}}\left(\frac{pN_j}{n}\right) - \left(\frac{pN_j}{n} - 1\right)\hat{\beta}_j$$

$$= \hat{\beta}_{j,\mathrm{LS}} + \lambda\left(p - \frac{n}{N_j}\right).$$

The noise level $\|X^T\epsilon\|_\infty/n$ is of order $\sqrt{\frac{\log p}{np}}$ in probability (or even in expectation). We have taken

$$\lambda \asymp \sqrt{\frac{\log p}{np}}.$$

(This is because the noise level $\|X^T\epsilon\|_\infty/n$ is of order $\sqrt{\frac{\log p}{np}}$ in probability (or even in expectation).) Moreover

$$\left(p - \frac{n}{N_j}\right) \asymp \frac{N_j/n - 1/p}{1/p^2} = \mathscr{O}_{\mathbb{P}}\left(\frac{p^2}{\sqrt{np}}\right).$$

Hence

$$\hat{b}_j = \hat{\beta}_{j,\text{LS}} + \mathscr{O}_{\mathbb{P}}\left(\frac{p\sqrt{\log p}}{n}\right).$$

- If $0 < \hat{\beta}_{j,\text{LS}} N_j/n < \lambda$ we have $\hat{\beta}_j = 0$ and find

$$\hat{b}_j = \hat{\beta}_{j,\text{LS}} + \hat{\beta}_{j,\text{LS}}\left(\frac{pN_j}{n} - 1\right)$$

and again

$$\left|\hat{\beta}_{j,\text{LS}}\left(\frac{pN_j/n}{n} - 1\right)\right| \le \lambda\left|p - \frac{n}{N_j}\right| = \mathscr{O}\left(\frac{p\sqrt{\log p}}{n}\right).$$

Note also that $|\hat{\beta}_{j,\text{LS}} - \beta_j^0| \asymp \sqrt{p/n}$ in probability. Hence, the de-sparsified estimator $\hat{b}_j$ is asymptotically equivalent to the standard least squares estimator $\hat{\beta}_{j,\text{LS}}$. □

Problems

5.1 Suppose in Theorem 5.1 one would have used the Lasso with different scaling $\{\tau_j\}_{j=1}^p$ than the one of the square-root Lasso. Give an expression for the remainder term $\text{rem}_j(\sigma)$ in terms of τ_j $(j = 1, \ldots, p)$.

5.2 The Proof of Lemma 5.3 is based on the inequality

$$a^T\bar{\tau}^{-1}\hat{R}\bar{\tau}^{-1}a \le \|\bar{\tau}^{-1}a\|_1\|\hat{R}\bar{\tau}^{-1}a\|_\infty \le \|\bar{\tau}^{-1}a\|_1^2$$

where $\hat{R}$ is defined there. Verify that indeed $\|\hat{R}\|_\infty \leq 1$. Write down the exact result when $a = e_j$, the jth unit vector ($j = 1, \ldots, p$).

5.3 Verify (5.9) in Example 5.1 (Proof of Lemma 5.4) of Sect. 5.5.

5.4 Suppose that $p \leq n$ and that $\hat{\Sigma}$ is non-singular. Consider the "de-sparsified" estimator

$$\hat{b} := \hat{\beta} + \hat{\Sigma}^{-1} X^T (Y - X\hat{\beta})/n.$$

Show that (no matter what initial estimator $\hat{\beta}$) this estimator $\hat{b}$ is equal to the least squares estimator $\hat{\beta}_{LS}$.

Chapter 6
Structured Sparsity

Abstract Oracle results are given for least squares and square-root loss with sparsity inducing norms Ω. A general class of sparsity inducing norms are those generated from cones. Examples are the group sparsity norm, the wedge norm, and the sorted ℓ_1-norm. Bounds for the error in dual norm Ω_* are given. De-sparsifying is discussed as well.

6.1 The Ω-Structured Sparsity Estimator

Like Chap. 2 this chapter studies the linear model with fixed design (or—only in Sect. 6.10—random design)

$$Y = X\beta^0 + \epsilon$$

where $Y \in \mathbb{R}^n$ is an observed response variable, X is a $n \times p$ input matrix, $\beta^0 \in \mathbb{R}^p$ is a vector of unknown coefficients and $\epsilon \in \mathbb{R}^n$ is unobservable noise. The Ω-*structured sparsity estimator* is

$$\hat{\beta} := \arg \min_{\beta \in \mathbb{R}^p} \left\{ \|Y - X\beta\|_n^2 + 2\lambda\Omega(\beta) \right\},$$

with Ω a given norm on $\mathbb{R}^p$. The reason for applying some other norm than the ℓ_1-norm depends on the particular application. In this chapter, we have in mind the situation of a sparsity inducing norm, which means roughly that it favours solutions $\hat{\beta}$ with many zeroes structured in a particular way.[1] Such generalizations of the Lasso are examined in Jenatton et al. (2011), Micchelli et al. (2010), Bach (2010), Bach et al. (2012), Maurer and Pontil (2012) for example. The norm Ω is constructed in such a way that the sparsity pattern in $\hat{\beta}$ follow a suitable structure. This may for example facilitate interpretation.

---

[1]For example the least-squares estimator with so-called *nuclear norm* penalization is formally also a structured sparsity estimator. This will be considered in Sect. 12.5. The topic of this chapter is rather norms which are weakly decomposable as defined in Definition 6.1.

© Springer International Publishing Switzerland 2016
S. van de Geer, *Estimation and Testing Under Sparsity*,
Lecture Notes in Mathematics 2159, DOI 10.1007/978-3-319-32774-7_6

This chapter largely follows van de Geer (2014) and Stucky and van de Geer (2015).

The question is now: can we prove oracle inequalities (as given in for example Theorem 2.1) for more general norms Ω than the ℓ_1-norm? To answer this question we first recall the ingredients of the proof Theorem 2.1.

- the two point margin
- the two point inequality
- the dual norm inequality
- the (ℓ_1-)triangle property
- (ℓ_1-)compatibility
- the convex conjugate inequality.

We will also need empirical process theory to bound certain functions of ϵ. This will be done in Chap. 8.

The convex conjugate inequality and two point margin have to do with the loss function and not with the penalty. Since our loss function is still least squares loss we can use these two ingredients as before. The other ingredients: two point inequality, dual norm inequality, Ω-triangle property and Ω-compatibility will be discussed in what follows.

6.2 Dual Norms and KKT-Conditions for Structured Sparsity

The dual norm of Ω is defined as

$$\Omega_*(w) := \max_{\Omega(\beta) \leq 1} |w^T \beta|, \ w \in \mathbb{R}^p.$$

Therefore the dual norm inequality holds by definition: for any two vectors w and β

$$|w^T \beta| \leq \Omega_*(w)\Omega(\beta).$$

The sub-differential of Ω is given by

$$\partial\Omega(\beta) = \begin{cases} \{z \in \mathbb{R}^p : \ \Omega_*(z) \leq 1\} & \text{if } \beta = 0 \\ \{z \in \mathbb{R}^p : \ \Omega_*(z) = 1, \ z^T\beta = \Omega(\beta)\} & \text{if } \beta \neq 0 \end{cases}$$

(Bach et al. 2012, Proposition 1.2). It follows that the Ω-structured sparsity estimator $\hat{\beta}$ satisfies

$$\Omega_*(X^T(Y - X\hat{\beta}))/n \leq \lambda,$$

and if $\hat{\beta} \neq 0$,

$$\Omega_*(X^T(Y - X\hat{\beta}))/n = \lambda, \ \hat{\beta}^T X^T(Y - X\hat{\beta})/n = \Omega(\hat{\beta}).$$

The KKT-conditions are

$$X^T(Y - X\hat{\beta})/n = \lambda\hat{z},$$

where $\Omega_*(\hat{z}) \leq 1$ and $\hat{z}^T\hat{\beta} = \Omega(\hat{\beta})$ (Problem 6.8).

6.3 Two Point Inequality

We call the result (6.1) below in Lemma 6.1 a *two point inequality*. See also Lemma 7.1 which treats more general loss functions.

Lemma 6.1 *Let $\hat{\beta}$ be the estimator*

$$\hat{\beta} := \arg\min_{\beta \in \mathbb{R}^p} \left\{ \|Y - X\beta\|_n^2 + 2\text{pen}(\beta) \right\},$$

where pen $: \ \mathbb{R}^p \to \mathbb{R}$ *is a convex penalty. Then for any $\beta \in \mathbb{R}^p$*

$$(\beta - \hat{\beta})^T X^T(Y - X\hat{\beta})/n \leq \text{pen}(\beta) - \text{pen}(\hat{\beta}) \tag{6.1}$$

Proof of Lemma 6.1 Fix a $\beta \in \mathbb{R}^p$ and define for $0 < t \leq 1$,

$$\hat{\beta}_t := (1 - t)\hat{\beta} + t\beta.$$

Since $\hat{\beta}$ is a minimizer

$$\|Y - X\hat{\beta}\|_n^2 + 2\text{pen}(\hat{\beta}) \leq \|Y - X\hat{\beta}_t\|_n^2 + 2\text{pen}(\hat{\beta}_t)$$

$$\leq \|Y - X\hat{\beta}_t\|_n^2 + 2(1 - t)\text{pen}(\hat{\beta}) + 2t\text{pen}(\beta)$$

where in the second step we used the convexity of the penalty. It follows that

$$\frac{\|Y - X\hat{\beta}\|_n^2 - \|Y - X\hat{\beta}_t\|_n^2}{t} + 2\text{pen}(\hat{\beta}) \leq 2\text{pen}(\beta).$$

But clearly

$$\lim_{t \downarrow 0} \frac{\|Y - X\hat{\beta}\|_n^2 - \|Y - X\hat{\beta}_t\|_n^2}{t} = 2(Y - X\hat{\beta})^T X(\beta - \hat{\beta})/n.$$

□

Note that the two point inequality (6.1) can be written in the form

$$(\hat{\beta} - \beta)^T \hat{\Sigma} (\hat{\beta} - \beta^0) \le \epsilon^T X(\hat{\beta} - \beta)/n + \text{pen}(\beta) - \text{pen}(\hat{\beta}).$$

For the case pen $= \lambda\Omega$ an alternative proof can be formulated from the KKT-conditions (Problem 6.8).

6.4 Weak Decomposability and Ω-Triangle Property

What we need is a more general version of the ℓ_1-triangle property given at the end of Sect. 2.5. This more general version will be termed the Ω-triangle property, and reads

$$\Omega(\beta) - \Omega(\beta') \le \Omega(\beta_S' - \beta_S) + \Omega(\beta_{-S}) - \Omega^{-S}(\beta_{-S}'), \ \forall \ \beta, \beta'.$$

Here Ω^{-S} is a norm defined on $\mathbb{R}^{p-|S|}$. This property holds if S is a *allowed* set which is defined as follows

Definition 6.1 The set S is called *(Ω-)allowed* if for a norm Ω^{-S} on $\mathbb{R}^{p-|S|}$ it holds that

$$\Omega(\beta) \ge \Omega(\beta_S) + \Omega^{-S}(\beta_{-S}), \ \forall \ \beta \in \mathbb{R}^p. \tag{6.2}$$

We call Ω *weakly decomposable* for the set S.

Clearly for the ℓ_1-norm $\| \cdot \|_1$ any subset S is allowed, Ω^{-S} is again the ℓ_1-norm and one has in fact equality: $\|\beta\|_1 = \|\beta_S\|_1 + \|\beta_{-S}\|_1$ for all β. More examples are in Sects. 6.9 and 6.12.2. Observe also that by the triangle inequality, for any vector β

$$\Omega(\beta) \le \Omega(\beta_S) + \Omega(\beta_{-S}).$$

For allowed sets, one thus in a sense also has the reverse inequality, albeit that $\Omega(\beta_{-S})$ is now replaced by an other norm [namely $\Omega^{-S}(\beta_{-S})$].

We introduce some further notation. If Ω and $\underline{\Omega}$ are two norms on Euclidean space, say $\mathbb{R}^p$, we write

$$\Omega \geq \underline{\Omega} \; \Leftrightarrow \; \Omega(\beta) \geq \underline{\Omega}(\beta) \; \forall \; \beta \in \mathbb{R}^p$$

and then say that Ω is a stronger norm than $\underline{\Omega}$ (where "stronger" need not be in strict sense, so Ω is actually at least as strong as $\underline{\Omega}$). Note that

$$\Omega \geq \underline{\Omega} \; \Rightarrow \; \Omega_* \leq \underline{\Omega}_*.$$

For an allowed set S, write (6.2) shorthand as

$$\Omega \geq \Omega(\cdot|S) + \Omega^{-S}$$

where for any set J and $\beta \in \mathbb{R}^p$, the notation $\Omega(\beta|J) = \Omega(\beta_J)$ is used. Define Ω^{-S} as the largest norm among the norms $\underline{\Omega}^{-S}$ for which

$$\Omega \geq \Omega(\cdot|S) + \underline{\Omega}^{-S}.$$

If $\Omega^{-S} = \Omega(\cdot | - S)$ we call Ω *decomposable* for the set S.

Let us compare the various norms.

Lemma 6.2 *Let S be an allowed set so that*

$$\Omega \geq \Omega(\cdot|S) + \Omega^{-S} =: \underline{\Omega}.$$

Then

$$\Omega_* \leq \underline{\Omega}_* = \max\{\Omega_*(\cdot|S), \Omega_*^{-S}\}$$

and

$$\Omega^{-S} \leq \Omega(\cdot| - S), \; \Omega_*^{-S} \geq \Omega_*(\cdot| - S).$$

We see that the original norm Ω is *stronger* than the decomposed version $\underline{\Omega} = \Omega(\cdot|S) + \Omega^{-S}$. As we will experience this means we will loose a certain amount of its strength by replacing Ω by $\underline{\Omega}$ at places.

Proof of Lemma 6.2 We first prove $\underline{\Omega}_* \leq \max\{\Omega_*(\cdot|S), \Omega_*^{-S}\}$. Clearly

$$\underline{\Omega}_*(w_S) = \Omega_*(w_S) = \max\{\Omega_*(w_S), \Omega_*^{-S}(0)\}$$

and

$$\underline{\Omega}_*(w_{-S}) = \Omega_*^{-S}(w_{-S}) = \max\{\Omega_*(0), \Omega_*^{-S}(w_{-S})\}.$$

So it suffices to consider vectors w with both $w_S \neq 0$ and $w_{-S} \neq 0$. By the definition of the dual norm $\underline{\Omega}_*$

$$\underline{\Omega}_*(w) = \max_{\underline{\Omega}(\beta) \leq 1} w^T \beta$$

$$= \max_{\Omega(\beta_S) + \Omega^{-S}(\beta_{-S}) \leq 1} \left\{ \frac{w_S^T \beta_S}{\Omega(\beta_S)} \Omega(\beta_S) + \frac{w_{-S}^T \beta_{-S}}{\Omega^{-S}(\beta_{-S})} \Omega^{-S}(\beta_{-S}) \right\}$$

$$\leq \max_{\Omega(\beta_S) + \Omega^{-S}(\beta_{-S}) \leq 1} \left\{ \Omega_*(w_S)\Omega(\beta_S) + \Omega_*^{-S}(w_{-S})\Omega^{-S}(\beta_{-S}) \right\}$$

$$\leq \max\{\Omega_*(w_S), \Omega_*^{-S}(w_{-S})\}.$$

The reverse inequality $\underline{\Omega}_* \geq \max\{\Omega_*(\cdot|S), \Omega_*^{-S}\}$ follows from

$$\underline{\Omega}_*(w) = \max_{\underline{\Omega}(\beta) \leq 1} w^T \beta \geq \max_{\underline{\Omega}(\beta) \leq 1,\ \beta_{-S} = 0} w^T \beta = \Omega_*(w_S)$$

and similarly $\underline{\Omega}(w) \geq \Omega_*^{-S}(w_{-S})$.

For the second result of the lemma, we use the triangle inequality

$$\Omega \leq \Omega(\cdot|S) + \Omega(\cdot|-S)$$

Since S is assumed to be allowed we also have

$$\Omega \geq \Omega(\cdot|S) + \Omega^{-S}.$$

So it must hold that $\Omega^{-S} \leq \Omega(\cdot|-S)$. Hence also $\Omega_*^{-S} \geq \Omega(\cdot|-S)$. $\qquad\square$

6.5 Ω-Compatibility

Just as for the Lasso, the results will depend on compatibility constants, which in the present setup are defined as follows.

Definition 6.2 (van de Geer 2014) Suppose S is an allowed set. Let $L > 0$ be some *stretching* factor. The Ω-*compatibility constant* (for S) is

$$\hat{\phi}_\Omega^2(L, S) := \min \left\{ |S| \|X\beta_S - X\beta_{-S}\|_n^2 :\ \Omega(\beta_S) = 1,\ \Omega^{-S}(\beta_{-S}) \leq L \right\}.$$

A comparison of this definition with its counterpart for the Lasso (Definition 2.1 in Sect. 2.6) we see that the ℓ_1-norm is merely replaced by a more general norm. The geometric interpretation is however less evident. On top of that the various Ω-compatibility constants do not allow a clear ordering, i.e., generally one cannot say that one norm gives smaller compatibility constants than another. Suppose for example that both $\Omega(\cdot|S)$ and Ω^{-S} are stronger than ℓ_1 (see also Sect. 6.7 for this terminology). Then clearly

$$(i) := \left\{\beta : \Omega(\beta_S) = 1, \ \Omega^{-S}(\beta_{-S})\|_1 \leq L\right\} \subset \left\{\beta : \|\beta_S\|_1 \leq 1, \ \|\beta_{-S} \leq L\right\} =: (ii).$$

But the latter set (ii) is *not* a subset of

$$(iii) := \left\{\beta : \|\beta_S\|_1 = 1, \ \|\beta_{-S}\|_1 \leq L\right\}.$$

Hence we cannot say whether the minimum over the first set (i) is larger (or smaller) than the minimum over the third set (iii). Section 15.3 presents for some bounds for compatibility constants (there $|S|/\hat{\phi}_\Omega^2(L, S)$ is called *effective sparsity*) when the design is formed by i.i.d. realizations of a random variable with favourable distributional properties.

6.6   A Sharp Oracle Inequality with Structured Sparsity

Let Ω be a norm on $\mathbb{R}^p$. We have in mind a sparsity inducing norm for which the collection of allowed sets (see Definition 6.1) does not consist of only the trivial sets $\emptyset$ and $\mathbb{R}^p$. Recall that the Ω-structured sparsity estimator $\hat{\beta}$ is defined as

$$\hat{\beta} := \arg\min_{\beta \in \mathbb{R}^p} \left\{ \|Y - X\beta\|_n^2 + 2\lambda\Omega(\beta) \right\}.$$

Theorem 6.1 *Consider an allowed set S. Let λ_S and λ^{-S} be constants such that*

$$\lambda_S \geq \Omega_*(X_S^T\epsilon)/n, \ \lambda^{-S} \geq \Omega_*^{-S}(X_{-S}^T\epsilon)/n.$$

Let $\delta_1 \geq 0$ and $0 \leq \delta_2 < 1$ be arbitrary. Take $\lambda > \lambda^{-S}$ and define

$$\underline{\lambda} := \lambda - \lambda^{-S}, \ \bar{\lambda} := \lambda + \lambda_S + \delta_1\underline{\lambda}$$

and stretching factor

$$L := \frac{\bar{\lambda}}{(1 - \delta_2)\underline{\lambda}}.$$

Then for any vector $\beta \in \mathbb{R}^p$ it holds that

$$2\delta_1 \underline{\lambda} \Omega(\hat{\beta}_S - \beta_S) + 2\delta_2 \underline{\lambda} \Omega^{-S}(\hat{\beta}_{-S} - \beta_{-S}) + \|X(\hat{\beta} - \beta^0)\|_n^2$$

$$\leq \|X(\beta - \beta^0)\|_n^2 + \frac{\bar{\lambda}^2 |S|}{\hat{\phi}_\Omega^2(L,S)} + 4\lambda \Omega(\beta_{-S}).$$

Theorem 6.1 is a special case of Theorem 7.1 in Sect. 7.5. We do however provide a direct proof in Sect. 6.12.3. The direct proof moreover facilitates the verification of the claims made in Proposition 6.1 which treats the case of square-root least squares loss with sparsity inducing norms.

One may minimize the result of Theorem 6.1 over all candidate oracles (β, S) with β a vector in $\mathbb{R}^p$ and S an allowed set. This then gives oracle values (β^*, S_*). Theorem 6.1 is a generalization of Theorem 2.2 in Sect. 2.8. As there, with the choice $\delta_1 = \delta_2 = 0$ it has no result for the $\Omega(\cdot|S)$ or Ω^{-S} estimation error. If we take these values strictly positive say $\delta_1 = \delta_1 = \delta > 0$ one obtains the following corollary.

Corollary 6.1 *Let S be an Ω-allowed set and define*

$$\underline{\Omega} = \Omega(\cdot|S) + \Omega^{-S}.$$

Then, using the notation of Theorem 6.1 with $\delta_1 = \delta_2 := \delta$, we have for any vector $\beta \in \mathbb{R}^p$

$$2\delta \underline{\lambda}\, \underline{\Omega}(\hat{\beta} - \beta^0) \leq 2\delta \underline{\lambda}\, \underline{\Omega}(\beta - \beta^0) + \|X(\beta - \beta^0)\|_n^2 + \frac{\bar{\lambda}^2 |S|}{\hat{\phi}_\Omega^2(L,S)} + 4\lambda \Omega(\beta_{-S}). \quad (6.3)$$

Remark 6.1 The good news is that the oracle inequalities thus hold for general norms. The bad news is that by the definition of an allowed set S

$$\Omega \geq \underline{\Omega},$$

where

$$\underline{\Omega} := \Omega(\cdot|S) + \Omega^{-S}.$$

Hence in general the bounds for $\underline{\Omega}$-estimation error of $\hat{\beta}$ (as given in Corollary 6.1) do not imply bounds for the Ω-estimation error of $\hat{\beta}$. As an illustration, we see in Example 6.2 ahead (Sect. 6.9) that Ω^{-S} can be very small when $|S|$ is large. Lemma 6.2 moreover shows that $\Omega_*^{-S} \geq \Omega_*(\cdot| - S)$, leading by the condition $\lambda > \Omega_*^{-S}(X_{-S}^T\epsilon)/n$ to a perhaps (somewhat) large tuning parameter.

6.7 Norms Stronger than ℓ_1

We say that the norm Ω is stronger than $\underline{\Omega}$ if $\Omega \geq \underline{\Omega}$. For such two norms the dual norm of Ω is weaker: $\Omega_* \leq \underline{\Omega}_*$. Thus, when Ω is stronger than the ℓ_1-norm $\| \cdot \|_1$, Theorem 6.1 gives stronger results than Theorem 2.2 and its bounds on the tuning parameter λ are weaker. (This is modulo the behaviour of the compatibility constants: $\hat{\phi}^2$ and $\hat{\phi}_\Omega^2$ are generally not directly comparable.) Section 6.9 considers a general class of norms that are stronger than the ℓ_1-norm.

With norms stronger than ℓ_1 one can apply the "conservative" ℓ_1-based choice of the tuning parameter. This is important for the following reason. In view of Corollary 6.1, one would like to choose S in (6.3) in an optimal "oracle" way trading off the terms involved. But such a choice depends on the unknown β^0. Hence we need to prove a bound for $\Omega_*^{-S}(X_{-S}^T \epsilon)/n$ which holds for all S which are allowed and which we want to include in our collection of candidate oracles. If the norm is stronger than the ℓ_1-norm a value $\lambda > \lambda_\epsilon$ with λ_ϵ at least $\|X^T \epsilon\|_\infty/n$ works. This "conservative" choice is a (slightly) too severe overruling of the noise and in that sense not optimal. There may be cases where one still can use smaller values. Perhaps by using cross-validation one can escape from this dilemma. On the other hand, maybe the only gain when using some "optimal" tuning parameter is in the logarithmic terms and constants. Chapter 8 further examines the situation for a general class of norms (see in particular Corollaries 12.4 and 12.5 in Sect. 12.2).

6.8 Structured Sparsity and Square-Root Loss

Let Ω be a norm on $\mathbb{R}^p$. The topic of this section is the *square-root Ω-structured* sparsity estimator

$$\hat{\beta} := \arg \min_{\beta \in \mathbb{R}^p} \left\{ \|Y - X\beta\|_n + \lambda_0 \Omega(\beta) \right\}.$$

The motivation for studying square-root quadratic loss is as before: it allows one to utilize a tuning parameter that does not depend on the scale of the noise. This motivation as perhaps somewhat troubled now, as we have seen in the previous sections (see also the discussion in Sect. 6.7) that the "good" (i.e. minimal effective) choice for the tuning parameter is more subtle as it may depend on the oracle. On the other hand, a conservative but scale-free choice may still be preferable to a choice that needs to know the scale. Moreover, for certain examples (for instance the group-Lasso example of Example 6.1 in the next section) dependence on the oracle is not an issue and square-root least squares loss gives a universal choice *and* not overly conservative choice for the tuning parameter.

Let the residuals be

$$\hat{\epsilon} := Y - X\hat{\beta}.$$

The idea in this section is as in Chap. 3 to first present an oracle inequality under the assumption that $\hat{\epsilon} \neq 0$, i.e, no overfitting. This is done in Sect. 6.8.1. Then Sect. 6.8.2 shows that indeed $\hat{\epsilon} \neq 0$ with high probability. Finally Sect. 6.8.3 combines the results. The arguments are completely parallel to those for the square-root Lasso as presented in Chap. 3.

6.8.1 Assuming There Is No Overfitting

In this subsection it is assumed that $\hat{\epsilon} \neq 0$. In the next section show this is shown to be the case with high probability if β^0 is Ω-sparse.

Proposition 6.1 *Suppose* $\|\hat{\epsilon}\|_n > 0$. *Consider an allowed set S. Let*

$$\lambda_{0,S} \geq \frac{\Omega_*(X_S^T \epsilon)}{n\|\epsilon\|_n}, \ \lambda_0^{-S} \geq \frac{\Omega_*^{-S}(X_{-S}^T \epsilon)}{n\|\epsilon\|_n}.$$

Let $\delta_1 \geq 0$ *and* $0 \leq \delta_2 < 1$ *be arbitrary. Take* $\lambda_0 \|\hat{\epsilon}\|_n > \lambda_0^{-S}\|\epsilon\|_n$ *and define*

$$\hat{\lambda}_L\|\epsilon\|_n := \lambda_0\|\hat{\epsilon}\|_n - \lambda_0^{-S}\|\epsilon\|_n, \ \hat{\lambda}_U\|\epsilon\|_n := \lambda_0\|\hat{\epsilon}\|_n + \lambda_{0,S}\|\epsilon\|_n + \delta_1\hat{\lambda}_L\|\epsilon\|_n$$

and

$$\hat{L} := \frac{\hat{\lambda}_U}{(1-\delta_2)\hat{\lambda}_L}.$$

Then for any β *we have*

$$2\delta_1\hat{\lambda}_L\|\epsilon\|_n\Omega(\hat{\beta}_S - \beta) + 2\delta_2\hat{\lambda}_L\|\epsilon\|_n\Omega^{-S}(\hat{\beta}_{-S}) + \|X(\hat{\beta} - \beta^0)\|_n^2$$

$$\leq \|X(\beta - \beta^0)\|_n^2 + \frac{\hat{\lambda}_U^2\|\epsilon\|_n^2|S|}{\hat{\phi}_\Omega^2(\hat{L}, S)} + 4\lambda_0\|\hat{\epsilon}\|_n\Omega(\beta_{-S}).$$

Proof of Proposition 6.1 This follows by the same arguments as for Theorem 6.1 (see Sect. 6.12.3) and using the two point inequality (6.10) from Problem 6.2. □

6.8.2 Showing There Is No Overfitting

Conditions that ensure that $\hat{\epsilon} \neq 0$, and in fact $\|\hat{\epsilon}\|_n$ is close to $\|\epsilon\|_n$, are of the same flavour as for the square-root Lasso in Lemma 3.1.

Lemma 6.3 *Suppose that for some $0 < \eta < 1$, some $\lambda_{0,\epsilon} > 0$ and some $\underline{\sigma} > 0$, we have*

$$\lambda_0(1 - \eta) \geq \lambda_{0,\epsilon}$$

and

$$\lambda_0 \Omega(\beta^0)/\underline{\sigma} \leq 2\left(\sqrt{1 + (\eta/2)^2} - 1\right). \tag{6.4}$$

Then on the set where $\Omega_*(X^T \epsilon)/(n\|\epsilon\|_n) \leq \lambda_{0,\epsilon}$ and $\|\epsilon\|_n \geq \underline{\sigma}$ we have

$$\left| \|\hat{\epsilon}\|_n/\|\epsilon\|_n - 1 \right| \leq \eta.$$

Proof of Lemma 6.3 This follows by exactly the same arguments as those used for Lemma 3.1.   □

6.8.3 A Sharp Oracle Inequality

Combining the previous two subsections yields the following oracle result.

Theorem 6.2 *Let S be an allowed set. Let for some positive constants $\lambda_{0,\epsilon}$, $\lambda_{0,S}$, λ_0^{-S}, $0 < \eta < 1$ and $\underline{\sigma}$, the Ω-sparsity (6.4) hold, and*

$$\lambda_{0,S} \geq \frac{\Omega_*(X_S^T \epsilon)}{n\|\epsilon\|_n}, \quad \lambda_0^{-S} \geq \frac{\Omega_*^{-S}(X_{-S}^T \epsilon)}{n\|\epsilon\|_n},$$

$$\lambda_{0,\epsilon} \geq \frac{\Omega_*(X^T \epsilon)}{n\|\epsilon\|_n}, \quad \|\epsilon\|_n \geq \underline{\sigma}$$

and finally

$$\lambda_0(1 - \eta) \geq \max\{\lambda_{0,c}, \lambda_0^{-S}\}.$$

Define

$$\underline{\lambda}_0 := \lambda_0(1 - \eta) - \lambda_0^{-S}, \quad \bar{\lambda}_0 := \lambda_0(1 + \eta)\lambda_{0,S} + \delta_1 \underline{\lambda}_0$$

and

$$L := \frac{\bar{\lambda}_0}{(1 - \delta_2)\underline{\lambda}_0}.$$

Then for any β we have

$$2\delta_1\underline{\lambda}_0\|\epsilon\|_n\Omega(\hat{\beta}_S - \beta) + 2\underline{\lambda}_0\|\epsilon\|_n\Omega_2(\hat{\beta}_{-S}) + \|X(\hat{\beta} - \beta^0)\|_n^2$$

$$\leq \|X(\beta - \beta^0)\|_n^2 + \frac{\bar{\lambda}_0^2\|\epsilon\|_n^2|S|}{\hat{\phi}_\Omega^2(L, S)} + 4\lambda_0(1 + \eta)\|\epsilon\|_n\Omega(\beta_{-S}).$$

Proof of Theorem 6.2 This follows from Proposition 6.1 combined with Lemma 6.3.   □

Note that since $\Omega(\cdot|S) + \Omega^{-S} \leq \Omega$, the inequality $\Omega_* \leq \max\{\Omega_*(\cdot|S), \Omega_*^{-S}\}$ holds true (Lemma 6.2). Thus, one may take $\lambda_{0,\epsilon} \leq \max\{\lambda_{0,S}, \lambda_0^{-S}\}$. Typically, high-probability bounds will hold with $\lambda_{0,S} \leq \lambda_0^{-S}$.

6.9 Norms Generated from Cones

This section introduces a general class of norms for which the weak decomposability property, as presented in Definition 6.1, holds. The corresponding allowed sets are the sets which one believes to be candidate active sets.

Let $\mathscr{A}$ be a convex cone in $\mathbb{R}_+^p =: [0, \infty)^p$. This cone is given beforehand and will describe the sparsity structure one believes is (approximately) valid for the underlying target β^0.

Definition 6.3 The norm Ω generated from the convex cone $\mathscr{A}$ is

$$\Omega(\beta) := \min_{a \in \mathscr{A}} \frac{1}{2} \sum_{j=1}^p \left[\frac{|\beta_j|^2}{a_j} + a_j \right], \quad \beta \in \mathbb{R}^p.$$

Here we use the convention $0/0 = 0$. If $\beta_j \neq 0$ one is forced to take $a_j \neq 0$ in the above minimum. It is shown in Micchelli et al. (2010) that Ω is indeed a norm. We present a proof for completeness.

Lemma 6.4 *The function Ω defined in Definition 6.3 above is a norm.*

Proof of Lemma 6.4 It is clear that $\Omega(\beta) \geq 0$ for all β and that it can only be zero when $\beta \equiv 0$. It is also immediate that the scaling property

$$\Omega(\lambda\beta) = \lambda\Omega(\beta), \ \forall \ \lambda > 0, \ \beta \in \mathbb{R}^p,$$

holds, where we use that $\mathscr{A}$ is a cone. The function $\beta \mapsto \Omega(\beta)$ is convex because $(a,b) \mapsto b^2/a$ and $a \mapsto a$ are convex functions and $\mathscr{A}$ is convex. The triangle inequality follows from this and from the scaling property. $\square$

The norm Ω is said to be generated from the convex cone $\mathscr{A}$. One may verify that a penalty proportional to Ω favours sparse vectors which lie in $\mathscr{A}$. It is easy to see that the ℓ_1-norm is a special case with $\mathscr{A} = \mathbb{R}^p_+$.

Having sparsity in mind, a minimal requirement seems to be that when coordinates are put to zero this does not increase the norm. This is indeed the case for a norm generated from a cone, as the following lemma shows.

Lemma 6.5 *For $J \subset \{1,\ldots,p\}$ we have*

$$\Omega(\cdot|J) \le \Omega(\cdot).$$

Proof of Lemma 6.5 Let $\beta \in \mathbb{R}^p$ be arbitrary. For all $a \in \mathscr{A}$

$$\frac{1}{2}\sum_{j\in J}\left[\frac{\beta_j^2}{a_j}+a_j\right] + \frac{1}{2}\sum_{j\notin J}a_j \le \frac{1}{2}\sum_{j=1}^p\left[\frac{\beta_j^2}{a_j}+a_j\right].$$

Hence also

$$\Omega(\beta_J) = \min_{a\in\mathscr{A}}\left\{\frac{1}{2}\sum_{j\in J}\left[\frac{\beta_j^2}{a_j}+a_j\right] + \frac{1}{2}\sum_{j\notin J}a_j\right\} \le \min_{a\in\mathscr{A}}\frac{1}{2}\sum_{j=1}^p\left[\frac{\beta_j^2}{a_j}+a_j\right].$$

$\square$

The rest of this section is organized as follows. First in Lemma 6.6 an alternative representation of the norm Ω generated from a cone is presented, and also the dual norm. Then Lemma 6.7 shows which sets S are allowed and the corresponding weak decomposability into $\Omega(\cdot|S)$ and Ω^{-S}. Then in Lemma 6.8 a bound for $\Omega(\cdot|-J)$ in terms of Ω^{-J} is given, for general sets J and hence in particular for allowed sets $J = S$. Lemma 6.9 states that $\underline{\Omega} := \Omega(\cdot|J) + \Omega^{-J}$ is stronger than the ℓ_1-norm. We end the section with some examples.

Lemma 6.6 *We have*

$$\Omega(\beta) = \min_{a\in\mathscr{A},\,\|a\|_1=1}\sqrt{\sum_{j=1}^p\frac{\beta_j^2}{a_j}} = \min_{a\in\mathscr{A},\,\|a\|_1\le1}\sqrt{\sum_{j=1}^p\frac{\beta_j^2}{a_j}} \tag{6.5}$$

and

$$\Omega_*(w) = \max_{a\in\mathscr{A},\,\|a\|_1=1}\sqrt{\sum_{j=1}^p a_j w_j^2} = \max_{a\in\mathscr{A},\,\|a\|_1\le1}\sqrt{\sum_{j=1}^p a_j w_j^2}. \tag{6.6}$$

Proof This is Problem 6.3.   □

For $J \subset \{1, \ldots, p\}$ we set

$$\mathscr{A}_J = \{a_J : a \in \mathscr{A}\}.$$

Note that $\mathscr{A}_J$ is a convex cone in $\mathbb{R}_+^{|J|}$ (whenever $\mathscr{A}$ is one in $\mathbb{R}_+^p$). Denote the norm on $\mathbb{R}^{|J|}$ generated by $\mathscr{A}_J$ as

$$\Omega^J(\beta_J) := \min_{a_J \in \mathscr{A}_J} \frac{1}{2} \sum_{j \in J} \left[\frac{\beta_j^2}{a_j} + a_j \right], \quad \beta \in \mathbb{R}^p.$$

Recall that a set S is called *allowed* if Ω is weakly decomposable for the set S.

Lemma 6.7 *If $\mathscr{A}_S$ considered as subset of $\mathbb{R}^p$ is a subset of $\mathscr{A}$ we have the weak decomposability*

$$\Omega \geq \Omega(\cdot | S) + \Omega^{-S}$$

so that S is allowed.

Proof Observe first that $\mathscr{A}_S \subset \mathscr{A}$ implies $\Omega(\beta_S) = \Omega^S(\beta_S)$. Moreover

$$\Omega(\beta) \geq \min_{a \in \mathscr{A}} \frac{1}{2} \sum_{j \in S} \left[\frac{\beta_j^2}{a_j} + a_j \right] + \min_{a \in \mathscr{A}} \frac{1}{2} \sum_{j \notin S} \left[\frac{\beta_j^2}{a_j} + a_j \right]$$

$$\geq \min_{a_S \in \mathscr{A}_S} \frac{1}{2} \sum_{j \in S} \left[\frac{\beta_j^2}{a_j} + a_j \right] + \min_{a_{-S} \in \mathscr{A}_{-S}} \frac{1}{2} \sum_{j \notin S} \left[\frac{\beta_j^2}{a_j} + a_j \right]$$

$$= \Omega_S(\beta_S) + \Omega^{-S}(\beta_{-S}).$$

□

Lemma 6.2 pointed out that in the case of an allowed set the Ω^{-S}-norm may be quite small. We now examine this for the special case of a norm generated from a cone.

Lemma 6.8 *Let $\mathscr{E}^{-J}$ be the extreme points of the Ω^{-J}-unit ball. Then*

$$\Omega(\cdot | - J) \leq \omega^{-J} \Omega^{-J}$$

where $\omega^{-J} = \max\{\Omega(e^{-J} | - J) : e^{-J} \in \mathscr{E}^{-J}\}$.

Proof Define $\omega := \max\{\Omega(\beta_{-J} | - J) : \Omega^{-J}(\beta_{-J}) = 1\}$. The maximum is attained in the extreme points of the Ω^{-J}-unit ball.   □

Recall the bad news in Remark 6.1 that the oracle results of Theorem 6.1 and its relatives in general do not imply bounds for the Ω-estimation error. However, there is some good news too: they do imply bounds for the ℓ_1-estimation error. This is clear from the next lemma.

Lemma 6.9 *For any set J,*

$$\|\beta_J\|_1 \leq \min\{\Omega(\beta_J|J), \Omega^J(\beta_J)\}, \forall \beta \in \mathbb{R}^p.$$

Proof of Lemma 6.9 We clearly have

$$\Omega(\beta) = \min_{a \in \mathscr{A}} \frac{1}{2} \sum_{j=1}^{p} \left[\frac{|\beta_j|^2}{a_j} + a_j \right] \geq \min_{a \in \mathbb{R}^p_+} \frac{1}{2} \sum_{j=1}^{p} \left[\frac{|\beta_j|^2}{a_j} + a_j \right].$$

But for each j the minimum of

$$\frac{1}{2} \sum_{j=1}^{p} \left[\frac{|\beta_j|^2}{a_j} + a_j \right]$$

over $a_j \geq 0$ is equal to $|\beta_j|$. We apply this argument with Ω respectively replaced by $\Omega(\cdot|J)$ and Ω^J. □

We give four examples from Micchelli et al. (2010).

Example 6.1 (Group Lasso Penalty) Let $\{G_j\}_{j=1}^m$ be a partition of $\{1, \ldots, p\}$ into m groups. The set $\mathscr{A}$ consists of all non-negative vectors which are constant within groups. This gives

$$\Omega(\beta) := \sum_{j=1}^{m} \sqrt{|G_j|} \|\beta_{G_j}\|_2.$$

With squared error loss a penalty proportional to this choice of Ω is called the *Group Lasso*. It is introduced in Yuan and Lin (2006). Oracle inequalities for the group Lasso have been derived in Lounici et al. (2011) for example. For the square-root version we refer to Bunea et al. (2014). The dual norm is

$$\Omega_*(w) = \max_{1 \leq j \leq m} \|w_{G_j}\|_2 / \sqrt{|G_j|}.$$

Any union of groups is an allowed set and we moreover have for any allowed set S

$$\Omega^{-S} = \Omega(\cdot| - S)$$

and

$$\Omega = \Omega(\cdot|S) + \Omega^{-S}.$$

In other words, this norm is decomposable which frees it from the concerns expressed in Remark 6.1.

Example 6.2 (Wedge Penalty) The norm corresponding to the wedge penalty is generated from the cone

$$\mathscr{A} := \{a_1 \geq a_2 \geq \cdots\}.$$

Let for some $s \in \mathbb{N}$, the set $S := \{1, \ldots, s\}$ be the first s indices. Then S is an allowed set. To see that Ω^{-S} can be much smaller than $\Omega(\cdot| - S)$, take the vector $\beta \in \mathbb{R}^p$ to be one in its $s + 1$th entry and zero elsewhere. Then $\Omega^{-S}(\beta) = 1$ but $\Omega(\beta| - S) = \sqrt{s+1}$.

Example 6.3 (DAG Penalty) Let $\mathscr{A} = \{Aa \geq 0\}$ where A is the incidence matrix of a directed acyclic graph (DAG) with nodes $\{1, \ldots, p\}$. Then removing orphans is allowed, i.e., successively removing nodes with only outgoing edges the remaining set is allowed at each stage.

Example 6.4 (Concavity Inducing Penalty) Let

$$\mathscr{A} := \{a_1 \geq \cdots \geq a_p,\ a_{k+2} - 2a_{k-1} + a_k \leq 0,\ k = 1, \ldots, p-2\}.$$

Then allowed sets are $S = \{1, \ldots, s\}$, $s = 1, \ldots, p$.

6.10 Ω_*-Bounds for the Ω-Structured Sparsity Estimator

Consider the Ω-structured sparsity estimator

$$\hat{\beta} := \arg \min_{\beta \in \mathbb{R}^p} \left\{ \|Y - X\beta\|_n^2 + 2\lambda \Omega(\beta) \right\}.$$

This section studies bounds for $\Omega_*(\hat{\beta} - \beta^0)$. Let us examine here (for the current chapter only in this section) the case of random design. Let $\hat{\Sigma} = X^T X/n$, $\Sigma_0 := \mathbb{E}\hat{\Sigma}$ and suppose $\Theta_0 := \Sigma_0^{-1}$ exists. We use for a $p \times q$ matrix $A = (A_1, \ldots, A_q)$ the notation

$$\|A\|_\Omega := \max_{\Omega_*(\beta) \leq 1} \Omega_*(A^T \beta), \quad \|A\|_\Omega := \max_{1 \leq k \leq q} \Omega(A_k).$$

Note that

$$\|A\|_{\|\cdot\|_1} \le \|A\|_1.$$

Further

$$\|A\|_{\|\cdot\|_\infty} = \|A\|_\infty.$$

If Ω is stronger than the ℓ_1-norm (see Sect. 6.7), its dual norm Ω_* is less strong than the ℓ_∞-norm. In Theorem 6.1 bounds for the $\underline{\Omega}$-error were presented, where $\underline{\Omega} \le \Omega$. We now consider bounds for the Ω_*-error in terms of the $\underline{\Omega}$-error. As in Lemma 4.1 in Sect. 4.4 we have the following result.

Lemma 6.10 *Let $\lambda_\epsilon \ge \Omega_*(X^T\epsilon)/n$. Consider some other norm $\underline{\Omega} \le \Omega$.*

- *If $\|\cdot\|_1 \le \underline{\Omega} \le \Omega$, then $\|\cdot\|_\infty$ is a stronger norm than Ω_* but we only obtain a bound for the Ω_*-error:*

$$\Omega_*(\hat{\beta} - \beta^0) \le \|\Theta_0\|_\Omega \left[\lambda + \lambda_\epsilon + \|\hat{\Sigma} - \Sigma^0\|_\infty \underline{\Omega}(\hat{\beta} - \beta^0)\right].$$

Norms stronger than $\|\cdot\|_1$ are perhaps the most interesting ones, but for completeness also the reverse situation is presented here:
- *If $\|\cdot\|_1 \ge \hat{\Omega}$, then $\|\cdot\|_\infty$ is a weaker norm than Ω_* but we only obtain a bound for the $\|\cdot\|_\infty$-error:*

$$\|\hat{\beta} - \beta_0\|_\infty \le \|\Theta_0\|_1 \left[\lambda + \lambda_\epsilon + \|\hat{\Sigma} - \Sigma^0\|_{\Omega_*} \underline{\Omega}(\hat{\beta} - \beta^0)\right].$$

Proof of Lemma 6.10 By the KKT-conditions

$$X^T(Y - X\hat{\beta})/n = \lambda\hat{z}, \ \Omega_*(\hat{z}) \le 1.$$

Therefore

$$\hat{\beta} - \beta^0 = \Theta_0 X^T\epsilon/n - \lambda\Theta_0\hat{z} - \Theta_0(\hat{\Sigma} - \Sigma_0)(\hat{\beta} - \beta^0).$$

- Suppose $\|\cdot\|_1 \le \underline{\Omega} \le \Omega$. We then bound the three terms in the above expression in Ω_*-norm. By the definition of $\|\cdot\|_\Omega$

$$\Omega_*(\Theta_0 X^T\epsilon)/n \le \|\Theta_0\|_\Omega \Omega_*(X^T\epsilon)/n, \ \Omega_*(\Theta_0\hat{z}) \le \|\Theta_0\|_\Omega$$

and

$$\Omega_*(\Theta_0(\hat{\Sigma} - \Sigma_0)(\hat{\beta} - \beta^0)) \leq \|\Theta_0\|_\Omega \Omega_*((\hat{\Sigma} - \Sigma_0)(\hat{\beta} - \beta^0))$$
$$\leq \|\Theta_0\|_\Omega \|(\hat{\Sigma} - \Sigma_0)(\hat{\beta} - \beta^0)\|_\infty$$
$$\leq \|\Theta_0\|_\Omega \|\hat{\Sigma} - \Sigma_0\|_\infty \|\hat{\beta} - \beta^0\|_1$$
$$\leq \|\Theta_0\|_\Omega \|\hat{\Sigma} - \Sigma_0\|_\infty \underline{\Omega}(\hat{\beta} - \beta^0).$$

- Suppose $\|\cdot\|_1 \geq \Omega$. We then bound the three terms in $\|\cdot\|_\infty$-norm:

$$\|\Theta_0 X^T \epsilon\|_\infty / n \leq \|\Theta_0\|_1 \|X^T \epsilon\|_\infty / n \leq \|\Theta_0\|_1 \lambda_\epsilon,$$

$$\|\Theta_0 \hat{z}\|_\infty \leq \|\Theta_0\|_1 \|\hat{z}\|_\infty \leq \|\Theta_0\|_1 \Omega_*(\hat{z}) \leq \|\Theta_0\|_1$$

and

$$\|\Theta_0(\hat{\Sigma} - \Sigma_0)(\hat{\beta} - \beta^0)\|_\infty \leq \|\Theta_0\|_1 \|(\hat{\Sigma} - \Sigma_0)(\hat{\beta} - \beta_0)\|_\infty$$
$$\leq \|\Theta_0\|_1 \|\hat{\Sigma} - \Sigma_0\|_{\underline{\Omega_*}} \Omega(\hat{\beta} - \beta^0).$$

□

6.11   Confidence Intervals Using Structured Sparsity

The same approach as in Sect. 3.8 is proposed, but now with penalization by a general norm. The structured sparsity estimator $\hat{\beta}$ can be de-sparsified using node-wise the multivariate square-root structured sparsity "estimators" for defining surrogate projections. It turns out that the ℓ_1-norm is generally a "safe" choice for getting surrogate projections (see Corollary 6.2).

Let $J \subset \{1, \ldots, p\}$ and Ω_{-J} be some norm on $(p - |J|)$-dimensional space.[2] Define the Ω_{-J}-structured sparsity square-root Lasso

$$\hat{\Gamma}_J := \arg\min_{\Gamma_J} \left\{ \|X_J - X_{-J}\Gamma_J\|_{\text{nuclear}} / \sqrt{n} + \lambda_\sharp \Omega_{-J}(\Gamma_J) \right\}.$$

where for $\Gamma_J = \{\gamma_j\}_{j \in J} = \{\gamma_{k,j}\}_{k \notin J, j \in J}$,

$$\Omega_{-J}(\Gamma_J) := \sum_{j \in J} \Omega_{-J}(\gamma_j).$$

---

[2]If Ω is a weakly decomposable norm on $\mathbb{R}^p$ and J is an allowed set, one may think of choosing $\Omega_{-J} = \Omega^{-J}$. Alternatively, if J is the complement of an allowed set, one might choose $\Omega_{-J} = \Omega(\cdot| - J)$.

The surrogate projection is

$$X_J \hat{P} X_{-J} := X_{-J} \hat{\Gamma}_J$$

with surrogate anti-projection

$$X_J \hat{A} X_{-J} = X_J - X_J \hat{P} X_{-J} =: X \hat{C}_J.$$

We let

$$\hat{T}_J := (X_J \hat{A} X_{-J})^T (X_J \hat{A} X_{-J})/n$$
$$\tilde{T}_J := (X_J \hat{A} X_{-J})^T X_J / n.$$

Let $\Omega_{*,-J}$ be the dual norm of Ω_{-J}:

$$\Omega_{*,-J}(w_{-J}) := \sup\{w_{-J}^T \beta_{-J} :\ \Omega_{-J}(\beta_{-J}) \leq 1\},\ w_{-J} \in \mathbb{R}^{p-|J|}.$$

Applying the KKT-conditions forming a counterpart of Lemma 3.4 in Sect. 3.8, we find (Problem 6.4)

$$\|X_{-J}^T (X_J \hat{A} X_{-J}) \hat{T}_J^{-1/2}\|_{\Omega_{*,-J}}/n \leq \lambda_\sharp \qquad (6.7)$$

where for a $(p - |J|) \times |J|$ matrix $W = \{W_j\}_{j \in J}$.

$$\|W\|_{\Omega_{*,-J}} := \max_{j \in J} \Omega_{*,-J}(W_j).$$

Definition 6.4 Let $\hat{\beta}$ be an initial estimator. The Ω_{-J}-de-sparsified estimator of β_J^0 is defined as a solution $\hat{b}_J$ of the equation

$$\tilde{T}_J \hat{b}_J = \tilde{T}_J \hat{\beta}_J + (X_J \hat{A} X_{-J})^T (Y - X\hat{\beta})/n.$$

Remark 6.2 If $\tilde{T}_J$ is invertible then obviously

$$\hat{b}_J = \hat{\beta}_J + \tilde{T}_J^{-1}(X_J \hat{A} X_{-J})^T (Y - X\hat{\beta})/n.$$

Lemma 6.11 *Assume $\hat{T}_J$ is invertible. Then*

$$\hat{T}_J^{-1/2} \tilde{T}_J (\hat{b}_J - \beta_J^0)/n = \hat{T}_J^{-1/2}(X_J \hat{A} X_{-J})^T \epsilon/n - \mathrm{rem}_J$$

where

$$\|\mathrm{rem}_J\|_\infty \leq \lambda_\sharp \Omega_{-J}(\hat{\beta}_{-J} - \beta_{-J}^0).$$

Proof of Lemma 6.11 Write

$$
\begin{aligned}
\hat{T}_J^{-1/2}\tilde{T}_J(\hat{b}_J - \beta_J^0) &= \hat{T}_J^{-1/2}\tilde{T}_J(\hat{\beta}_J - \beta_J^0) + \hat{T}_J^{-1/2}(X_J\hat{A}X_{-J})^T(Y - X\hat{\beta})/n \\
&= \hat{T}_J^{-1/2}\tilde{T}_J(\hat{\beta}_J - \beta_J^0) + \hat{T}_J^{-1/2}(X_J\hat{A}X_{-J})^T\epsilon/n \\
&\quad -\hat{T}_J^{-1/2}(X_J\hat{A}X_{-J})^T X(\hat{\beta} - \beta^0)/n \\
&= \hat{T}_J^{-1/2}(X_J\hat{A}X_{-J})^T\epsilon/n - \underbrace{\hat{T}_J^{-1/2}(X_J\hat{A}X_{-J})^T X_{-J}(\hat{\beta}_{-J} - \beta_{-J}^0)/n}_{:=\mathrm{rem}_J}.
\end{aligned}
$$

But, invoking (6.7),

$$
\begin{aligned}
|\mathrm{rem}_{j,J}| &\leq \Omega_{-J}(\hat{\beta}_{-J} - \beta_{-J}^0)\left\| (\hat{T}_J^{-1/2}(X_J\hat{A}X_{-J})^T X_{-J})_j \right\|_{\Omega_*}/n \\
&\leq \lambda_\sharp \Omega_{-J}(\hat{\beta}_{-J} - \beta_{-J}^0).
\end{aligned}
$$

$\square$

The point is now that "good" bounds for the Ω_{-J}-estimation error are not so obvious, even when the initial estimator is the Ω-structured sparsity estimator with a weakly decomposable Ω. Let S_* be the allowed "oracle" set obtained from Theorem 6.1 in Sect. 6.6. Then we may have good bounds for the $\underline{\Omega}$-error of the Ω-structured sparsity estimator with $\underline{\Omega} = \Omega(\cdot|S_*) + \Omega^{-S_*}$. However, since S_* is unknown it may not be an option to use Ω^{-S^*} as the norm Ω_{-J} used for getting surrogate projections. If one applies the square-root Lasso for calculating the surrogate projections (i.e., $\Omega_{-J} = \|\cdot\|_1$), then one obtains by the above result a bound for the remainder in terms of the ℓ_1-error of $\hat{\beta}_{-J}$. Thus, if the initial estimator was $\hat{\beta}$ is the Ω-structured sparsity estimator with a norm stronger than ℓ_1 we may invoke the oracle inequality of Sect. 6.6. This is why the following corollary is formulated.

Corollary 6.2 *Take $\Omega_{-J} = \|\cdot\|_1$ and let $\underline{\Omega}$ be a norm stronger than ℓ_1. Then the remainder in Lemma 6.11 satisfies*

$$
\|\mathrm{rem}_J\|_\infty \leq \lambda_\sharp \|\hat{\beta}_{-J} - \beta_{-J}^0\|_1 \leq \lambda_\sharp \underline{\Omega}(\hat{\beta} - \beta^0).
$$

If $\hat{\beta}$ is the Ω-structured sparsity estimator one may bound $\underline{\Omega}(\hat{\beta} - \beta^0)$ for suitable $\underline{\Omega}$ by applying the oracle inequality of Sect. 6.6. As a special case, one may invoke Lemma 6.9 which says that when Ω is a norm generated from a convex cone, then with the notation used there, for any set J

$$
\|\cdot\|_1 \leq \Omega^{-J}.
$$

Thus, as explained in Corollary 6.2, if Ω is generated from a convex cone, the Ω-structured sparsity estimator $\hat{\beta}$ may be de-sparsified using $\Omega_{-J} = \| \cdot \|_1$. We discuss two examples where another choice for Ω_{-J} appears to be more natural.

Example 6.5 (Continuation of Example 6.1: Group Lasso Penalty) Let $\{G_j\}_{j=1}^m$ be the groups, which form a partition of $\{1, \ldots, p\}$. The penalty is

$$\Omega(\beta) := \sum_{j=1}^{m} \sqrt{|G_j|} \|\beta_{G_j}\|_2.$$

In this setup, it may be natural to de-sparsity all the coefficients within a whole groups at once, and to apply the group Lasso for the de-sparsifying step. In that case, one takes $J = G_j$ for some fixed j (the group of interest for which one wants to construct confidence intervals) and $\Omega_{-J} = \Omega(\cdot| - G_j) = \Omega^{-G_j}$.

Example 6.6 (Continuation of Example 6.2: Wedge Penalty)

Case 1 Suppose we want a confidence set for the first s variables (with typically the largest in absolute value coefficients). Let $S := \{1, \ldots, s\}$. Then S is an allowed set. It makes sense to use the penalty $\Omega_{-J} = \Omega^{-S}$, because it means that, in the chain graph from $\{p\}$ to $\{1\}$, for $j \leq s$ and $k > s$ the regression of X_j on X_k is judged as decreasing in k in importance. Let $S_0 = S_{\beta^0} = \{1, \ldots, s_0\}$. Then with the bound

$$\Omega(\hat{\beta}_{-S_0}) \leq \sqrt{s_0 + 1}\, \Omega^{-S_0}(\hat{\beta}_{-S_0}),$$

we have

$$\Omega^{-S}(\hat{\beta}_{-S} - \beta^0_{-S}) \leq \Omega(\hat{\beta} - \beta^0) \leq \sqrt{s_0 + 1}\, \Omega^{-S_0}(\hat{\beta}_{-S_0}) + \Omega(\hat{\beta}_{S_0} - \beta^0).$$

Another choice for Ω_{-J} would be $\Omega_{-J} = \Omega(\cdot| - S)$.

Case 2 Suppose we want a confidence set for the last m variables, or we want to test whether $\beta^0_{p-m} = \cdots = \beta^0_p = 0$. Then we take $J = S^c$ where $S = \{1, \ldots, p - m\}$ is an allowed set. A possible choice for Ω_{-J} is now $\Omega_{-J} = \Omega(\cdot|S)$ but this seems to make little sense (it favours the largest coefficients for the variables that are the furthest away). The "all purposes" choice $\Omega_{-J} = \| \cdot \|_1$ seems to be the most appropriate here.

6.12   Complements

6.12.1   The Case Where Some Coefficients Are Not Penalized

Suppose the coefficients with index set $U \subset \{1, \ldots, p\}$ are not penalized. The Ω-structured sparsity estimator is then

$$\hat{\beta} := \arg \min_{\beta \in \mathbb{R}^p} \left\{ \|Y - X\beta\|_n^2 + 2\lambda\Omega(\beta| - U) \right\}$$

where $\Omega(\beta| - U) := \Omega(\beta_{-U})$, $\beta \in \mathbb{R}^p$. We need the following result.

Lemma 6.12 *Suppose that* $\Omega(\cdot| - U) \leq \Omega$. *Then for all* $z_{-U} \in \mathbb{R}^p$

$$\Omega_*(z_{-U}| - U) = \Omega_*(z_{-U}).$$

Proof By the definition of Ω_*

$$\Omega_*(z_{-U}) = \max_{\Omega(\beta)\leq 1} \beta^T z_{-U}.$$

Hence

$$\Omega_*(z_{-U}) \geq \max_{\Omega(\beta)\leq 1, \; \beta = \beta_{-U}} \beta^T z_{-U} = \max_{\Omega(\beta_{-U})\leq 1} \beta_{-U}^T z_{-U} = \Omega_*(z_{-U}| - U).$$

On the other hand, the condition $\Omega(\cdot| - U) \leq \Omega$ implies

$$\Omega(\beta) \leq 1 \Rightarrow \Omega(\beta_{-U}) \leq 1$$

and therefore

$$\Omega_*(z_{-U}) \leq \max_{\Omega(\beta_{-U})\leq 1} \beta_{-U}^T z_{-U} = \Omega_*(z_{-U}| - U).$$

$\square$

When $\Omega(\cdot| - U) \leq \Omega$ the KKT-conditions are (see Problem 6.6)

$$X^T(Y - X\hat{\beta})/n + \lambda\hat{z}_{-U} = 0, \Omega_*(\hat{z}_{-U}) \leq 1, \hat{z}_{-U}^T \hat{\beta}_{-U} = \Omega(\hat{\beta}_{-U}).$$

6.12.2   The Sorted ℓ_1-Norm

Let $\lambda_1 \geq \lambda_2 \geq \cdots \geq \lambda_p$ be a given increasing sequence. For $\beta = (\beta_1, \ldots, \beta_p)^T \in \mathbb{R}^p$ we define the vector of absolute values in increasing order

$|\beta|_{(1)} \geq |\beta|_{(2)} \geq \cdots \geq |\beta|_{(p)}$. The sorted ℓ_1-norm is

$$\Omega(\beta) = \sum_{j=1}^{p} \lambda_j |\beta|_{(j)}.$$

It was introduced in Bogdan et al. (2013). In Zeng and Mario (2014) it is shown that this is indeed a norm and its dual norm is provided. We now show that this norm is weakly decomposable.

Lemma 6.13 *Let*

$$\Omega(\beta) = \sum_{j=1}^{p} \lambda_j |\beta|_{(j)},$$

and

$$\Omega^{-S}(\beta_{-S}) = \sum_{l=1}^{r} \lambda_{p-r+l} |\beta|_{(l,-S)},$$

where $r = p - s$ and $|\beta|_{(1,-S)} \geq \cdots \geq |\beta|_{(r,-S)}$ is the ordered sequence in β_{-S}. Then $\Omega(\beta) \geq \Omega(\beta_S) + \Omega^{-S}(\beta_{-S})$. Moreover Ω^{-S} is the strongest norm among all $\underline{\Omega}^{-S}$ for which $\Omega(\beta) \geq \Omega(\beta_S) + \underline{\Omega}^{-S}(\beta_{-S})$

Proof of Lemma 6.13 Without loss of generality assume $\beta_1 \geq \cdots \geq \beta_p \geq 0$. We have

$$\Omega(\beta^S) + \Omega^{-S}(\beta_{-S}) = \sum_{j=1}^{p} \lambda_j \beta_{\pi(j)}$$

for a suitable permutation π. It follows that (Problem 6.7)

$$\Omega(\beta_S) + \Omega^{-S}(\beta_{-S}) \leq \Omega(\beta).$$

To show Ω^{-S} is the strongest norm it is clear we need only to search among candidates of the form

$$\underline{\Omega}^{-S}(\beta_{-S}) = \sum_{l=1}^{r} \underline{\lambda}_{p-r+l} \beta_{\pi^{-S}(l)}$$

where $\{\underline{\lambda}_{p-r+l}\}$ is a decreasing positive sequence and where $\pi^{-S}(1), \ldots, \pi^{-S}(r)$ is a permutation of indices in S^c. This is then maximized by ordering the indices in S^c in decreasing order (see Problem 6.7). But then it follows that the largest norm is obtained by taking $\underline{\lambda}_{p-r+l} = \lambda_{p-r+l}$ for all $l = 1, \ldots, r$. □

6.12.3 A Direct Proof of Theorem 6.1

Instead of checking the conditions of the more general Theorem 7.1 we give here a direct proof. This also helps to follow the assertion of Proposition 6.1. We simplify the notation somewhat by writing $\Omega_2 := \Omega^{-S}$, $\lambda_1 := \lambda_S$ and $\lambda_2 := \lambda^{-S}$.

- If

$$\delta_1 \underline{\lambda} \Omega(\hat{\beta}_S - \beta_S) + \delta_2 \underline{\lambda} \Omega_2(\hat{\beta}_{-S} - \beta_{-S}) + (\hat{\beta} - \beta)^T \hat{\Sigma}(\hat{\beta} - \beta^0) \leq 2\lambda\Omega(\beta_{-S})$$

we know from the two point margin that

$$2\delta_1 \underline{\lambda} \Omega(\hat{\beta}_S - \beta_S) + 2\delta_2 \underline{\lambda} \Omega_2(\hat{\beta}_{-S} - \beta_{-S}) + \|X(\hat{\beta} - \beta^0)\|_n^2$$
$$\leq \|X(\beta - \beta^0)\|_n^2 + 4\lambda\Omega(\beta_{-S}).$$

- Suppose now that

$$\delta_1 \underline{\lambda} \Omega(\hat{\beta}_S - \beta_S) + \delta_2 \underline{\lambda} \Omega_2(\hat{\beta}_{-S} - \beta_{-S}) + (\hat{\beta} - \beta)^T \hat{\Sigma}(\hat{\beta} - \beta^0) \geq 2\lambda\Omega(\beta_{-S}). \qquad (6.8)$$

By Lemma 6.1

$$(\hat{\beta} - \beta)^T \hat{\Sigma}(\hat{\beta} - \beta^0) \leq (\hat{\beta} - \beta)^T X^T \epsilon / n + \lambda\Omega(\beta) - \lambda\Omega(\hat{\beta})$$
$$\leq \lambda_1 \Omega(\hat{\beta}_S - \beta_S) + \lambda_2 \Omega_2(\hat{\beta}_{-S}) + (\lambda + \lambda_2)\Omega(\beta_{-S}) + \lambda\Omega(\beta_S) - \lambda\Omega(\hat{\beta})$$
$$\leq (\lambda + \lambda_1)\Omega(\hat{\beta}_S - \beta_S) - \underline{\lambda}\Omega_2(\hat{\beta}_{-S} - \beta_{-S}) + 2\lambda\Omega(\beta_{-S}).$$

We summarize this and give the inequality a number for reference:

$$(\hat{\beta} - \beta)^T \hat{\Sigma}(\hat{\beta} - \beta^0) \leq (\lambda + \lambda_1)\Omega(\hat{\beta}_S - \beta_S) - \underline{\lambda}\Omega_2(\hat{\beta}_{-S} - \beta_{-S}) + 2\lambda\Omega(\beta_{-S}).$$
$$(6.9)$$

From (6.8) we see that

$$(1 - \delta_2)\underline{\lambda}\Omega_2(\hat{\beta}_{-S} - \beta_{-S}) \leq \bar{\lambda}\Omega(\hat{\beta}_S - \beta_S)$$

or

$$\Omega_2(\hat{\beta}_{-S} - \beta_{-S}) \leq L\Omega(\hat{\beta}_S - \beta_S).$$

It follows that

$$\Omega(\hat{\beta}_S - \beta_S) \leq \sqrt{|S|}\|X(\hat{\beta} - \beta)\|_n / \hat{\phi}_\Omega(L, S).$$

But then, inserting (6.9),

$$(\hat{\beta} - \beta)^T \hat{\Sigma} (\hat{\beta} - \beta^0) + \delta_1 \underline{\lambda} \Omega (\hat{\beta}_S - \beta_S) + \delta_2 \underline{\lambda} \Omega_2 (\hat{\beta}_{-S} - \beta_{-S})$$

$$\leq \bar{\lambda} \Omega (\hat{\beta}_S - \beta_S) + 2\lambda \Omega (\beta_{-S})$$

$$\leq \bar{\lambda} \sqrt{|S|} \|X(\hat{\beta} - \beta)\|_n / \hat{\phi}_\Omega (L, S) + 2\lambda \Omega (\beta_{-S})$$

$$\leq \frac{1}{2} \frac{\bar{\lambda}^2 |S|}{\hat{\phi}_\Omega^2 (L, S)} + \frac{1}{2} \|X(\hat{\beta} - \beta)\|_n^2 + 2\lambda \Omega (\beta_{-S}).$$

By the two point margin this gives

$$\|X(\hat{\beta} - \beta^0)\|_n^2 + 2\delta_1 \underline{\lambda} \Omega (\hat{\beta}_S - \beta_S) + 2\delta_2 \underline{\lambda} \Omega_2 (\hat{\beta}_{-S} - \beta_{-S})$$

$$\leq \|X(\beta - \beta^0)\|_n^2 + \frac{\bar{\lambda}^2 |S|}{\hat{\phi}^2 (L, S)} + 4\lambda \Omega (\beta_{-S}).$$

□

Problems

6.1 Show that Lemma 6.1 remains valid if we minimize over a convex subset of $\mathbb{R}^p$ instead of the whole $\mathbb{R}^p$.

6.2 Check that Lemma 6.1 for the square-root case reads as follows. Let for a convex penalty pen($\cdot$)

$$\hat{\beta} := \arg\min_{\beta \in \mathbb{R}^p} \left\{ \|Y - X\beta\|_n + \text{pen}(\beta) \right\}.$$

Assume that $\hat{\epsilon} := Y - X\hat{\beta}$ is not the zero-vector. Then for all β

$$(\beta - \hat{\beta})^T X^T (Y - X\hat{\beta})/n \leq \|\hat{\epsilon}\|_n \text{pen}(\beta) - \|\hat{\epsilon}\|_n \text{pen}(\hat{\beta}). \tag{6.10}$$

6.3 Prove the first part of the first result (6.5) in Lemma 6.6 (Sect. 6.9) by writing

$$\Omega (\beta) = \min_{a \in \mathscr{A}, \, \|a\|_1 = 1} \min_{c > 0} \frac{1}{2} \sum_{j=1}^{p} \left[\frac{|\beta_j|^2}{ca_j} + ca_j \right].$$

The second part of the first result (6.5) can be obtained by normalizing: for $0 < \|a\|_1 \leq 1$

$$\sum_{j=1}^{p} \frac{\beta_j^2}{a_j} = \sum_{j=1}^{p} \frac{\beta_j^2}{a_j/\|a\|_1} \frac{1}{\|a\|_1} \geq \sum_{j=1}^{p} \frac{\beta_j^2}{a_j/\|a\|_1}.$$

To prove the first part of the second result (6.6) in Lemma 6.6, define for $\beta \in \mathbb{R}^p$, $a(\beta) \in \mathscr{A} \cup \{a : \|a\|_1 = 1\}$ such that

$$\Omega(\beta) = \sqrt{\sum_{j=1}^{p} \frac{\beta_j^2}{a_j(\beta)}}$$

and use the Cauchy-Schwarz inequality to see that

$$|w^T \beta| \leq \Omega(\beta) \sqrt{\sum_{j=1}^{p} a_j(\beta) w_j^2}.$$

Similarly, take $\bar{a}(w) \in \mathscr{A} \cup \{a : \|a\|_1 = 1\}$ such that

$$\max_{a \in \mathscr{A}, \|a\|_1 = 1} \sqrt{\sum_{j=1}^{p} a_j w_j^2} = \sqrt{\sum_{j=1}^{p} \bar{a}_j(w) w_j^2}$$

and take $\beta_j = \bar{a}_j(w) w_j$. Then show that

$$|w^T \beta| \geq \Omega(\beta) \sqrt{\sum_{j=1}^{p} \bar{a}_j(w) w_j^2}.$$

To prove the second part of the second result (6.6) write

$$\max_{a \in \mathscr{A}, \|a\|_1 \leq 1} \sqrt{\sum_{j=1}^{p} a_j w_j^2} = \max_{a \in \mathscr{A}, \|a\|_1 = 1} \max_{c \leq 1} c \sqrt{\sum_{j=1}^{p} a_j w_j^2}.$$

6.4 Use the same arguments as in Lemma 3.4 and use the KKT-conditions given in Sect. 6.2 to show (6.7) in Sect. 6.11.

6.5 The multiple regression model is

$$Y_{i,t} = X_{i,t} \beta_t^0 + \epsilon_{i,t}, \ i = 1, \ldots, n, \ t = 1, \ldots, T,$$

where $\{\epsilon_{i,t} : i = 1, \ldots, n, \ t = 1, \ldots, T\}$ are i.i.d. zero mean noise variables. For each i and t, the input $X_{i,t}$ is a given p-vector and for each t, the vector $\beta_t^0 \in \mathbb{R}^p$ is unknown. The index t refers to "time". Let us say we believe that over time the coefficients are either all zero or all non-zero. A group Lasso penalty that takes such structure into account is

$$\Omega(\beta) := \sum_{j=1}^{p} \left[\sum_{t=1}^{T} \beta_{j,t}^2 \right]^{1/2},$$

for $\beta^T = (\beta_1^T, \cdots, \beta_t^T) \in \mathbb{R}^{pT}$. Let the estimator be

$$\hat{\beta} := \arg \min_{\beta \in \mathbb{R}^{pT}} \left\{ \sum_{t=1}^{T} \sum_{i=1}^{n} |Y_{i,t} - X_{i,t}\beta_t|^2/(nT) + 2\lambda\Omega(\beta) \right\}.$$

By applying Theorem 6.1 in Sect. 6.6, formulate a sharp oracle inequality for

$$\sum_{t=1}^{T} \sum_{i=1}^{n} |X_{i,t}(\hat{\beta}_t - \beta_t^0)|^2/(nT).$$

Let now $j \in \{1, \ldots, p\}$ be fixed (the variable of interest). How would you construct a confidence set for $\{\beta_{j,t}^0 : t = 1, \ldots, T\}$?

6.6 Consider the situation of Sect. 6.12.1 where coefficients in the set U are left unpenalized. Let $\Omega(\cdot | -U) \leq \Omega$. Check that the KKT-conditions for the case of unpenalized coefficients with indices U are as given in Sect. 6.12.1. Then verify that Theorem 6.1 is true for candidate oracles (β, S) with allowed sets S containing U.

6.7 Let $\lambda_1 \geq \cdots \geq \lambda_p$ be a decreasing sequence of positive numbers. Let moreover π be a permutation of $\{1, \ldots, p\}$. The identity permutation is denoted by id. Let $\beta_1 \geq \cdots \geq \beta_p$ be a decreasing sequence of non-negative numbers. Show that the sum $\sum_{j=1}^{p} \lambda_j \beta_{\pi(j)}$ is maximized over all permutations π at $\pi = id$.

6.8 Let $\hat{\beta}$ be the structured sparsity estimator

$$\hat{\beta} := \arg \min_{\beta \in \mathbb{R}^p} \left\{ \|Y - X\beta\|_n^2 + 2\lambda\Omega(\beta) \right\}.$$

Show the KKT-conditions

$$X^T(Y - X\hat{\beta})/n = \lambda\hat{z}, \ \hat{z} \in \partial\Omega(\hat{\beta}).$$

Multiply with $(\beta - \hat{\beta})^T$ to get the two point inequality

$$(\beta - \hat{\beta})^T X^T(Y - X\hat{\beta})/n \leq \lambda\Omega(\beta) - \lambda\Omega(\hat{\beta}).$$

Chapter 7
General Loss with Norm-Penalty

Abstract Results obtained for least-squares loss are now extended to general convex loss. The regularization penalty is again a norm, and the concept of weak decomposability and "allowedness" is extended as well: the norm used in the penalty is required to have the *triangle property*. The generalized notion of compatibility will be *effective sparsity*. The oracle results obtained require almost-differentiability of the loss. It is moreover assumed that the population version of the problem, the theoretical risk function has strict convexity properties. This will be called the (two point) *margin condition*.

7.1 Introduction

Let $X_1, \ldots, X_n$ be independent observations with values in some observation space $\mathscr{X}$ and let for β in a space $\bar{\mathscr{B}} \subset \mathbb{R}^p$ be given a loss function $\rho_\beta : \mathscr{X} \to \mathbb{R}$. The parameter space—or model space—$\mathscr{B}$ is some given subset of $\bar{\mathscr{B}}$. It is potentially high-dimensional, so that possibly $p \gg n$. We require throughout convexity of parameter space and loss function. That is, we require Condition 7.1.1 without further explicit mentioning.

Condition 7.1.1 *The parameter space $\mathscr{B}$ as well as the map $\beta \mapsto \rho_\beta$, $\beta \in \mathscr{B}$ is convex.*

Define for all β in the extended space $\bar{\mathscr{B}}$ the empirical risk

$$R_n(\beta) := P_n \rho_\beta := \frac{1}{n} \sum_{i=1}^{n} \rho_\beta(X_i)$$

and the theoretical risk

$$R(\beta) := P\rho_\beta := \mathbb{E}R_n(\beta).$$

© Springer International Publishing Switzerland 2016
S. van de Geer, *Estimation and Testing Under Sparsity*,
Lecture Notes in Mathematics 2159, DOI 10.1007/978-3-319-32774-7_7

Let Ω be a norm on $\mathbb{R}^p$. This chapter studies the Ω-*structured sparsity M-estimator*

$$\hat{\beta} := \arg\min_{\beta \in \mathscr{B}} \left\{ R_n(\beta) + \lambda\Omega(\beta) \right\}$$

with $\lambda > 0$ a tuning parameter. The "M" stands for "minimization". We alternatively term $\hat{\beta}$ the Ω-regularized empirical risk minimizer.

The "true" parameter or "target" is defined as the minimizer of the theoretical risk over the extended space $\bar{\mathscr{B}}$

$$\beta^0 := \arg\min_{\beta \in \bar{\mathscr{B}}} R(\beta)$$

(where uniqueness is not required without expressing this in the notation). In many cases one simply is interested in the target with $\mathscr{B} = \bar{\mathscr{B}}$.[1] On the other hand β^0 may be some more general reference value, see Problem 7.6. As a look-ahead, the main result, Theorem 7.1 in Sect. 7.1, makes no explicit mention of any target β^0 (as it should be from a learning point of view). However, there is a mention of a local set $\mathscr{B}_{\text{local}}$. This generally points to a neighbourhood of some target β^0.

7.2 Two Point Inequality, Convex Conjugate and Two Point Margin

We first need to introduce a "local" set $\mathscr{B}_{\text{local}}$. Without further explicit mentioning, we require:

Condition 7.2.1 *The set $\mathscr{B}_{\text{local}}$ is a convex subset of $\mathscr{B}$.*

The set $\mathscr{B}_{\text{local}}$ is commonly a neighbourhood of β^0 (for some suitable topology). The reason is that typically the conditions we will impose (to be precise, Condition 7.2.2) only hold locally. One then needs to prove that the estimator is in the local neighbourhood. Here one may exploit the assumed convexity of the loss. Section 7.6 illustrates how this works. There $\mathscr{B}_{\text{local}}$ is a subset of the collection of $\beta' \in \mathscr{B}$ which are in a suitable Ω-norm close to β^0. In the case of quadratic loss, one generally does not need to localize, i.e, then one can take $\mathscr{B}_{\text{local}} = \mathscr{B}$. For the moment we leave the form of the local set unspecified (but we do require its convexity).

In what follows we will use parameter values β and β'. The value β will represent a "candidate oracle", that is, one should think of it as some fixed vector. The assumption $\beta \in \mathscr{B}_{\text{local}}$ is thus reasonable: candidate oracles are supposed to know

---

[1] An example where this is not the case is where $\mathscr{B}$ is a lower-dimensional subspace of $\bar{\mathscr{B}}$. This is comparable to the situation where one approximates a function (an ∞-dimensional object) by a p-dimensional linear function (with p large).

how to get close to the target β^0. The value β' as a rule represents the estimator $\hat{\beta}$. Thus the assumption $\beta' \in \mathscr{B}_{local}$ may mean that some work is to be done here.

Definition 7.1 We call R_n *almost-differentiable* if for all $\beta', \beta \in \mathscr{B}_{local}$

$$\lim_{t \downarrow 0} \frac{R_n((1-t)\beta' + t\beta) - R_n(\beta')}{t} \leq \dot{R}_n(\beta')^T(\beta - \beta')$$

for some $\dot{R}_n(\beta') \in \mathbb{R}^p$. We call $\dot{R}_n(\beta')$ an *almost-derivative* of R_n at β'.

The reason for not working with the standard differentiability concept is that then also loss functions that are differentiable up to null-sets can be included. This monograph however does not treat sharp oracle results for non-differentiable loss such as least absolute deviations loss (but we do give a non-sharp one in Sect. 12.5). Note that almost-differentiability has as side effect that it also allows β' to be at the border of parameter space.

Lemma 7.1 (Two Point Inequality) *Suppose R_n is almost-differentiable and that $\hat{\beta} \in \mathscr{B}_{local}$. Then for all $\beta \in \mathscr{B}_{local}$*

$$-\dot{R}_n(\hat{\beta})^T(\beta - \hat{\beta}) \leq \lambda\Omega(\beta) - \lambda\Omega(\hat{\beta}).$$

Proof of Lemma 7.1 Let $\beta \in \mathscr{B}$ and define for $0 < t < 1$,

$$\hat{\beta}_t := (1-t)\hat{\beta} + t\beta.$$

Recall that we require $\mathscr{B}_{local}$ to be convex, so $\hat{\beta}_t \in \mathscr{B}_{local}$ for all $0 < t < 1$. We have for pen $:= \lambda\Omega$

$$R_n(\hat{\beta}) + \text{pen}(\hat{\beta}) \leq R_n(\hat{\beta}_t) + \text{pen}(\hat{\beta}_t) \leq R_n(\hat{\beta}_t) + (1-t)\text{pen}(\hat{\beta}) + t\text{pen}(\beta).$$

Hence

$$\frac{R_n(\hat{\beta}) - R_n(\hat{\beta}_t)}{t} \leq \text{pen}(\beta) - \text{pen}(\hat{\beta}).$$

The results now follows by sending $t \downarrow 0$.   □

Definition 7.2 (Convex Conjugate) Let G be an increasing strictly convex non-negative function on $[0, \infty)$ with $G(0) = 0$. The *convex conjugate* of G is

$$H(v) := \sup_{u \geq 0}\left\{uv - G(u)\right\}, \ v \geq 0.$$

For example, the convex conjugate of the function $u \mapsto u^2/2$ is $v \mapsto v^2/2$.

Clearly, if H is the convex conjugate of G one has for all positive u and v

$$uv \le G(u) + H(v).$$

This is the one-dimensional version of the so-called Fenchel-Young inequality.

Condition 7.2.2 (Two Point Margin Condition) *There is an increasing strictly convex non-negative function G with $G(0) = 0$ and a semi-norm τ on $\mathscr{B}$ such that for all β and β' in $\mathscr{B}_{\text{local}}$ we have*

$$R(\beta) - R(\beta') \ge \dot{R}(\beta')^T (\beta - \beta') + G(\tau(\beta - \beta'))$$

for some $\dot{R}(\beta') \in \mathbb{R}^p$.

Note that $R(\cdot)$ is in view of our assumptions a convex function. One calls

$$B_R(\beta, \beta') := R(\beta) - R(\beta') - \dot{R}(\beta')^T (\beta - \beta'), \beta, \ \beta' \in \mathscr{B}_{\text{local}}$$

the *Bregman divergence*. Convexity implies that

$$B_R(\beta, \beta') \ge 0, \ \forall \ \beta, \ \beta' \in \mathscr{B}_{\text{local}}.$$

But the Bregman divergence is not symmetric in β and β' (nor does it satisfy the triangle inequality). The two point margin condition thus assumes that the Bregman divergence is lower bounded by a symmetric convex function. We present examples in Chap. 11.

We have in mind applying the two point margin condition at $\beta' = \hat{\beta}$ and $\beta = \beta^*$ where β^* is some "oracle" which trades off approximation error, effective sparsity and part of the vector β^* where the Ω-norm is smallish. Important to realize here is that the oracle β^* is a fixed vector. We note now that in the two point margin condition we assume the margin function G and the semi-norm τ not to depend on β' and β. The first (no dependence on β') is important, the last (no dependence on β) can be omitted (because we only need our conditions at a fixed value β^*). For ease of interpretation we refrain from the more general formulation.

7.3   Triangle Property and Effective Sparsity

In this section we introduce the *triangle property* for general norms Ω. The triangle property is a major ingredient for proving sharp oracle inequalities, see Theorem 7.1 in Sect. 7.5. Section 7.4 shows that the triangle property holds for certain vectors which are either *allowed* or *allowed** (or both). Examples can be found in Chap. 12.

Definition 7.3 Let Ω^+ and Ω^- be two semi-norms. We call them a *complete pair* if $\Omega^+ + \Omega^-$ is a norm.

Definition 7.4 Let Ω be a norm on $\mathbb{R}^p$. We say that the *triangle property* holds at β if for a complete pair of semi-norms Ω_β^+ and Ω_β^- and $\Omega_\beta^- \not\equiv 0$ one has

$$\Omega(\beta) - \Omega(\beta') \leq \Omega_\beta^+(\beta' - \beta) - \Omega_\beta^-(\beta'), \ \forall \ \beta' \in \mathbb{R}^p.$$

Note that in this definition one may choose for Ω_β^+ a very strong norm. This has its advantages (Theorem 7.1 then gives bounds for estimation error in a strong norm) but also a major disadvantage as for stronger norms Ω_β^+ the effective sparsity defined below might be larger.

In the next lemma, a vector β is written as the sum of two terms:

$$\beta = \beta^+ + \beta^-.$$

The situation we have in mind is the following. The vector β represents a candidate oracle. It may have a "relevant" sparsity-like part β^+ and a "irrelevant" smallish-like part β^-. For the "relevant" part, the triangle property is assumed. The "irrelevant" part of a candidate oracle better have small Ω-norm, otherwise this candidate oracle fails, i.e., it will not pass the test of being oracle. So we think of the situation where $\Omega(\beta^-)$ is small. The term $\Omega(\beta^-)$ is carried around in all the calculations: it is simply there without playing a very active role in the derivations.

Lemma 7.2 *Let* $\beta = \beta^+ + \beta^-$ *where* β^+ *has the triangle property and where* $\Omega_{\beta^+}^+(\beta^-) = 0$. *Then for any* $\beta' \in \mathbb{R}^p$

$$\Omega(\beta) - \Omega(\beta') \leq \Omega^+(\beta' - \beta) - \Omega^-(\beta' - \beta) + 2\Omega(\beta^-)$$

with $\Omega^+ =: \Omega_{\beta^+}^+$ *and* $\Omega_{\beta^+}^- =: \Omega^-$.

Proof of Lemma 7.2 We will first show that $\Omega^-(\beta^-) \leq \Omega(\beta^-)$. By applying the triangle property at $\beta' := \beta^+$ we obtain $0 \leq -\Omega^-(\beta^+)$. Hence $\Omega^-(\beta^+) = 0$. We next apply the triangle property at $\beta' := \beta^+ + \beta^-$. This gives

$$\Omega(\beta^+) - \Omega(\beta^+ + \beta^-) \leq \Omega^+(\beta^-) - \Omega^-(\beta^+ + \beta^-) = -\Omega^-(\beta^+ + \beta^-)$$

since by assumption $\Omega^+(\beta^-) = 0$. By the triangle inequality

$$\Omega^-(\beta^+ + \beta^-) \geq \Omega^-(\beta^-) - \Omega^-(\beta^+) = \Omega^-(\beta^-)$$

since we just showed that $\Omega^-(\beta^+) = 0$. Thus we have

$$\Omega(\beta^+) - \Omega(\beta^+ + \beta^-) \leq -\Omega^-(\beta^-).$$

On the other hand, by the triangle inequality

$$\Omega(\beta^+) - \Omega(\beta^+ + \beta^-) \geq -\Omega(\beta^-).$$

Combining the two gives indeed $\Omega^-(\beta^-) \le \Omega(\beta^-)$.

Let now β' be arbitrary. By the triangle inequality

$$\Omega(\beta) - \Omega(\beta') \le \Omega(\beta^+) + \Omega(\beta^-) - \Omega(\beta').$$

Apply the triangle property to find

$$\Omega(\beta) - \Omega(\beta') \le \Omega^+(\beta^+ - \beta') - \Omega^-(\beta') + \Omega(\beta^-).$$

Then apply twice the triangle inequality to get

$$\Omega(\beta) - \Omega(\beta') \le \Omega^+(\beta - \beta') + \Omega^+(\beta^-) - \Omega^-(\beta - \beta') + \Omega^-(\beta) + \Omega(\beta^-)$$

$$\le \Omega^+(\beta - \beta') - \Omega^-(\beta - \beta') + 2\Omega(\beta^-),$$

where in the last step we used that $\Omega^+(\beta^-) = 0$ and $\Omega^-(\beta) \le \Omega^-(\beta^-) \le \Omega(\beta^-)$.
$\square$

Definition 7.5 Let β have the triangle property. For τ a semi-norm on $\mathbb{R}^p$ and for a stretching factor $L > 0$, we define

$$\Gamma_\Omega(L, \beta, \tau) := \left(\min\left\{ \tau(\tilde\beta) : \ \tilde\beta \in \mathbb{R}^p, \ \Omega_\beta^+(\tilde\beta) = 1, \ \Omega_\beta^-(\tilde\beta) \le L \right\} \right)^{-1}.$$

We call $\Gamma_\Omega^2(L, \beta, \tau)$ the *effective sparsity* (for the norm Ω, the vector β, the stretching factor L and the semi-norm τ).

Effective sparsity is a generalization of compatibility, see Problem 7.1. The reason for the (somewhat) new terminology is because the scaling by the size of some active set is no longer defined in this general context.

7.4 Two Versions of Weak Decomposability

Having defined the triangle property, the next question is for which norms this property is true. In Problem 7.2 one sees that it is a generalization of the earlier introduced weak decomposability (Definition 6.1 in Sect. 6.4). This will be the first version considered in this section. A second version will be called weak decomposablity*, and holds for example for the nuclear norm (see Sect. 12.5).

Definition 7.6 We call a vector β *allowed* if for a complete pair of semi-norms Ω_β^+ and Ω_β^- with $\Omega_\beta^+(\beta) = \Omega(\beta)$, $\Omega_\beta^- \not\equiv 0$ and $\Omega_\beta^-(\beta) = 0$, one has

$$\Omega \ge \Omega_\beta^+ + \Omega_\beta^-.$$

We then call Ω *weakly decomposable* at β. If in fact we have equality: $\Omega = \Omega_\beta^+ + \Omega_\beta^-$, we call Ω *decomposable* at β.

Recall that for $\beta \neq 0$

$$\partial\Omega(\beta) = \{z \in \mathbb{R}^p : \Omega_*(z) = 1,\ z^T\beta = \Omega(\beta)\}.$$

Definition 7.7 We call a vector β *allowed** if for a complete pair of semi-norms Ω_β^+ and Ω_β^- with $\Omega_\beta^- \not\equiv 0$ one has for all $\beta' \in \mathbb{R}^p$

$$\min_{z \in \partial\Omega(\beta)} z^T(\beta - \beta') \leq \Omega_\beta^+(\beta' - \beta) - \Omega_\beta^-(\beta').$$

We then call Ω *weakly decomposable** at β.

Lemma 7.3 *Suppose β is an allowed or an allowed* vector. Then the triangle property holds at β:*

$$\Omega(\beta) - \Omega(\beta') \leq \Omega_\beta^+(\beta' - \beta) - \Omega_\beta^-(\beta').$$

Proof of Lemma 7.3

- If β is an allowed vector we have for any β' the inequality

$$\Omega(\beta) - \Omega(\beta') \leq \Omega(\beta) - \Omega_\beta^+(\beta') - \Omega_\beta^-(\beta') \leq \Omega_\beta^+(\beta' - \beta) - \Omega_\beta^-(\beta').$$

- If β is an allowed* vector we have for any $z \in \partial\Omega(\beta)$

$$\Omega(\beta) - \Omega(\beta') \leq z^T(\beta - \beta').$$

Hence

$$\Omega(\beta) - \Omega(\beta') \leq \min_{z \in \partial\Omega(\beta)} z^T(\beta - \beta') \leq \Omega_\beta^+(\beta' - \beta) - \Omega_\beta^-(\beta').$$

$\square$

If there is a "relevant" part β^+ and an "irrelevant" part β^- in the vector β we get:

Corollary 7.1 *Let $\beta = \beta^+ + \beta^-$ where β^+ is allowed or allowed* and where $\Omega_{\beta^+}^+(\beta^-) = 0$. Then by Lemma 7.2 combined with Lemma 7.3 we have for any $\beta' \in \mathbb{R}^p$*

$$\Omega(\beta) - \Omega(\beta') \leq \Omega^+(\beta' - \beta) - \Omega^-(\beta' - \beta) + 2\Omega(\beta^-)$$

with $\Omega^+ = \Omega_{\beta^+}^+$ and $\Omega_{\beta^+}^- = \Omega^-$.

We note that β allowed* does not imply β allowed (nor the other way around). In fact there are norms Ω where for all allowed* β

$$\Omega \leq \Omega_\beta^+ + \Omega_\beta^-$$

i.e. $\leq$ instead of $\geq$ as is per definition the case for allowed vectors. Lemma 12.5 in Sect. 12.5.2 shows an example. Here Ω is the nuclear norm as defined there (Sect. 12.5).

7.5   A Sharp Oracle Inequality

Notation for the Candidate Oracle In the next theorem we fix some $\beta \in \mathscr{B}_{\text{local}}$, a "candidate oracle". We assume β to be the sum of two vectors $\beta = \beta^+ + \beta^-$ where Ω has the triangle property at β^+ and where $\Omega_{\beta+}^+(\beta^-) = 0$. Write then $\Omega^+ := \Omega_{\beta+}^+$ and $\Omega^- := \Omega_{\beta+}^-$ We let

$$\underline{\Omega} := \gamma_\beta \Omega_\beta^+ + (1 - \gamma_\beta)\Omega_\beta^- =: \underline{\Omega}_{\beta+}$$

be the strongest norm among all convex combinations $\gamma\Omega_\beta^+ + (1-\gamma)\Omega_\beta^-, \gamma \in [0, 1]$.

Theorem 7.1 *Assume R_n is almost-differentiable and that Condition 7.2.2 (the two point margin condition) holds. Let H be the convex conjugate of G. Let λ_ϵ be a constant satisfying*

$$\lambda_\epsilon \geq \underline{\Omega}_* \left(\dot{R}_n(\hat{\beta}) - \dot{R}(\hat{\beta}) \right). \tag{7.1}$$

Set $\lambda_1 := \lambda_\epsilon \gamma_{\beta+}$ and $\lambda_2 := \lambda_\epsilon(1-\gamma_{\beta+})$. Take the tuning parameter λ large enough, so that $\lambda > \lambda_2$. Let $\delta_1 \geq 0$ and $0 \leq \delta_2 < 1$ be arbitrary and define

$$\underline{\lambda} := \lambda - \lambda_2, \ \bar{\lambda} := \lambda + \lambda_1 + \delta_1\underline{\lambda}$$

and stretching factor

$$L := \frac{\bar{\lambda}}{(1 - \delta_2)\underline{\lambda}}.$$

Then, when $\hat{\beta} \in \mathscr{B}_{\text{local}}$,

$$\delta_1\underline{\lambda}\Omega^+(\hat{\beta} - \beta) + \delta_2\underline{\lambda}\Omega^-(\hat{\beta} - \beta) + R(\hat{\beta})$$

$$\leq R(\beta) + H\left(\bar{\lambda}\Gamma_\Omega(L, \beta^+, \tau) \right) + 2\lambda\Omega(\beta^-).$$

Note that it is assumed that $\hat{\beta} \in \mathscr{B}_{\text{local}}$. Theorem 7.2 gives an illustration how this can be established. Note also that no reference is made to the target β^0. However, in Theorem 7.2 $\mathscr{B}_{\text{local}}$ as some local neighbourhood of β^0, so in the end the target *does* play a prominent role.

Chapter 10 studies (probability) inequalities for $\underline{\Omega}_*(\dot{R}_n(\hat{\beta}) - \dot{R}(\hat{\beta}))$. This term occurs because in the proof of the theorem the dual norm inequality is applied:

$$(\dot{R}_n(\hat{\beta}) - \dot{R}(\hat{\beta}))^T(\hat{\beta} - \beta) \le \underline{\Omega}_*(\dot{R}_n(\hat{\beta}) - \dot{R}(\hat{\beta}))\underline{\Omega}(\hat{\beta} - \beta).$$

This is in some cases too rough. An alternative route can be found in Problem 10.3.

We refer the a vector $\beta^* = \beta^{*+} + \beta^{*-}$ which trades off approximation error, estimation error (the term involving $H(\cdot)$ in Theorem 7.1) and Ω-smallish coefficients, as the oracle.

Typically, the margin function G is quadratic, say $G(u) = u^2/2$, $u \ge 0$. Then its convex conjugate $H(v) = v^2/2$, $v \ge 0$ is quadratic as well. The estimation error is then

$$H\left(\bar{\lambda}\Gamma_\Omega(L, \beta^+, \tau)\right) = \bar{\lambda}^2\Gamma_\Omega^2(L, \beta^+, \tau)/2.$$

Proof of Theorem 7.1 Define

$$\text{Rem}(\hat{\beta}, \beta) := R(\beta) - R(\hat{\beta}) - \dot{R}(\hat{\beta})^T(\beta - \hat{\beta}).$$

Then we have

$$R(\hat{\beta}) - R(\beta) + \text{Rem}(\hat{\beta}, \beta) = -\dot{R}(\hat{\beta})^T(\beta - \hat{\beta}).$$

- So if

$$\dot{R}(\hat{\beta})^T(\beta - \hat{\beta}) \ge \delta_1\underline{\lambda}\Omega^+(\hat{\beta} - \beta) + \delta_2\underline{\lambda}\Omega^-(\hat{\beta} - \beta) - 2\lambda\Omega(\beta^-)$$

we find from Condition 7.2.2

$$\delta_1\underline{\lambda}\Omega^+(\hat{\beta} - \beta) + \delta_2\underline{\lambda}\Omega^-(\hat{\beta} - \beta) + R(\hat{\beta}) \le R(\beta) + 2\lambda\Omega(\beta^-)$$

(as $\text{Rem}(\hat{\beta}, \beta) \ge 0$). So then we are done.
- Assume now in the rest of the proof that

$$\dot{R}(\hat{\beta})^T(\beta - \hat{\beta}) \le \delta_1\underline{\lambda}\Omega^+(\hat{\beta} - \beta) + \delta_2\underline{\lambda}\Omega^-(\hat{\beta} - \beta) - 2\lambda\Omega(\beta^-).$$

From Lemma 7.1

$$-\dot{R}_n(\hat{\beta})^T(\beta - \hat{\beta}) \le \lambda\Omega(\beta) - \lambda\Omega(\hat{\beta}).$$

Hence by the dual norm inequality

$$-\dot{R}(\hat{\beta})^T(\beta - \hat{\beta}) + \delta_1\underline{\lambda}\Omega^+(\hat{\beta} - \beta) + \delta_2\underline{\lambda}\Omega^-(\hat{\beta} - \beta)$$

$$\le (\dot{R}_n(\hat{\beta}) - \dot{R}(\hat{\beta}))^T(\beta - \hat{\beta}) + \delta_1\underline{\lambda}\Omega^+(\hat{\beta} - \beta) + \delta_2\underline{\lambda}\Omega^-(\hat{\beta} - \beta)$$

$$+ \lambda\Omega(\beta) - \lambda\Omega(\hat{\beta})$$

$$\le \lambda_\epsilon\underline{\Omega}(\hat{\beta} - \beta) + \delta_1\underline{\lambda}\Omega^+(\hat{\beta} - \beta) + \delta_2\underline{\lambda}\Omega^-(\hat{\beta} - \beta) + \lambda\Omega(\beta) - \lambda\Omega(\hat{\beta})$$

$$\le \lambda_1\gamma_{\beta^+}\Omega^+(\hat{\beta} - \beta) + \lambda_2(1 - \gamma_{\beta^+})\Omega^-(\hat{\beta} - \beta) + \delta_1\underline{\lambda}\Omega^+(\hat{\beta} - \beta)$$

$$+ \delta_2\underline{\lambda}\Omega^-(\hat{\beta} - \beta) + \lambda\Omega^+(\hat{\beta} - \beta) - \lambda\Omega^-(\hat{\beta} - \beta) + 2\lambda\Omega(\beta^-)$$

$$= \bar{\lambda}\Omega^+(\hat{\beta} - \beta) - (1 - \delta_2)\underline{\lambda}\Omega^-(\hat{\beta} - \beta) + 2\lambda\Omega(\beta^-)$$

(here we applied Corollary 7.1). In summary

$$-\dot{R}(\hat{\beta})^T(\beta - \hat{\beta}) + \delta_1\underline{\lambda}\Omega^+(\hat{\beta} - \beta) + \delta_2\underline{\lambda}\Omega^-(\hat{\beta} - \beta)$$

$$\le \bar{\lambda}\Omega^+(\hat{\beta} - \beta) - (1 - \delta_2)\underline{\lambda}\Omega^-(\hat{\beta} - \beta) + 2\lambda\Omega(\beta^-) \qquad (7.2)$$

But then

$$(1 - \delta_2)\underline{\lambda}\Omega^-(\beta - \hat{\beta}) \le \bar{\lambda}\Omega^+(\hat{\beta} - \beta)$$

or

$$\Omega^-(\hat{\beta} - \beta) \le L\Omega^+(\hat{\beta} - \beta).$$

The implies by the definition of the effective sparsity $\Gamma_\Omega(L, \beta^+, \tau)$

$$\Omega^+(\hat{\beta} - \beta) \le \tau(\hat{\beta} - \beta)\Gamma_\Omega(L, \beta^+, \tau).$$

Continuing with (7.2), we find

$$-\dot{R}(\hat{\beta})^T(\beta - \hat{\beta}) + \underline{\lambda}\Omega^-(\hat{\beta} - \beta) + \delta_1\underline{\lambda}\Omega^+(\hat{\beta} - \beta)$$

$$\le \bar{\lambda}\Omega^+(\hat{\beta} - \beta) + 2\lambda\Omega(\beta^-)$$

$$\le \bar{\lambda}\Gamma_\Omega(L, \beta^+, \tau)\tau(\hat{\beta} - \beta) + 2\lambda\Omega(\beta^-)$$

or

$$R(\hat{\beta}) - R(\beta) + \text{Rem}(\hat{\beta}, \beta) + \underline{\lambda}\Omega^-(\hat{\beta} - \beta) + \delta_1\underline{\lambda}\Omega^+(\hat{\beta} - \beta)$$

$$\leq \bar{\lambda}\Gamma_{\Omega}(L, \beta^+, \tau)\tau(\hat{\beta} - \beta) + 2\lambda\Omega(\beta^-)$$

$$\leq H\left(\bar{\lambda}\Gamma_{\Omega}(L, \beta^+, \tau)\right) + G(\tau(\hat{\beta} - \beta)) + 2\lambda\Omega(\beta^-)$$

$$\leq H\left(\bar{\lambda}\Gamma_{\Omega}(L, \beta^+, \tau)\right) + \text{Rem}(\hat{\beta}, \beta) + 2\lambda\Omega(\beta^-).$$

$\square$

7.6 Localizing (or a Non-sharp Oracle Inequality)

This section considers the situation where one stirs towards showing that $\hat{\beta}$ is consistent in Ω-norm. The local set $\mathscr{B}_{\text{local}}$ is taken as a convex subset of the collection $\{\beta' : \Omega(\beta' - \beta) \leq M_{\beta}\}$ with β a candidate oracle and M_{β} a constant depending on how well this candidate oracle is trading off approximation error, estimation error and smallish coefficients. For further discussions, see Sect. 11.1 and see Problem 11.1 for an alternative approach to localization.

Theorem 7.2 below does not require almost-differentiability of R_n and only needs Condition 7.2.2 at β' equal to β^0 and with $\dot{R}(\beta^0) = 0$. We call this the *one point margin condition*.

Condition 7.6.1 (One Point Margin Condition) *There is an increasing strictly convex function G with $G(0) = 0$ and a semi-norm τ on $\mathscr{B}$ such that for all $\tilde{\beta} \in \mathscr{B}_{\text{local}}$*

$$R(\tilde{\beta}) - R(\beta^0) \geq G(\tau(\tilde{\beta} - \beta^0)).$$

Notation for the Candidate Oracle Fix a candidate oracle $\beta \in \mathscr{B}_{\text{local}}$ which is the sum $\beta = \beta^+ + \beta^-$ of two vectors β^+ and β^- with β^+ having the triangle property and with $\Omega^+_{\beta+}(\beta^-) = 0$. Then write $\Omega^+ := \Omega^+_{\beta+}$, $\Omega^- := \Omega^-_{\beta+}$ and (for simplicity) $\Omega := \Omega^+ + \Omega^-$.

Theorem 7.2 *Assume Condition 7.6.1 and let H be the convex conjugate of G. Suppose that for some constants $0 < M_{\max} \leq \infty$ and λ_{ϵ}, and for all $0 < M \leq M_{\max}$*

$$\sup_{\beta' \in \mathscr{B}: \; \Omega(\beta'-\beta) \leq M} \left| [R_n(\beta') - R(\beta')] - [R_n(\beta) - R(\beta)] \right| \leq \lambda_{\epsilon}M. \qquad (7.3)$$

Let $0 < \delta < 1$, take $\lambda \geq 8\lambda_\epsilon/\delta$ and define M_β by

$$\delta\lambda M_\beta := 8H\left(\lambda(1+\delta)\Gamma_\Omega\left(\frac{1}{1-\delta}, \beta^+, \tau\right)\right) + 8\left(R(\beta) - R(\beta^0)\right) + 16\lambda\Omega(\beta^-).$$

Assume that $M_\beta \leq M_{\max}$ and that $\{\beta' \in \mathscr{B} : \underline{\Omega}(\beta' - \beta) \leq M_\beta\} \subset \mathscr{B}_{\text{local}}$. Then $\underline{\Omega}(\hat{\beta} - \beta) \leq M_\beta$ and hence $\hat{\beta} \in \mathscr{B}_{\text{local}}$. Moreover, it holds that

$$R(\hat{\beta}) - R(\beta) \leq (\lambda_\epsilon + \lambda)M_\beta + \lambda\Omega^-(\beta).$$

Probability inequalities for the empirical process

$$\left\{[R_n(\beta') - R(\beta')] - [R_n(\beta) - R(\beta)] : \underline{\Omega}(\beta' - \beta) \leq M, \beta' \in \mathscr{B}\right\}$$

(with $\beta \in \mathscr{B}$ and $M > 0$ fixed but arbitrary) are studied in Chap. 10. Generally, the constant $M_{\max}$ plays no role and can be taken as $M_{\max} = \infty$.

We note that—unlike Theorem 7.1—Theorem 7.2 involves the approximation error $R(\beta) - R(\beta^0)$ and hence it only gives "good" results if the approximation error $R(\beta) - R(\beta^0)$ is "small". Perhaps in contrast to general learning contexts, this is not too much of a restriction in certain cases. For example in linear regression with fixed design we have seen in Sect. 2.2 that high-dimensionality implies that the model is not misspecified.

If $\mathscr{B} = \bar{\mathscr{B}}$, then the target $\beta^0 = \arg\min_{\beta \in \mathscr{B}} R(\beta)$ is by definition in the class $\mathscr{B}$. If one is actually interested in a target $\beta^0 = \min_{\beta \in \bar{\mathscr{B}}} R(\beta)$ outside the class $\mathscr{B}$, this target may generally have margin behaviour different from the minimizer within $\mathscr{B}$.

We remark here that we did not try to optimize the constants in Theorem 7.2.

Some explanation of the oracle we are trying to mimic here is in place. The oracle is some fixed vector $\beta^* = \beta^{*+} + \beta^{*-}$ satisfying the conditions as stated with $\Omega^+ := \Omega^+_{\beta^{*+}}$ and $\Omega^- := \Omega^-_{\beta^{*+}}$. We take β^* in a favourable way, that is, in such a way that $M_* := M_{\beta^*}$ is the smallest value among all β's satisfying the conditions as stated and such that in addition $\underline{\Omega}(\beta^* - \beta^0) \leq M_*$ where $\underline{\Omega} = \Omega^+ + \Omega^-$, i.e. the oracle is in a suitable $\underline{\Omega}$-neighbourhood of the target (note that $\underline{\Omega}$ depends on β^*). We define $\mathscr{B}_{\text{local}}$ as $\mathscr{B}_{\text{local}} := \mathscr{B} \cap \{\beta' : \underline{\Omega}(\beta' - \beta^0) \leq 2M_*\}$. Then obviously $\beta^* \in \mathscr{B}_{\text{local}}$ and by the triangle inequality $\{\beta' \in \mathscr{B} : \underline{\Omega}(\beta' - \beta^*) \leq M_*\} \subset \mathscr{B}_{\text{local}}$. Hence, then we may apply the above theorem with $\beta = \beta^*$. See for example Sect. 12.4.1 for an illustration how Theorem 7.2 can be applied.

The situation simplifies substantially if one chooses β^0 itself as candidate oracle, see Problem 7.3. However, one then needs to assume that β^0 is has "enough" structured sparsity.

Proof of Theorem 7.2 To simplify the notation somewhat we write $M := M_\beta$. Define $\tilde{\beta} := t\hat{\beta} + (1-t)\beta$, where

$$t := \frac{M}{M + \underline{\Omega}(\hat{\beta} - \beta)}.$$

Then

$$\underline{\Omega}(\tilde{\beta} - \beta) = t\underline{\Omega}(\hat{\beta} - \beta) = \frac{M\underline{\Omega}(\hat{\beta} - \beta)}{M + \underline{\Omega}(\hat{\beta} - \beta)} \leq M.$$

Therefore $\tilde{\beta} \in \mathscr{B}_{\text{local}}$. Moreover, by the convexity of $R_n + \lambda\Omega$

$$R_n(\tilde{\beta}) + \lambda\Omega(\tilde{\beta}) \leq tR_n(\hat{\beta}) + t\lambda\Omega(\hat{\beta}) + (1-t)R_n(\beta) + (1-t)\lambda\Omega(\beta)$$
$$\leq R_n(\beta) + \lambda\Omega(\beta).$$

Rewrite this and apply the assumption (7.3):

$$R(\tilde{\beta}) - R(\beta) \leq -\Big[[R_n(\tilde{\beta}) - R(\tilde{\beta})] - [R_n(\beta) - R(\beta)] \Big] + \lambda\Omega(\beta) - \lambda\Omega(\tilde{\beta})$$

$$\leq \lambda_\epsilon M + \lambda\Omega(\beta) - \lambda\Omega(\tilde{\beta})$$

$$\leq \lambda_\epsilon M + \lambda\Omega^+(\tilde{\beta} - \beta) - \lambda\Omega^-(\tilde{\beta} - \beta) + 2\lambda\Omega^-(\beta),$$

where we invoked Lemma 7.2.

- If $\lambda\Omega^+(\tilde{\beta} - \beta) \leq (1-\delta)[\lambda_\epsilon M + R(\beta) - R(\beta^0) + 2\lambda\Omega(\beta^-)]/\delta$, we obtain

$$\delta\lambda\Omega^+(\tilde{\beta} - \beta) \leq \lambda_\epsilon M + [R(\beta) - R(\beta^0)] + 2\lambda\Omega(\beta^-)$$

  as well as

$$\delta\lambda\Omega^-(\tilde{\beta} - \beta) \leq \lambda_\epsilon M + [R(\beta) - R(\beta^0)] + 2\lambda\Omega(\beta^-).$$

  So then

$$\delta\lambda(\Omega^+ + \Omega^-)(\tilde{\beta} - \beta) \leq 2\lambda_\epsilon M + 2[R(\beta) - R(\beta^0)] + 4\lambda\Omega(\beta^-).$$

- If $\lambda\Omega^+(\tilde{\beta} - \beta) \geq (1-\delta)[\lambda_\epsilon M + R(\beta) - R(\beta^0) + 2\lambda\Omega(\beta^-)]/\delta$ we obtain

$$[R(\tilde{\beta}) - R(\beta^0)] + \lambda\Omega^-(\tilde{\beta} - \beta) \leq \lambda\Omega^+(\tilde{\beta} - \beta)/(1-\delta).$$

So then we may apply effective sparsity with stretching factor $L = 1/(1 - \delta)$. Hence

$$
[R(\tilde{\beta}) - R(\beta^0)] + \lambda\Omega^-(\tilde{\beta} - \beta) + \delta\lambda\Omega^+(\tilde{\beta} - \beta)
$$
$$
\leq \lambda(1 + \delta)\Omega^+(\tilde{\beta} - \beta) + \lambda_\epsilon M + [R(\beta) - R(\beta^0)] + 2\lambda\Omega(\beta^-)
$$
$$
\leq \lambda(1 + \delta)\tau(\tilde{\beta} - \beta)\Gamma_\Omega(1/(1 - \delta), \beta^+, \tau) + \lambda_\epsilon M
$$
$$
+ [R(\beta) - R(\beta^0)] + 2\lambda\Omega(\beta^-)
$$
$$
\leq 2H\left(\lambda(1 + \delta)\Gamma_\Omega(1/(1 - \delta), \beta^+, \tau)\right) + \lambda_\epsilon M + 2[R(\beta) - R(\beta^0)] + 2\lambda\Omega(\beta^-).
$$
$$
+ R(\tilde{\beta}) - R(\beta^0)
$$

It follows hat

$$
\delta\lambda(\Omega^+ + \Omega^-)(\tilde{\beta} - \beta) \leq \lambda\Omega^-(\tilde{\beta} - \beta) + \delta\underline{\lambda}\Omega^+(\tilde{\beta} - \beta)
$$
$$
\leq 2H\left(\lambda\Gamma_\Omega(1/(1 - \delta), \beta^+, \tau)\right)
$$
$$
+ \lambda_\epsilon M + 2[R(\beta) - R(\beta^0)] + 2\lambda\Omega(\beta^-).
$$

Hence, we have shown in both cases that

$$
\delta\lambda(\Omega^+ + \Omega^-)(\tilde{\beta} - \beta) \leq 2H\left(\lambda(1 + \delta)\Gamma_\Omega(1/(1 - \delta), \beta^+, \tau)\right)
$$
$$
+ 2[R(\beta) - R(\beta^0)] + 2\lambda_\epsilon M + 4\lambda\Omega(\beta^-)
$$
$$
= \delta\lambda M/4 + 2\lambda_\epsilon M \leq \delta\lambda M/2
$$

where we used the definition of M and that $\lambda \geq 8\lambda_\epsilon/\delta$. In turn, this implies that

$$
(\Omega^+ + \Omega^-)(\hat{\beta} - \beta) \leq M.
$$

For the second result of the theorem we apply the formula

$$
R(\hat{\beta}) - R(\beta) \leq -\left[[R_n(\hat{\beta}) - R(\hat{\beta})] - [R_n(\beta) - R(\beta)]\right] + \lambda\Omega(\beta) - \lambda\Omega(\hat{\beta})
$$
$$
\leq \lambda_\epsilon M + \lambda\Omega^+(\hat{\beta} - \beta) + 2\lambda\Omega^-(\beta)
$$
$$
\leq (\lambda_\epsilon + \lambda)M + 2\lambda\Omega^-(\beta).
$$

□

Problems

7.1 Let $\hat{\phi}^2(L, S)$ be the compatibility constant defined in Definition 2.1 (Sect. 2.6). Let moreover for a vector $\beta \in \mathbb{R}^p$ and a set $S \subset \{1, \ldots, p\}$

$$\beta^+ := \beta_S, \ \beta^- := \beta_{-S}.$$

Take $\Omega = \Omega^+ = \Omega^- = \|\cdot\|_1$. Now verify that

$$\Gamma_{\Omega}^2(L, \beta, \tau) = |S|/\hat{\phi}^2(L, S),$$

with $\tau(\tilde{\beta}) = \|X\tilde{\beta}\|_n$, $\tilde{\beta} \in \mathbb{R}^p$. Here, $\Gamma_{\Omega}^2(L, \beta, \tau)$ is the effective sparsity given in Definition 7.3 of Sect. 7.5.

7.2 Consider the triangle property as introduced in Definition 7.4 in Sect. 7.3. Let $S \subset \{1, \ldots, p\}$ be some set and $\beta \in \mathbb{R}^p$ be some vector. Show that if the weak-decomposablity

$$\Omega \geq \Omega(\cdot|S) + \Omega^{-S}$$

given in Definition 6.1 (Sect. 6.4) holds, then the triangle property is true at β_S, with $\Omega_{\beta_S}^+ = \Omega(\cdot|S)$ and $\Omega_{\beta_S}^- = \Omega^{-S}$. Note that the norms in the weak decomposition in this case only depend on β via the set S. In other cases (e.g. for the nuclear norm Sect. 12.5) the weak decomposition depends on the vector β in a more involved way,

7.3 Verify that Theorem 7.2 with the candidate oracle chosen to be the target itself implies the following clearer version. Suppose first of all that the target β^0 is in the parameter space $\mathscr{B}$ and that β^0 has the triangle property. Write $\Omega^+ := \Omega_{\beta^0}^+$, $\Omega^- := \Omega_{\beta^0}^-$ and $\underline{\Omega} := \Omega^+ + \Omega^-$. Assume that for some constant M_0

$$R(\beta) - R(\beta^0) \geq G(\tau(\beta - \beta^0))$$

for all $\beta \in \mathscr{B}$ with $\underline{\Omega}(\beta - \beta^0) \leq M_0$. Assume moreover that for all M

$$\sup_{\beta' \in \mathscr{B}: \ \underline{\Omega}(\beta' - \beta^0) \leq M} \left| [R_n(\beta') - R(\beta')] - [R_n(\beta) - R(\beta)] \right| \leq \lambda_{\epsilon} M.$$

Then for $\lambda \geq 16\lambda_{\epsilon}$

$$\underline{\Omega}(\hat{\beta} - \beta^0) \leq 4H\left(3\lambda \Gamma_{\Omega}(2, \beta^0, \tau)/2\right)$$

provided that the right-hand side is at most M_0.

7.4 Let $z \in \partial\Omega(\beta)$ and $z^* \in \partial\Omega(\beta^*)$. Show that $(\beta - \beta^*)^T(z - z^*) \leq 0$.

7.5 Here is how one could go about in the case of possibly non-convex loss. In that case we assume $\mathscr{B}_{\text{local}} = \mathscr{B}$.

As before fix some candidate oracle $\beta \in \mathscr{B}_{\text{local}} = \mathscr{B}$ which is the sum $\beta = \beta^+ + \beta^-$ of two vectors β^+ and β^- with β^+ having the triangle property and with $\Omega_{\beta^+}^+(\beta^-) = 0$. Write then $\Omega^+ := \Omega_{\beta^+}^+$, $\Omega^- := \Omega_{\beta^+}^-$ and $\underline{\Omega} := \Omega^+ + \Omega^-$.

Theorem 7.3 *Suppose that the loss function is possibly not convex but does satisfy Condition 7.6.1 with $\mathscr{B}_{\text{local}} = \mathscr{B}$. Let H be the convex conjugate of G. Suppose that for some constants $M_{\min}$ and λ_ϵ*

$$\sup_{\beta' \in \mathscr{B}} \frac{\left|[R_n(\beta') - R(\beta')] - [R_n(\beta) - R(\beta)]\right|}{\underline{\Omega}(\beta' - \beta) + M_{\min}} \leq \lambda_\epsilon. \tag{7.4}$$

Then for $\lambda \geq \lambda_\epsilon$ and $L = (2\lambda + \lambda_\epsilon)/(\lambda - \lambda_\epsilon)$ it holds that for any $0 < \eta < 1$

$$(\lambda - \lambda_\epsilon)\underline{\Omega}(\hat{\beta} - \beta) + (1 - \eta)(R(\hat{\beta}) - R(\beta^0))$$
$$\leq 3(R(\beta) - R(\beta^0)) + 2\eta H(3\lambda\Gamma_\Omega(L, \beta, \tau)/\eta) + 6\lambda\Omega(\beta^-) + 3\lambda_\epsilon M_{\min}.$$

This is a non-sharp oracle inequality. The constants in this result can be improved but with the proof technique suggested below there remains a constant larger than one in front of the approximation error $R(\beta) - R(\beta^0)$. Comparing the condition of the above theorem with condition (7.3) in Theorem 7.2, one sees that one now requires "uniformity" in M over all $\underline{\Omega}(\beta' - \beta) = M$. This can be dealt with invoking the peeling technique (a terminology from van de Geer 2000).

Observe that Condition 7.6.1 per se does not imply convexity of $\beta' \mapsto R(\beta')$ within $\mathscr{B}_{\text{local}}$.

Here are a few hints for the proof. First, show that

$$R(\hat{\beta}) - R(\beta) + \lambda\Omega(\hat{\beta}) \leq \lambda_\epsilon\underline{\Omega}(\hat{\beta} - \beta) + \lambda_\epsilon M_{\min} + \lambda\Omega(\beta).$$

Then employ the triangle property, in fact, Lemma 7.2:

$$R(\hat{\beta}) - R(\beta) \leq \lambda_\epsilon\underline{\Omega}(\hat{\beta} - \beta) + \lambda_\epsilon M_{\min} + \lambda\Omega^+(\hat{\beta} - \beta) - \lambda\Omega^-(\hat{\beta} - \beta) + 2\lambda\Omega^-(\beta).$$

If $\lambda\Omega^+(\hat{\beta} - \beta) \geq \lambda_\epsilon M_{\min} + R(\beta) - R(\beta^0) + 2\lambda\Omega^-(\beta)$ one gets

$$R(\hat{\beta}) - R(\beta^0) + (\lambda - \lambda_\epsilon)\underline{\Omega}(\hat{\beta} - \beta) \leq 3\lambda\Omega^+(\hat{\beta} - \beta)$$
$$\leq 3\lambda\Gamma_\Omega(L, \beta, \tau)\tau(\hat{\beta} - \beta).$$

Then, writing $\Gamma := \Gamma_\Omega(L, \beta, \tau)$, by the triangle inequality and Condition 7.6.1,

$$3\lambda\Gamma\tau(\hat{\beta} - \beta) \leq 3\lambda\Gamma\left(\tau(\hat{\beta} - \beta^0) + \tau(\beta - \beta^0)\right)$$

$$\leq \eta\left(G(\tau(\hat{\beta} - \beta^0)) + G(\tau(\beta - \beta^0))\right) + 2\eta H(3\lambda\Gamma/\eta).$$

7.6 Here is a simple instance where the target β^0 is not necessarily the "true parameter": there may be misspecification. Consider i.i.d. observations $\{(X_i, Y_i)\}_{i=1}^n$ with distribution P and with $X_i \in \mathcal{X} \subset \mathbb{R}^p$ the input and $Y_i \in \mathbb{R}$ the response $(i = 1, \ldots, n)$. We look at the generalized linear model. Let $f_\beta(x) := x\beta$, for all $\beta \in \mathbb{R}^p$ and all $x \in \mathcal{X}$ and $\rho_\beta(x, y) := \rho(f_\beta(x), y)$ where ρ is a given function. We are interested in the regression

$$f^0(x) := \arg\min_{\xi \in \mathbb{R}} \mathbb{E}(\rho(\xi, Y_1)|X_1 = x), \; x \in \mathcal{X}.$$

Let then f_{β^0} be a linear approximation of f^0. What properties would you require for this approximation? How does misspecification effect the (one or two point) margin condition (see also Sect. 11.6)? (You may answer: "it depends". These are points of discussion rather than mathematical questions.)

Chapter 8
Empirical Process Theory for Dual Norms

Abstract This chapter presents probability inequalities for the (dual) norm of a Gaussian vector in $\mathbb{R}^p$. For Gaussian vectors there are ready-to-use concentration inequalities (e.g. Borell, 1975). Here however, results are derived using direct arguments. The extension to for example sub-Gaussian vectors is then easier to read off. Bounds for the supremum are given, for the ℓ_2-norm, and more generally for dual norms of norms generated from a convex cone.

8.1 Introduction

Consider a vector $\epsilon \in \mathbb{R}^n$ with independent entries with mean zero and variance σ_0^2. We let X be a given $n \times p$ matrix. We are interested in the behaviour of $\Omega_*(X^T\epsilon)/n$ where Ω_* is the dual norm of Ω. Note that $X^T\epsilon$ is a p-dimensional random vector with components $X_j^T\epsilon$ where X_j is the jth column of X ($j = 1, \ldots, p$). For each j the random variable $W_j := X_j^T\epsilon/n$ is an average of n independent random variables with mean zero and variance $\sigma_0^2\|X_j\|_n^2/n$. Under suitable conditions, W_j has "Gaussian-type" behaviour. In this chapter, we assume for simplicity throughout that ϵ is Gaussian:

Condition 8.1.1 *The vector $\epsilon \in \mathbb{R}^n$ has a $\mathcal{N}_n(0, \sigma_0^2)$-distribution.*

Then, for all j, $W_j := X_j^T\epsilon/n$ is Gaussian as well and derivations are simpler than for more general distributions. Although the Gaussianity assumption is not crucial for the main picture, it does make a difference (see for example Problem 8.4).

8.2 The Dual Norm of ℓ_1 and the Scaled Version

The dual norm of $\|\cdot\|_1$ is $\|\cdot\|_\infty$. As a direct consequence of Lemma 17.5 in Sect. 17.3 we find the following corollary.

© Springer International Publishing Switzerland 2016   121
S. van de Geer, *Estimation and Testing Under Sparsity*,
Lecture Notes in Mathematics 2159, DOI 10.1007/978-3-319-32774-7_8

Corollary 8.1 *Let $\epsilon \sim \mathcal{N}_n(0, \sigma_0^2 I)$ and let X be a fixed $n \times p$ matrix with* $\mathrm{diag}(X^T X)/n = I$. *Let $0 < \alpha < 1$ be a given error level. Then for*

$$\lambda_\epsilon := \sigma_0 \sqrt{\frac{2 \log(2p/\alpha)}{n}},$$

we have

$$\mathbb{P}\left(\|X^T \epsilon\|_\infty/n \geq \lambda_\epsilon \right) \leq \alpha.$$

The scaled (or scale free) version of $\|X^T \epsilon\|_\infty/n$ is $\|X^T \epsilon\|_\infty/(n\|\epsilon\|_n)$. We now consider this latter quantity. First we present a probability inequality for the angle between a fixed and a random vector on the sphere in $\mathbb{R}^n$.

Lemma 8.1 *Let $\epsilon \sim \mathcal{N}_n(0, \sigma_0^2)$ where $n \geq 2$. Then for any $u \in \mathbb{R}^n$ with $\|u\|_n = 1$ and for all $0 < t < (n-1)/2$ we have*

$$\mathbb{P}\left(\frac{|u^T \epsilon|}{n\|\epsilon\|_n} > \sqrt{\frac{2t}{n-1}} \right) \leq 2 \exp[-t].$$

Proof of Lemma 8.1 Without loss of generality we may assume $\sigma_0 = 1$. Because $\epsilon/\|\epsilon\|_n$ is uniformly distributed on the sphere with radius $\sqrt{n}$ in $\mathbb{R}^n$, we may without loss of generality assume that $u = \sqrt{n}e_1$, the first unit vector scaled with $\sqrt{n}$. Then $u^T \epsilon/(n\|\epsilon\|_n) = \epsilon_1/(\sqrt{n}\|\epsilon\|_n) = \epsilon_1/\sqrt{\sum_{i=1}^n \epsilon_i^2}$. It follows that for $0 < t < n/2$

$$\mathbb{P}\left(\frac{|u^T \epsilon|}{n\|\epsilon\|_n} \geq \sqrt{2t/n} \right) = \mathbb{P}\left(\epsilon_1^2 \geq \frac{2t}{n} \sum_{i=1}^n \epsilon_i^2 \right)$$

$$= \mathbb{P}\left(\left(1 - \frac{2t}{n}\right)\epsilon_1^2 \geq \frac{2t}{n} \sum_{i=2}^n \epsilon_i^2 \right)$$

$$= \mathbb{P}\left(\epsilon_1^2 \geq \left(\frac{2t}{n-2t}\right) \sum_{i=2}^n \epsilon_i^2 \right).$$

The random variable $Z := \sum_{i=2}^n \epsilon_i^2$ has a chi-squared distribution with $n-1$ degrees of freedom. It follows that (Problem 8.1) for $v > 0$

$$\mathbb{E}e^{-vZ/2} = \left(\frac{1}{1+v}\right)^{\frac{n-1}{2}}.$$

We moreover have that for all $a > 0$,

$$\mathbb{P}(\epsilon_1^2 \geq 2a) \leq 2 \exp[-a].$$

So we find, with f_Z being the density of $Z = \sum_{i=2}^{n} \epsilon_i^2$,

$$\mathbb{P}\left(\epsilon_1^2 \geq \left(\frac{2t}{n-2t}\right)\sum_{i=2}^{n}\epsilon_i^2\right) = \int_0^{\infty}\mathbb{P}\left(\epsilon_1^2 \geq \left(\frac{2tz}{n-2t}\right)\right)f_Z(z)dz$$

$$\leq 2\int_0^{\infty}\exp\left[-\frac{tz}{n-2t}\right]f_Z(z)dz$$

$$= 2\left(\frac{1}{1+\frac{2t}{n-2t}}\right)^{\frac{n-1}{2}} = 2\left(\frac{n-2t}{n}\right)^{\frac{n-1}{2}}$$

$$\leq 2\exp\left[-t\left(\frac{n-1}{n}\right)\right].$$

Finalize the proof by replacing t by $tn/(n-1)$.   □

Lemma 8.2 *Let* $\epsilon \sim \mathcal{N}_n(0,\sigma_0^2 I)$ *and let* X *be a fixed* $n \times p$ *matrix with* $\mathrm{diag}(X^T X)/n = I$. *Let* α, $\underline{\alpha}$ *and* $\bar{\alpha}$ *be given positive error levels. Define*

$$\underline{\sigma}^2 := \sigma_0^2\left(1 - 2\sqrt{\frac{\log(1/\underline{\alpha})}{n}}\right),$$

$$\bar{\sigma}^2 := \sigma_0^2\left(1 + 2\sqrt{\frac{\log(1/\bar{\alpha})}{n}} + \frac{2\log(1/\bar{\alpha})}{n}\right)$$

and

$$\lambda_{0,\epsilon} := \sqrt{\frac{2\log(2p/\alpha)}{n-1}}.$$

We have

$$\mathbb{P}(\|\epsilon\|_n \leq \underline{\sigma}) \leq \underline{\alpha}, \ \mathbb{P}(\|\epsilon\|_n \geq \bar{\sigma}) \leq \bar{\alpha}$$

and

$$\mathbb{P}(\|X^T\epsilon\|_{\infty}/(n\|\epsilon\|_n) \geq \lambda_{0,\epsilon}) \leq \alpha.$$

Proof of Lemma 8.2 Without loss of generality we can assume $\sigma_0^2 = 1$. From Laurent and Massart (2000) we know that for all $t > 0$

$$\mathbb{P}\left(\|\epsilon\|_n^2 \leq 1 - 2\sqrt{t/n}\right) \leq \exp[-t]$$

and

$$\mathbb{P}\left(\|\epsilon\|_n^2 \geq 1 + 2\sqrt{t/n} + 2t/n \right) \leq \exp[-t].$$

A proof of the latter can also be found in Lemma 8.6.

Apply this with $t = \log(1/\underline{\alpha})$ and $t = \log(1/\bar{\alpha})$ respectively. Lemma 8.1 and the union bound yield the probability inequality for $\|X^T\epsilon\|_\infty/(n\|\epsilon\|_n)$.   □

8.3 Dual Norms Generated from Cones

In Maurer and Pontil (2012) one can find first moment inequalities for a general class of dual norms. Here, we consider only a special case and we establish probability inequalities directly (i.e. not via concentration inequalities).

Let Ω be the norm generated from a given convex cone $\mathscr{A}$:

$$\Omega(\beta) := \min_{a \in \mathscr{A}} \frac{1}{2} \sum_{j=1}^{p} \left[\frac{\beta_j^2}{a_j} + a_j \right], \beta \in \mathbb{R}^p.$$

(see Sect. 6.9). Lemma 6.6 expresses the dual norm as

$$\Omega_*(w) = \max_{a \in \mathscr{A}, \|a\|_1 = 1} \sqrt{\sum_{j=1}^{p} a_j w_j^2}, \ w \in \mathbb{R}^p.$$

Aim of the rest of this chapter is to bound $\Omega_*(W)$, with $W_1, \ldots, W_p$ random variables (in our setup, $W_j = X_j^T\epsilon/n, j = 1, \ldots, p$). Recall that in order to simplify the exposition its is assumed that these are Gaussian random variables. The results can be extended to sub-Gaussian ones.

It is easy to see that $\Omega \geq \|\cdot\|_1$ and hence we have $\Omega_* \leq \|\cdot\|_\infty$. However, in some instances this bound can be improved (in terms of log-factors). This is for example the case for the group Lasso, as we show below.

8.4 A Generalized Bernstein Inequality

Bernstein's inequality can be found in Theorem 9.2 in Sect. 9.1. In this section it is shown that under a condition on the moment generating function of a non-negative random variable Z one has a Bernstein-like inequality involving a sub-Gaussian part and a sub-exponential part. We apply this in the next section to squared Gaussians.

The following result can be deduced from Birgé and Massart (1998, Lemma 8 and its proof) or Bühlmann and van de Geer (2011, Lemma 14.9 and its proof).

Lemma 8.3 *Let* $Z \in \mathbb{R}$ *be a random variable that satisfies for some K and c and for all* $L > K$

$$\mathbb{E} \exp[Z/L] \le \exp\left[\frac{c}{(L^2 - LK)}\right].$$

Then for all $t > 0$

$$\mathbb{P}\left(Z \ge 2\sqrt{tc} + Kt\right) \le \exp[-t].$$

Proof of Lemma 8.3 Let $a > 0$ be arbitrary and take

$$K/L = 1 - (1 + aK/c)^{-1/2},$$

apply Chebyshev's inequality to obtain

$$\mathbb{P}\left(Z \ge a\right) \le \exp\left[-\frac{a^2}{aK + 2c + 2\sqrt{acK + c^2}}\right].$$

Now, choose $a = Kt + 2\sqrt{tc}$ to get

$$\mathbb{P}\left(Z \ge 2\sqrt{tc} + Kt\right) \le \exp[-t].$$

$\square$

Lemma 8.4 *Let* $Z \in \mathbb{R}$ *be a random variable that satisfies for a constant* L_0

$$C_0^2 := \mathbb{E} \exp[|Z|/L_0] < \infty.$$

Then for $L > 2L_0$

$$\mathbb{E} \exp[(Z - \mathbb{E}Z)/L] \le \exp\left[\frac{2L_0^2 C_0^2}{L^2 - 2LL_0}\right].$$

Proof of Lemma 8.4 We have for $m \in \{1, 2, \ldots\}$

$$\mathbb{E}|Z|^m \le m! L_0^m C_0^2.$$

Hence

$$\mathbb{E}|Z - \mathbb{E}Z|^m \le m!(2L_0)^m C_0^2.$$

So for $L < 2L_0$

$$\mathbb{E}\exp[(Z - \mathbb{E}Z)/L] \leq 1 + \sum_{m=2}^{\infty} \frac{1}{m!L^m}\mathbb{E}|Z - \mathbb{E}Z|^m \leq 1 + \sum_{m=2}^{\infty}\left(\frac{2L_0}{L}\right)^m C_0^2$$

$$= 1 + \frac{2L_0^2 C_0^2}{L^2 - 2LL_0} \leq \exp\left[\frac{2L_0^2 C_0^2}{L^2 - 2LL_0}\right].$$

□

Combining Lemma 8.3 with Lemma 8.4 returns us the following form of Bernstein's inequality.

Corollary 8.2 *Let* $Z_1, \ldots, Z_n$ *be independent random variables in* $\mathbb{R}$ *that satisfy for some constant* L_0

$$C_0^2 := \max_{1 \leq i \leq n} \mathbb{E}\exp[|Z_i|/L_0] < \infty.$$

Then one can apply Lemma 8.3 with $K = 2L_0$ *and* $c = 2nL_0^2 C_0^2$ *to find that for all* $t > 0$

$$\mathbb{P}\left(\frac{1}{n}\sum_{i=1}^{n}(Z_i - \mathbb{E}Z_i) \geq 2L_0\left(C_0\sqrt{2t/n} + t/n\right)\right) \leq \exp[-t].$$

8.5   Bounds for Weighted Sums of Squared Gaussians

Consider p normally distributed random variables $W_1, \ldots, W_p$, with mean zero and variance σ_0^2/n. Let $W := (W_1, \ldots, W_p)^T$ be the p-dimensional vector collecting the $W_j, j = 1, \ldots, p$. Let $a_1, \ldots, a_m$ be m given vectors in $\mathbb{R}^p$, with $\|a_k\|_1 = 1$ for $k = 1, \ldots, m$.

Key ingredient of the proof of the next lemma is that for a $\mathcal{N}(0, 1)$-distributed random variable V, the conditions of Lemma 8.3 hold with $K = 2$ if we take $Z = V^2 - 1$, see Laurent and Massart (2000, Lemma 1 and its proof).

Lemma 8.5 *Let* $0 < \alpha < 1$ *be a given error level. Then for*

$$\lambda_\epsilon^2 := \frac{\sigma_0^2}{n}\left(1 + 2\sqrt{\log(m/\alpha)} + 2\log(m/\alpha)\right)$$

we have

$$\mathbb{P}\left(\max_{1 \leq k \leq m}\sum_{j=1}^{p} a_{j,k} W_j^2 \geq \lambda_\epsilon^2\right) \leq \alpha.$$

Lemma 8.5 is somewhat a quick and dirty lemma, although the bound is "reasonable". As a special case, suppose that $a_k = e_k$, the kth unit vector, $k = 1, \ldots, m$, and $m = p$. Then we see that the bound of Corollary 8.1 in Sect. 8.2 is generally better than the one of the above lemma (Problem 8.3). Thus, since we know that the dual norm of a norm Ω generated by a convex cone is weaker than the $\|\cdot\|_\infty$-norm, Lemma 8.5 is in general somewhat too rough.

Proof of Lemma 8.5 Write $V_j := \sqrt{n}W_j/\sigma_0$. First check that for all $L > 2$

$$\mathbb{E}\exp\left[(V_j^2 - 1)/L\right] \le \exp\left[\frac{1}{L^2 - 2L}\right]$$

(Problem 8.2, see also Laurent and Massart (2000, Lemma 1 and its proof). We moreover have for all k

$$\mathbb{E}\exp\left[\sum_{j=1}^p a_{j,k}(V_j^2 - 1)/L\right] = \mathbb{E}\left(\prod_{j=1}^p \exp\left[a_{j,k}(V_j^2 - 1)/L\right]\right).$$

Now employ Hölder's inequality, which says that for two random variables X and Y in $\mathbb{R}$, and for $0 < \gamma < 1$

$$\mathbb{E}|X|^\gamma|Y|^{1-\gamma} \le (\mathbb{E}|X|)^\gamma(\mathbb{E}|Y|)^{1-\gamma}.$$

Hence also

$$\mathbb{E}\left(\prod_{j=1}^p \exp\left[a_{j,k}(V_j^2 - 1)/L\right]\right) \le \prod_{j=1}^p\left(\mathbb{E}\exp\left[(V_j^2 - 1)/L\right]\right)^{a_{j,k}}$$

$$\le \prod_{j=1}^p\left(\exp\left[\frac{1}{L^2 - 2L}\right]\right)^{a_{j,k}} = \exp\left[\frac{1}{L^2 - 2L}\right].$$

Therefore by Lemma 8.3, for all $t > 0$

$$\mathbb{P}\left(\sum_{j=1}^p a_{j,k}(V_j^2 - 1) > 2t + 2\sqrt{t}\right) \le \exp[-t].$$

Apply the union bound to find that for all $t > 0$

$$\mathbb{P}\left(\max_{1 \le k \le m}\sum_{j=1}^p a_{j,k}(V_j^2 - 1) \ge 2\sqrt{t + \log(m)} + 2(t + \log m)\right) \le \exp[-t].$$

Finally, take $t = \log(1/\alpha)$.   □

8.6 The Special Case of Chi-Squared Random Variables

We now reprove part of Lemma 1 in Laurent and Massart (2000) (see also Problem 8.6). This allows us a comparison with the results of the previous section.

Lemma 8.6 *Let χ_T^2 be a chi-squared distributed with T degrees of freedom. Then for all $t > 0$*

$$\mathbb{P}\left(\chi_T^2 \geq T + 2\sqrt{tT} + 2t\right) \leq \exp[-t].$$

Proof of Lemma 8.6 Let $V_1, \ldots, V_T$ be i.i.d. $\mathcal{N}(0,1)$. Then (see the Proof of Lemma 8.5)

$$\mathbb{E}\exp\left[(V_j^2 - 1)/L\right] \leq \exp\left[\frac{1}{L^2 - 2L}\right].$$

Hence, by the independence of the V_j,

$$\mathbb{E}\exp\left[\sum_{j=1}^{T}(V_j^2 - 1)/L\right] \leq \exp\left[\frac{T}{L^2 - 2L}\right].$$

The result now follows from Lemma 8.3 (with $K = 2$ and $c = T$).   □

As a consequence, when one considers the maximum of a collection of chi-squared random variables, each with a relatively large number of degrees of freedom, one finds that the log-term in the bound becomes negligible.

Corollary 8.3 *Let, for $j = 1, \ldots, m$, the random variables $\chi_{T_j}^2$ be chi-square distributed with T_j degrees of freedom. Define $T_{\min} := \min\{T_j : j = 1, \ldots, m\}$. Let $0 < \alpha < 1$ be a given error level. Then for*

$$\lambda_0^2 := \frac{1}{n}\left(1 + 2\sqrt{\frac{\log(m/\alpha)}{T_{\min}}} + \frac{2\log(m/\alpha)}{T_{\min}}\right),$$

we have

$$\mathbb{P}\left(\max_{1 \leq j \leq m} \chi_{T_j}^2/T_j \geq n\lambda_0^2\right) \leq \alpha.$$

8.7 The Wedge Dual Norm

The wedge penalty is proportional to the norm

$$\Omega(\beta) = \min_{a \in \mathscr{A},\, \|a\|_1 = 1} \sqrt{\sum_{j=1}^{p} \frac{\beta_j^2}{a_j}}, \quad \beta \in \mathbb{R}^p,$$

with $\mathscr{A} := \{a_1 \geq \cdots \geq a_p\}$ (see Example 6.2 in Sect. 6.9). Its dual norm is

$$\Omega_*(w) = \max_{a \in \mathscr{A},\, \|a\|_1 = 1} \sqrt{\sum_{j=1}^{p} a_j w_j^2}, \quad w \in \mathbb{R}^p.$$

The maximum is attained in the extreme points of $\mathscr{A} \cap \{\|a\|_1 = 1\}$ so

$$\Omega_*(w) = \max_{1 \leq k \leq p} \sqrt{\sum_{j=1}^{k} \frac{w_j^2}{k}}, \quad w \in \mathbb{R}^p.$$

Lemma 8.7 *Let* $V_1, \ldots, V_p$ *be i.i.d.* $\mathscr{N}(0, 1)$. *Then for all* $t > 0$

$$\mathbb{P}\left(\max_{1 \leq k \leq p} \frac{1}{k} \sum_{j=1}^{k} V_j^2 \geq 1 + 2\sqrt{t} + 2t \right) \leq \frac{e^{-t}}{1 - e^{-t}}.$$

Proof of Lemma 8.7 By Lemma 8.6, for all $k \in \{1, \ldots, p\}$ and all $t > 0$

$$\mathbb{P}\left(\frac{1}{k} \sum_{j=1}^{k} V_j^2 \geq 1 + 2\sqrt{t} + 2t \right) \leq \exp[-kt].$$

Hence

$$\mathbb{P}\left(\max_{1 \leq k \leq p} \frac{1}{k} \sum_{j=1}^{k} V_j^2 \geq 1 + 2\sqrt{t} + 2t \right) \leq \sum_{k=1}^{p} \exp[-kt] \leq \frac{e^{-t}}{1 - e^{-t}}.$$

$\square$

Problems

8.1 Let Z have a chi-squared distribution with T degrees of freedom. It has density

$$f_Z(z) = \frac{z^{\frac{T}{2}-1} e^{-z/2}}{2^{\frac{T}{2}} \Gamma(\frac{T}{2})}, \ z > 0.$$

Show that for $v > 0$

$$\mathbb{E} e^{-vZ/2} = \left(\frac{1}{1+v} \right)^{\frac{T}{2}}.$$

(This is used in Lemma 8.1 in Sect. 8.2.)

8.2 Let Z be a standard normal random variable. Show that for all $L > 2$

$$\mathbb{E} \exp \left[(Z^2 - 1)/L \right] \leq \exp \left[\frac{1}{L^2 - 2L} \right].$$

(This is applied in the Proof of Lemma 8.5 in Sect. 8.5.)

8.3 Compare the result of Lemma 8.5 in Sect. 8.5 with that of Corollary 8.1 in Sect. 8.2 for the special case where $a_k = e_k$, the kth unit vector ($k = 1, \ldots, m$) and $m = p$. Show that Corollary 8.1 wins.

8.4 Let $\epsilon \in \mathbb{R}^n$, $\epsilon \sim \mathcal{N}_n(0, \sigma_0^2 I)$ and $X_G := \{X_j\}_{j \in G}$, where $G \subset \{1, \ldots, p\}$ has cardinality $|G| =: T$ and where $\|X_j\|_n = 1$ for all j. Show that for all $t > 0$

$$\mathbb{P} \left(\frac{\|X_G^T \epsilon\|_2^2}{T} \geq \sigma_0^2 (1 + 2\sqrt{t} + 2t) \right) \leq \exp[-t].$$

Let us now drop the assumption of normality of ϵ. Assume its entries are independent and sub-Gaussian with constants L_0 and C_0^2:

$$\max_{1 \leq i \leq n} \mathbb{E} \exp[\epsilon_i^2 / L_0] = C_0^2.$$

Can you prove a similar result as above but now for the sub-Gaussian case instead of the Gaussian case? Hint: apply (for all j) Lemma 8.4 with the variable Z there chosen to be $Z := (X_j^T \epsilon)^2$.

8.5 This is a continuation of Problem 6.5 which concerns the multiple regression model. Consider i.i.d. random variables $\{\epsilon_{i,t} : i = 1, \ldots, n, \ t = 1, \ldots, T\}$. We assume sub-Gaussianity: for a at least some constant a_0 and τ_0

$$\mathbb{P}(|\epsilon_{i,t}| \geq \tau_0 \sqrt{a}) \leq \exp[-a], \ \forall \, j, t.$$

Let $\{X_{i,j,t} : i = 1,\ldots,n,\ j = 1,\ldots,p,\ t = 1,\ldots,T\}$ be fixed input. Define $W_{j,t} := \sum_{i=1}^{n} \epsilon_{i,t} X_{i,j,t}/n$. Derive a probability inequality for

$$\max_{1 \le j \le p} \left[\sum_{t=1}^{T} W_{j,t}^2 \right]^{1/2}$$

exploiting the independence of the entries of the vector $\{W_{j,1},\ldots,W_{j,T}\}$ $(j = 1,\ldots,p)$ (see also Lounici et al. 2011).

8.6 One may apply results of Borell (1975) which show that the square root of a chi-squared random variable χ_T^2 concentrates around its expectation:

$$\mathbb{P}(|\chi_T - \mathbb{E}\chi_T| > \sqrt{2t}) \le 2\exp[-t], \ \forall\, t > 0.$$

Compare with Lemma 8.6 in Sect. 8.6.

Chapter 9
Probability Inequalities for Matrices

Abstract A selection of probability inequalities from the literature for sums of mean zero matrices is presented. The chapter starts out with recalling Hoeffding's inequality and Bernstein's inequality. Then symmetric matrices and rectangular matrices are considered (the latter mostly in Sects. 9.3 and 9.4). The paper Tropp (2012) serves as main reference and in Sect. 9.4 result from Lounici (2011). See also Tropp (2015) for further results.

9.1 Hoeffding and Bernstein

Theorem 9.1 (Hoeffding's Inequality) *Let $X_1, \ldots, X_n$ be independent mean zero random variables that satisfy $|X_i| \leq c_i$ $(i = 1, \ldots, n)$ for positive constants $c_1, \ldots, c_n$. Then for all $t > 0$*

$$\mathbb{P}\left(\frac{1}{n} \sum_{i=1}^{n} X_i \geq \|c\|_n \sqrt{\frac{2t}{n}} \right) \leq \exp[-t].$$

Theorem 9.1 is Corollary 17.1 in Sect. 17.3. The proof is prepared there, with its final step in Problem 17.1.

Theorem 9.2 (Bernstein's Inequality) *Let $X_1, \ldots, X_n$ be independent mean zero random variables that satisfy*

$$\frac{1}{n} \sum_{i=1}^{n} \mathbb{E}|X_i|^m \leq \frac{m!}{2} K^{m-2} R^2, \forall\, m \in \{2, 3, \ldots\}$$

for certain positive constants K and R. Then for all $t > 0$

$$\mathbb{P}\left(\frac{1}{n} \sum_{i=1}^{n} X_i \geq R \sqrt{\frac{2t}{n}} + \frac{Kt}{n} \right) \leq \exp[-t].$$

Theorem 9.2 follows from Lemma 8.3 in Sect. 8.4.

© Springer International Publishing Switzerland 2016
S. van de Geer, *Estimation and Testing Under Sparsity*,
Lecture Notes in Mathematics 2159, DOI 10.1007/978-3-319-32774-7_9

9.2   Matrix Hoeffding and Matrix Bernstein

Recall the notation: for a matrix A, $\Lambda_{\max}^2(A)$ is the largest eigenvalue of $A^T A$. Note that thus $\Lambda_{\max}(A^T A) = \Lambda_{\max}^2(A)$.

We cite a Hoeffding inequality from Tropp (2012).

Theorem 9.3 *Let $X_1, \ldots, X_n$ be random symmetric $p \times p$ matrices. Suppose that for some symmetric matrices C_i, $X_i^2 - C_i^2$ is positive semi-definite. Define*

$$R_n^2 := \Lambda_{\max}\left(\frac{1}{n}\sum_{i=1}^{n} C_i^2\right).$$

Let $\epsilon_1, \ldots, \epsilon_n$ be a Rademacher sequence.
Then for all $t > 0$

$$\mathbb{P}\left(\Lambda_{\max}\left(\frac{1}{n}\sum_{i=1}^{n} \epsilon_i X_i\right) \geq R_n \sqrt{\frac{8(t + \log(2p))}{n}}\right) \leq \exp[-t].$$

The next result (a matrix Bernstein's inequality) also follows from Tropp (2012) (it is a slight variant of Theorem 6.2 in that paper). It is based on a generalization of Theorem 17.1.

Theorem 9.4 *Let $\{X_i\}_{i=1}^n$ be symmetric $p \times p$ matrices satisfying for a constant K and symmetric matrices $\{\Sigma_i\}_{i=1}^n$*

$$\mathbb{E}X_i = 0, \mathbb{E}X_i^m \leq \frac{m!}{2}K^{m-2}\Sigma_i, \ m = 2, 3, \ldots, \ i = 1, \ldots, n.$$

Define

$$R^2 := \Lambda_{\max}\left(\frac{1}{n}\sum_{i=1}^{n} \Sigma_i\right).$$

Then for all $t > 0$

$$\mathbb{P}\left(\Lambda_{\max}\left(\frac{1}{n}\sum_{i=1}^{n} X_i\right) \geq R\sqrt{\frac{2(t + \log(2p))}{n}} + K\frac{(t + \log(2p))}{n}\right) \leq \exp[-t].$$

For the bounded case we cite the following result from Tropp (2012), Theorem 1.6 in that paper.

Theorem 9.5 *Let $\{X_i\}_{i=1}^n$ be $q \times p$ matrices satisfying for a constant K*

$$\mathbb{E}X_i = 0, \ \max_{1 \leq i \leq n} \Lambda_{\max}(X_i) \leq K.$$

Define

$$R^2 := \max\left\{ \Lambda_{\max}\left(\frac{1}{n}\sum_{i=1}^{n}\mathbb{E}X_iX_i^T\right), \Lambda_{\max}\left(\frac{1}{n}\sum_{i=1}^{n}\mathbb{E}X_i^TX_i\right)\right\}.$$

Then for all $t > 0$

$$\mathbb{P}\left(\Lambda_{\max}\left(\frac{1}{n}\sum_{i=1}^{n}X_i\right) \geq R\sqrt{\frac{2(t+\log(p+q))}{n}} + \frac{K(t+\log(p+q))}{3n}\right) \leq \exp[-t].$$

9.3 Matrix Inequalities for Gaussian and Rademacher Sums

A Rademacher sequence is a sequence of i.i.d. random variables $\varepsilon_1,\ldots,\varepsilon_n$ with $\mathbb{P}(\varepsilon_i = 1) = \mathbb{P}(\varepsilon_i = -1) = 1/2$ $(i = 1,\ldots,n)$.

The following theorem is from Tropp (2012) as well (see also Oliveira (2010) for the case of symmetric matrices).

Theorem 9.6 *Let $\epsilon_1,\ldots,\epsilon_n$ be a Rademacher sequence or alternatively a sequence of i.i.d. standard normal random variables. Let $X_1,\ldots,X_n$ be fixed $q \times p$ matrices. Define*

$$R_n^2 := \max\left\{ \Lambda_{\max}\left(\frac{1}{n}\sum_{i=1}^{n}X_iX_i^T\right), \Lambda_{\max}\left(\frac{1}{n}\sum_{i=1}^{n}X_i^TX_i\right)\right\}.$$

Then for all $t > 0$

$$\mathbb{P}\left(\Lambda_{\max}\left(\frac{1}{n}\sum_{i=1}^{n}\epsilon_iX_i\right) \geq R_n\sqrt{\frac{2(t+\log(p+q))}{n}}\right) \leq \exp[-t].$$

9.4 Matrices with a Finite Orlicz Norm

For a random variable $Z \in \mathbb{R}$ and constant $\alpha \geq 1$ the Ψ_α-Orlicz norm is defined as

$$\|Z\|_{\Psi_\alpha} := \inf\{c > 0 : \mathbb{E}\exp[|Z|^\alpha]/c^\alpha] \leq 2\}.$$

In Koltchinskii et al. (2011) one can find the following result.

Theorem 9.7 *Let $X_1,\ldots,X_n$ be i.i.d. $q \times p$ matrices that satisfy for some $\alpha \geq 1$ and all i*

$$\mathbb{E}X_i = 0, \quad K := \|\Lambda_{\max}(X_i)\|_{\Psi_\alpha} < \infty.$$

Define

$$R^2 := \max \left\{ \Lambda_{\max} \left(\frac{1}{n} \sum_{i=1}^{n} \mathbb{E} X_i X_i^T \right), \Lambda_{\max} \left(\frac{1}{n} \sum_{i=1}^{n} \mathbb{E} X_i^T X_i \right) \right\}.$$

Then for a constant C and for all t > 0

$$\mathbb{P} \left(\Lambda_{\max} \left(\frac{1}{n} \sum_{i=1}^{n} X_i \right) \geq CR \sqrt{\frac{t + \log(p+q)}{n}} \right.$$

$$\left. + C \log^{1/\alpha} \left(\frac{K}{R} \right) \left(\frac{t + \log(p+q)}{n} \right) \right)$$

$$\leq \exp[-t].$$

9.5   An Example from Matrix Completion

This example will be applied in Sect. 12.5 where the nuclear norm regularization penalty is invoked. Let $p \geq q$ and $\mathscr{X}$ be the space of all $q \times p$ matrices X of the form

$$X = \begin{pmatrix} 0 & \cdots & 0 & \cdots & 0 & 0 \\ \vdots & & \vdots & & \vdots & \vdots \\ 0 & \cdots & 1 & \cdots & 0 & 0 \\ 0 & \cdots & 0 & \cdots & 0 & 0 \\ \vdots & & \vdots & & \vdots & \vdots \\ 0 & \cdots & 0 & \cdots & 0 & 0 \end{pmatrix},$$

that is matrices—termed masks—X consisting of all 0's except for a single 1 at some entry. The number of masks is thus $|\mathscr{X}| = pq$. Note that XX^T is a $q \times q$ matrix with all 0's except for a single 1 and $X^T X$ is a $p \times p$ matrix with all 0's except a single 1. Let $X_1, \ldots, X_n$ be i.i.d. and uniformly distributed on $\mathscr{X}$. One sees that for all i

$$\mathbb{E} X_i X_i^T = \frac{1}{pq} \iota \iota^T$$

where ι is a q-vector consisting of only 1's. It follows that for all i

$$\Lambda_{\max}(\mathbb{E} X_i X_i^T) = \frac{1}{p}.$$

In the same way one sees that $\Lambda_{\max}(\mathbb{E}X_i^T X_i) = 1/q$ for all i. Hence, since we assumed $p \geq q$,

$$\max\left\{\Lambda_{\max}\left(\frac{1}{n}\sum_{i=1}^n \mathbb{E}X_i X_i^T\right), \Lambda_{\max}\left(\frac{1}{n}\sum_{i=1}^n \mathbb{E}X_i^T X_i\right)\right\} = \frac{1}{q}.$$

Let now $\epsilon_1, \ldots, \epsilon_n$ be i.i.d $\mathcal{N}(0, 1)$ and independent of $X_1, \ldots, X_n$. Then the matrix $\epsilon_i X_i$ has mean zero and for all i

$$\mathbb{E}\epsilon_i^2 X_i X_i^T = \mathbb{E}X_i X_i^T, \ \mathbb{E}\epsilon_i^2 X_i^T X_i = \mathbb{E}X_i^T X_i.$$

Moreover

$$\Lambda_{\max}(\epsilon_i X_i) = |\epsilon_i|\Lambda_{\max}(X_i) \leq |\epsilon_i|, \forall \, i.$$

Hence

$$\|\Lambda_{\max}^2(\epsilon_i X_i)\|_{\psi_2} \leq \|\epsilon_i\|_{\psi_2} = \sqrt{8/3}, \ \forall i$$

(Problem 9.1). It follows from Theorem 9.7 that for all $t > 0$

$$\mathbb{P}\left(\Lambda_{\max}\left(\frac{1}{n}\sum_{i=1}^n \epsilon_i X_i\right) \geq C\sqrt{\frac{t + \log(p + q)}{nq}}\right.$$

$$\left. + C\sqrt{\frac{1}{2}\log\left(\frac{8q}{3}\right)\left(\frac{t + \log(p + q)}{n}\right)}\right)$$

$$\leq \exp[-t].$$

Problems

9.1 Let $Z \sim \mathcal{N}(0, 1)$. Show that

$$\|Z\|_{\psi_2} = \sqrt{8/3},$$

where $\|\cdot\|_{\psi_2}$ is the Orlicz-norm defined in Sect. 9.4 (with $\alpha = 2$).

9.2 Suppose the positive random variable Z satisfies for some $(a, b, c) \in \mathbb{R}_+^3$,

$$\mathbb{P}(Z \geq a + 2b\sqrt{t} + ct) \leq \exp[-t], \forall \, t \geq 0.$$

Show that

$$\mathbb{E}Z \leq a + \sqrt{\pi}b + c.$$

Chapter 10
Inequalities for the Centred Empirical Risk and Its Derivative

Abstract In Theorems 7.1 and 7.2 oracle inequalities for norm-penalized empirical risk minimizers (or M-estimators) were shown. These inequalities require a probability bound for the dual norm of the empirical process. Here such bounds are provided for. The models and methods studied are exponential families, projection estimators and generalized linear models.

10.1 Introduction

Let $\beta \in \mathscr{B} \subset \mathbb{R}^p$ and, for each $\beta \in \mathscr{B}$, let ρ_β be a given real-valued loss function on $\mathscr{X}$. Consider independent observations $X_1, \ldots, X_n$ with values in $\mathscr{X}$ and define the empirical risk $R_n(\beta) := \sum_{i=1}^n \rho_\beta(X_i)/n$ and the theoretical risk $R(\beta) := \mathbb{E}R_n(\beta)$. The empirical process is the process $R_n(\beta') - R(\beta')$ as β' ranges over a subset of $\mathscr{B}$. Let Ω be a norm on $\mathbb{R}^p$ with dual norm Ω_*. In our context, we need results uniformly in β' so that we can apply them with β' equal to some random estimator $\hat{\beta}$. We are furthermore dealing with a candidate oracle β, which is a fixed, nonrandom, vector. We will examine for a positive constant M and fixed $\beta \in \mathbb{R}^p$ the quantity

$$\mathbf{Z}_M := \sup_{\beta' \in \mathscr{B}:\ \Omega(\beta'-\beta) \le M} \left| \left[R_n(\beta') - R(\beta') \right] - \left[R_n(\beta) - R(\beta) \right] \right|.$$

For the case where $\beta' \mapsto R_n(\beta')$ is differentiable we write $\dot{R}_n(\cdot)$ and $\dot{R}(\cdot)$ for the derivative of R_n and R respectively. Uniform-in-β'-inequalities are needed in Theorem 7.1 for $\dot{R}_n(\beta') - \dot{R}(\beta')$. Write the latter as

$$\dot{R}_n(\beta') - \dot{R}(\beta') = \dot{R}_n(\beta) - \dot{R}(\beta) + \left(\left[\dot{R}_n(\beta') - \dot{R}(\beta') \right] - \left[\dot{R}_n(\beta) - \dot{R}(\beta) \right] \right).$$

The first term is not too difficult to deal with as β is fixed. To handle the second term involving the increments, we study

$$\dot{\mathbf{Z}}_M := \sup_{\beta' \in \mathscr{B}:\ \Omega(\beta'-\beta) \le M} \Omega_* \left(\left[\dot{R}_n(\beta') - \dot{R}(\beta') \right] - \left[\dot{R}_n(\beta) - \dot{R}(\beta) \right] \right).$$

© Springer International Publishing Switzerland 2016   139
S. van de Geer, *Estimation and Testing Under Sparsity*,
Lecture Notes in Mathematics 2159, DOI 10.1007/978-3-319-32774-7_10

The tentative conclusion from this chapter is that the empirical process condition (7.1) of Theorem 7.1 (the sharp oracle inequality) can be dealt with

o   in the case where the derivative $\dot{R}_n(\beta') - \dot{R}(\beta')$ does not depend on β' (as is the case in Sects. 10.2 and 10.3 and in Sect. 10.4.1); in other words for the case where $R_n(\beta') - R(\beta')$ is linear in β'
o   or in the case where Ω is at least as strong as the ℓ_1-norm.

In other cases one may want to go back to the original object $(\dot{R}(\hat{\beta}) - \dot{R}(\hat{\beta}))^T(\hat{\beta} - \beta)$. Theorem 7.1 bounds this by applying the dual norm inequality. A direct argument, as outlined in Problem 10.3, may lead to better results. Otherwise, one may want to stay with Theorem 7.2 (instead of trying to use this non-sharp oracle inequality as initial localization step towards a sharp oracle inequality).

In order to be able to apply the results of this chapter to Theorem 7.1 or Theorem 7.2 one needs to replace Ω by the corresponding $\underline{\Omega}$ (which generally depends on β) coming from the triangle property.

In the regression examples, $X_i \in \mathscr{X}$ is replaced by (X_i, Y_i), where $X_i \in \mathscr{X}$ is the co-variable and $Y_i \in \mathbb{R}$ is the response variable $(i = 1, \ldots, n)$.

10.2   Exponential Families

In the exponential family case, $R_n(\beta') - R(\beta')$ is linear in β'. Hence, then its derivative $\dot{R}_n(\beta') - \dot{R}(\beta')$ does not depend on β' so that $\dot{\mathbf{Z}}_M = 0$.

In the next two subsections, we will look at density estimation and regression problems.

10.2.1   Density Estimation

Suppose that $\mathscr{X} \subset \mathbb{R}^p$ (here and more generally one may first want to apply a change of dictionary—or feature mapping—as in Sect. 10.3) and

$$\rho_\beta(x) = -x\beta + d(\beta), \ \beta \in \mathscr{B}$$

where $d : \mathscr{B} \to \mathbb{R}$ is a given function. In the exponential family case, $\rho_\beta(x)$ is minus a log-density with respect to some dominating measure ν, and d is the normalization constant that takes care that the density integrates to one:

$$d(\beta) = \log\left(\int \exp[x\beta]d\nu(x)\right).$$

The parameter space $\mathscr{B}$ is now generally taken as (some convex subset of) $\{\beta : d(\beta) < \infty\}$. Note that the latter is a convex set.

Using the notation

$$W := \frac{1}{n} \sum_{i=1}^{n} (X_i^T - \mathbb{E} X_i^T) \in \mathbb{R}^p,$$

one clearly has for all β and β' in the parameter space $\mathscr{B}$

$$\left| \left[R_n(\beta') - R(\beta') \right] - \left[R_n(\beta) - R(\beta) \right] \right| = \left| W^T (\beta' - \beta) \right|$$

and

$$\dot{R}_n(\beta') - \dot{R}(\beta') = -W.$$

It follows that

$$\mathbf{Z}_M \leq M \Omega_*(W)$$

and

$$\Omega_* \left(\dot{R}_n(\beta) - \dot{R}(\beta) \right) = \Omega_*(W), \quad \dot{\mathbf{Z}}_M = 0.$$

10.2.2 Regression with Fixed Design

In the regression context one has

$$\rho_\beta(x, y) = -yx\beta + d(x\beta),$$

where

$$d(\xi) = \log(\int \exp[y\xi] dv(y)), \xi \in \mathscr{M} \subset \mathbb{R}$$

with $\mathscr{M}$ (a convex subset of) the set $\{\xi : \int \exp[\xi y] dv(y) < \infty\}$.
   Some examples are:

- least squares regression: $\rho_\beta(x, y) = -yx\beta + (x\beta)^2/2$,
- logistic regression: $\rho_\beta(x, y) = -yx\beta + \log(1 + e^{x\beta}), y \in \{0, 1\}$,
- Poisson regression: $\rho_\beta(x, y) = -yx\beta + e^{x\beta}$.

In the fixed design case, the inputs $X_1, \ldots, X_n$ are non-random and hence

$$\left\| \left[R_n(\beta') - R(\beta') \right] - \left[R_n(\beta) - R(\beta) \right] \right\| = \left| W^T (\beta' - \beta) \right|$$

with now

$$W := \frac{1}{n} \sum_{i=1}^{n} (Y_i - \mathbb{E} Y_i) X_i^T.$$

Thus

$$\mathbf{Z}_M \leq M \Omega_*(W)$$

and

$$\Omega_* \left(\dot{R}_n(\beta) - \dot{R}(\beta) \right) = \Omega_*(W), \ \dot{\mathbf{Z}}_M = 0.$$

10.3   Projection Estimators

Let $X_1, \ldots, X_n$ be i.i.d. random variables with values in $\mathscr{X}$ and with density p^0 with respect to some sigma-finite measure ν. Let $\{\psi_j\}_{j=1}^{p} \subset L_2(\nu)$ be a given collection of functions (the dictionary) and define $\psi := (\psi_1, \ldots, \psi_p)$. Write the $L_2(\nu)$-norm as $\| \cdot \|_\nu$. The projection estimator of p^0 is based on the loss function

$$\rho_\beta(x) = -\psi(x)\beta + \frac{1}{2} \| \psi \beta \|_\nu^2.$$

A sharp oracle inequality is derived in Sect. 12.3. Here we study only the empirical process. We have

$$\left\| \left[R_n(\beta') - R(\beta') \right] - \left[R_n(\beta) - R(\beta) \right] \right\| = \left| W^T (\beta' - \beta) \right|,$$

where this time

$$W := \frac{1}{n} \sum_{i=1}^{n} (\psi^T (X_i) - \mathbb{E} \psi^T (X_i)).$$

Therefore

$$\mathbf{Z}_M \leq M \Omega_*(W)$$

and

$$\Omega_* \left(\dot{R}_n(\beta) - \dot{R}(\beta) \right) = \Omega_*(W), \quad \dot{\mathbf{Z}}_M = 0.$$

10.4 The Linear Model with Random Design

The regression model is

$$Y = X\beta^0 + \epsilon,$$

where ϵ is mean zero noise independent of X. Recall the Gram matrix $\hat{\Sigma} := X^T X / n$, $X^T = (X_1^T, \ldots, X_n^T)$. Suppose the inputs $X_1, \ldots, X_n$ are independent mean zero random row-vectors and let the co-variance (or inner-product) matrix be $\Sigma_0 := \mathbb{E}\hat{\Sigma}$.

10.4.1 Known Co-variance Matrix

Suppose Σ_0 is known. Then one can take as loss function "linearized" least squares loss

$$\rho_\beta(x, y) = -yx\beta + \frac{1}{2}\beta^T \Sigma_0 \beta.$$

Write

$$W := X^T \epsilon / n + (\hat{\Sigma} - \Sigma_0)\beta^0.$$

Then

$$\left| \left[R_n(\beta') - R(\beta') \right] - \left[R_n(\beta) - R(\beta) \right] \right| = \left| W^T (\beta' - \beta) \right|.$$

Thus

$$\mathbf{Z}_M \leq M\Omega_*(W)$$

$$\leq M\Omega_*(X^T \epsilon)/n + M\Omega_* \left((\hat{\Sigma} - \Sigma_0)\beta^0 \right)$$

and

$$\Omega_*\left(\dot{R}_n(\beta) - \dot{R}(\beta)\right) = \Omega_*(W)$$

$$\leq \Omega_*(X^T\epsilon)/n + \Omega_*\left((\hat{\Sigma} - \Sigma_0)\beta^0\right), \ \dot{Z}_M = 0.$$

Problem 14.2 examines bounds for $(\hat{\Sigma} - \Sigma_0)\beta^0$.

10.4.2   Unknown Co-variance Matrix

In this case we take least squares loss

$$\rho_\beta(x, y) = \frac{1}{2}(y - x\beta)^2.$$

End-goal is to prove oracle inequalities for the theoretical squared norm $\|X(\hat{\beta} - \beta^0)\|^2$ where $\|X\tilde{\beta}\|^2 := \mathbb{E}\|X\tilde{\beta}\|_n^2$, $\tilde{\beta} \in \mathbb{R}^p$. There are (at least) two ways to proceed, depending on the assumptions one is willing to take. Very roughly, these two approaches are:

○ one assumes that $X_1, \ldots, X_n$ are i.i.d. copies of a sub-Gaussian vector X_0 for example (see Definition 15.2 in Sect. 15.1). Then use the approach as outlined in Problem 10.2.
○ one assumes that $\Omega_*(X_i^T) \leq K_\Omega$ for all $i = 1, \ldots, n$, where K_Ω is a constant under control and then stirs towards showing that $\Omega(\hat{\beta} - \beta)$ is small.

We have the second approach in mind in this section.
Let

$$W_\beta := X^T\epsilon/n + (\hat{\Sigma} - \Sigma_0)(\beta - \beta^0).$$

We have

$$\left[R_n(\beta') - R(\beta')\right] - \left[R_n(\beta) - R(\beta)\right]$$

$$= \epsilon^T X(\beta' - \beta)/n$$

$$+ \frac{1}{2}(\beta' - \beta^0)^T(\hat{\Sigma} - \Sigma_0)(\beta' - \beta^0) - \frac{1}{2}(\beta - \beta^0)^T(\hat{\Sigma} - \Sigma_0)(\beta - \beta^0)$$

$$= W_\beta^T(\beta' - \beta) + (\beta' - \beta)^T(\hat{\Sigma} - \Sigma_0)(\beta - \beta^0)$$

where we inserted a variant of the two point inequality (see Sect. 2.4). The first term $W_\beta^T(\beta' - \beta)$ is linear in β' and as such can be dealt with using the dual norm argument: for $\Omega(\beta' - \beta) \le M$

$$\left| W_\beta^T(\beta - \beta^0) \right| \le M\Omega_*(W_\beta)$$

$$\le M\Omega_*(\epsilon^T X)/n + \Omega_*\left(\hat{\Sigma} - \Sigma_0)(\beta - \beta^0) \right).$$

The quadratic term is more delicate (which is why an alternative approach is indicated in the beginning of this subsection). Clearly, for $\Omega(\beta' - \beta) \le M$ it holds that

$$(\beta' - \beta)^T(\hat{\Sigma} - \Sigma_0)(\beta' - \beta) \le M^2 \Lambda_{\Omega*}(\hat{\Sigma} - \Sigma_0)$$

where, for symmetric a $p \times p$ matrix $A = (A_1, \cdots A_p)$, the notation $\Lambda_{\Omega_*}(A) := \max_{\Omega(\tilde{\beta}) \le 1} \tilde{\beta}^T A \tilde{\beta}$ is implemented[1]. When M is small, the quadratic term is of smaller order than the linear term.

Consider now (for completeness as it does not bring much news) the derivatives. We have

$$\dot{R}_n(\beta') - \dot{R}(\beta') = \epsilon^T X/n + (\hat{\Sigma} - \Sigma_0)(\beta' - \beta^0)$$

and so

$$\dot{R}_n(\beta) - \dot{R}(\beta) = \epsilon^T X/n + (\hat{\Sigma} - \Sigma_0)(\beta - \beta^0) = W_\beta.$$

Moreover

$$\left[\dot{R}_n(\beta') - \dot{R}(\beta') \right] - \left[\dot{R}_n(\beta) - \dot{R}(\beta) \right] = (\hat{\Sigma} - \Sigma^0)(\beta' - \beta).$$

Therefore

$$\Omega_*\left(\dot{R}_n(\beta) - \dot{R}(\beta) \right) \le \Omega_*(W_\beta)$$

$$\le \Omega_*(\epsilon^T X)/n + \Omega_*\left((\hat{\Sigma} - \Sigma^0)(\beta - \beta^0) \right)$$

---

[1]Note that for $A = (A_1, \ldots, A_p)$, it holds true that $\Lambda_{\Omega_*}(A) \le \|\|A\|\|_{\Omega_*} =: \max_k \Omega_*(A_k)$.

and

$$\dot{\mathbf{Z}}_M = \sup_{\beta' \in \mathcal{B}, \, \Omega(\beta'-\beta) \le M} \Omega_* \left((\hat{\Sigma} - \Sigma^0)(\beta' - \beta) \right) \le M \| \hat{\Sigma} - \Sigma^0 \|_{\Omega_*}$$

where for a $p \times q$ matrix $A = (A_1, \ldots A_q)$, $\|A\|_{\Omega_*} := \max_{1 \le k \le q} \Omega_*(A_k)$.

10.5   Generalized Linear Models

Consider "robust" loss of the form $\rho_\beta(x, y) = \rho(x\beta, y)$, where $\rho(\cdot, y)$ is a 1-Lipschitz function on $\mathbb{R}$:

$$|\rho(\xi, y) - \rho(\xi', y)| \le |\xi - \xi'|, \ \forall \ \xi, \ \xi', \ y. \tag{10.1}$$

Examples are:

- quantile regression: $\rho(\xi, y) = \rho(y - \xi)$, $\rho(z) = \alpha|z|\mathbf{1}_{\{z>0\}} + (1 - \alpha)|z|\mathbf{1}_{\{z<0\}}$, $z \in \mathbb{R}$, with $0 < \alpha < 1$ given. This includes least absolute deviation regression ($\alpha = 1/2$) as a special case.
- Huber loss: $\rho(\xi, y) = \rho(y - \xi)$, $\rho(z) = [z^2\mathbf{1}_{\{|z| \le K\}} + K(2|z| - K)\mathbf{1}_{\{|z|>K\}}]/(2K)$, $z \in \mathbb{R}$, with $K > 0$ a given constant.

We use a Rademacher sequence which is a sequence of i.i.d. random variables taking the values $+1$ or -1 each with probability $\frac{1}{2}$.

Lemma 10.1 *Suppose the Lipschitz condition (10.1) on the loss function ρ. Assume moreover that for some constant K_Ω*

$$\max_{1 \le i \le n} \Omega_*(X_i^T) \le K_\Omega.$$

Then for all $t > 0$

$$\mathbb{P}\left(\mathbf{Z}_M \ge M\left[4\mathbb{E}\Omega_*(X^T\tilde{\epsilon})/n + K_\Omega \sqrt{8t/n} \right] \right) \le \exp[-t]$$

where $\tilde{\epsilon}_1, \ldots, \tilde{\epsilon}_n$ is a Rademacher sequence independent of $\{(X_i, Y_i)\}_{i=1}^n$.

Proof of Lemma 10.1 In view of the Symmetrization Theorem (see Theorem 16.1 in Sect. 16.1) combined with the Contraction Theorem (see Theorem 16.2 in Sect. 16.1) we have by the Lipschitz condition (10.1)

$$\mathbb{E}\mathbf{Z}_M \le 4\mathbb{E}\left\{ \sup_{\beta' \in \mathcal{B}: \, \Omega(\beta'-\beta) \le M} \frac{1}{n}\left| \tilde{\epsilon}^T X(\beta' - \beta) \right| \right\}$$

so by the dual norm inequality

$$\mathbb{E}\mathbf{Z}_M \leq 4M\mathbb{E}\Omega_*(X^T\epsilon)/n.$$

Again by the Lipschitz condition (10.1), it further holds that for all i

$$|\rho_{\beta'}(X_i, Y_i) - \rho_\beta(X_i, Y_i)| \leq |X_i(\beta' - \beta)| \leq \Omega_*(X_i^T)\Omega(\beta' - \beta) \leq K_\Omega\Omega(\beta' - \beta).$$

The result now follows from Massart's Concentration Inequality (see Theorem 16.4 in Sect. 16.2).   □

Next, suppose that $\xi \mapsto \rho(\xi, y)$ is differentiable. Write its derivative as

$$\dot{\rho}(\xi, y) := \partial\rho(\xi, y)/\partial\xi.$$

Then

$$\dot{R}_n(\beta') = \frac{1}{n}\sum_{i=1}^{n}X_i^T\dot{\rho}(X_i\beta', Y_i), \ \beta' \in \mathcal{B}.$$

Because in this case $R_n(\beta') - R(\beta')$ is not linear in β', the centered derivative $\dot{R}_n(\beta') - \dot{R}(\beta')$ depends on β':

$$\dot{R}_n(\beta) - \dot{R}(\beta) = \frac{1}{n}\sum_{i=1}^{n}[X_i^T\dot{\rho}(X_i\beta, Y_i) - \mathbb{E}X_i^T\dot{\rho}(X_i\beta, Y_i)].$$

We will now show how one can deal with the dependence on β' when the norm Ω is at least as strong as $\|\cdot\|_1$, see the second point of the tentative conclusion at the end of Sect. 10.1. Otherwise, see Problem 10.3.

We assume that

$$|\dot{\rho}(\xi, y) - \dot{\rho}(\xi', y)| \leq |\xi - \xi'|, \ \forall \ \xi, \ \xi', \ y. \tag{10.2}$$

This is for example the case for the Huber loss given above.

Lemma 10.2 *Suppose the Lipschitz condition (10.2) on the derivative $\dot{\rho}$. Assume moreover that for some constants K_1 and K_Ω*

$$\max_{1\leq i\leq n}\|X_i^T\|_\infty \leq K_1, \ \max_{1\leq i\leq n}\Omega_*(X_i^T) \leq K_\Omega.$$

Then for all $t > 0$ we have with probability at least $1 - \exp[-t]$

$$\sup_{\beta' \in \mathscr{B}:\ \Omega(\beta'-\beta) \leq M} \left\| \left[\dot{R}_n(\beta') - \dot{R}(\beta') \right] - \left[\dot{R}_n(\beta) - \dot{R}(\beta) \right] \right\|_\infty$$

$$\leq K_1 M \left[4\mathbb{E}\Omega_*(X^T \tilde{\epsilon})/n + K_\Omega \sqrt{\frac{8(t + \log p)}{n}} \right]$$

where $\tilde{\epsilon}_1, \ldots, \tilde{\epsilon}_n$ is a Rademacher sequence independent of $\{(X_i, Y_i)\}_{i=1}^n$.

Proof of Lemma 10.2 By the Symmetrization Theorem (see Theorem 16.1 in Sect. 16.1) combined with the Contraction Theorem (see Theorem 16.2 in Sect. 16.1) and invoking the Lipschitz condition (10.2), one sees that for each component j

$$\mathbb{E} \left\{ \sup_{\beta' \in \mathscr{B}:\ \Omega(\beta'-\beta) \leq M} \left| \left[\dot{R}_n(\beta') - \dot{R}(\beta') \right]_j - \left[\dot{R}_n(\beta) - \dot{R}(\beta) \right]_j \right| \right\}$$

$$\leq 4K_1 \mathbb{E} \left\{ \sup_{\beta' \in \mathscr{B}:\ \Omega(\beta'-\beta) \leq M} \frac{1}{n} |\tilde{\epsilon}^T X(\beta' - \beta)| \right\} \leq 4K_1 M \mathbb{E}\Omega_*(X^T\epsilon)/n.$$

Moreover for $\Omega(\beta - \beta) \leq M$ it holds that for all i and j

$$|X_{i,j}(\dot{\rho}(X_i\beta', Y_i) - \dot{\rho}(X_i\beta, Y_i))| \leq K_1|X_i(\beta' - \beta)| \leq K_1 K_\Omega M$$

again by the Lipschitz condition (10.2) The result now follows from Massart's Concentration Inequality (see Theorem 16.4 in Sect. 16.2) and the union bound. $\square$

Recall that if Ω is at least as strong as the ℓ_1-norm, then $\Omega_* \leq \| \cdot \|_\infty$. This can be invoked to deduce from Lemma 10.2 a bound for $\dot{Z}_M$.

10.6   Other Models

In the models considered so far the part of the loss depending on the data x depends on β through a linear function, say $\psi(x)\beta$. An example of a model where the situation is more complicated is the Gaussian mixture model. For simplicity, let us assume there are two components, that the mixing coefficients are $1/2$ and that the variance in each component is one. The mean of each component is described by a high-dimensional linear model. We have then (up to constants) minus log-likelihood

$$\rho_\beta(x, y) = \log(\phi(y - x\beta^{(1)}) + \phi(y - x\beta^{(2)})), \ x \in \mathbb{R}^p, \ y \in \mathbb{R}.$$

Here ϕ is the standard normal density. In this setup, one may apply the two-dimensional Contraction Theorem (Theorem 16.3 in Sect. 16.1), assuming that the means in each component stay within a bounded set. Admittedly, the minus log-likelihood function for the mixture model is not convex so that our theory (Theorems 7.1 and 7.2) cannot be applied. However, one can establish a non-sharp oracle result if one assumes a priori that $\sup_x |x\beta^{(k)}|$, $k = 1, 2$, remains bounded for all $\beta = (\beta^{(1)}, \beta^{(2)}) \in \mathscr{B}$ (Problem 7.5).

Problems

10.1 In Sect. 10.2.1, assume i.i.d. observations $X_1, \ldots, X_n$. Show that without loss of generality one may assume that $\mathbb{E}X_1 = 0$.

10.2 This is a rudimentary outline of an approach alternative to the one of Sect. 10.4.2 (concerning least squares loss with random design). To avoid too may digressions we take β^0 as candidate oracle. First, write the basic inequality

$$\|Y - X\hat{\beta}\|_n^2 + 2\lambda\Omega(\hat{\beta}) \le \|Y - X\beta^0\|_n^2 + 2\lambda\Omega(\beta^0)$$

as

$$\|X(\hat{\beta} - \beta^0)\|_n^2 + 2\lambda\Omega(\hat{\beta}) \le 2\epsilon^T X(\hat{\beta} - \beta^0)/n + 2\lambda\Omega(\beta^0).$$

Assume next that $X_1, \ldots, X_n$ are i.i.d. copies of a sub-Gaussian vector X_0 (see Definition 15.2 in Sect. 15.1, one may also relax this to weak isotropy). Then use similar arguments as in Chap. 15 (see Problem 15.1) to conclude that with large probability for suitable constant C

$$\inf_{\Omega^-(\beta'-\beta^0)\le L\Omega^+(\beta'-\beta^0)} \|X(\beta' - \beta^0)\|_n^2 \ge \|X(\beta' - \beta^0)\|^2(1 - \eta)$$

with $\eta := C(L + 1)\Gamma_\Omega\lambda_{\tilde{\epsilon}}$, where $\Gamma_\Omega = \Gamma_\Omega(L, \beta^0, \|X \cdot \|)$ is the effective sparsity and $\lambda_{\tilde{\epsilon}} := \mathbb{E}\Omega_*(X^T\tilde{\epsilon})/n$ with $\tilde{\epsilon} \in \mathbb{R}^n$ a Rademacher sequence independent of X. Conclude that when $\Omega^-(\hat{\beta} - \beta^0) \le L\Omega^+(\hat{\beta} - \beta^0)$, then with large probability

$$(1 - \eta)\|X(\hat{\beta} - \beta^0)\|^2 + 2\lambda\Omega(\hat{\beta}) \le 2\epsilon^T X(\hat{\beta} - \beta^0)/n + 2\lambda\Omega(\beta^0).$$

Now proceed using arguments as in Problem 7.5 for example.

10.3 Consider the situation of Sect. 10.5. In this problem, the aim is to bound $(\dot{R}_n(\hat{\beta}) - \dot{R}(\hat{\beta}))^T(\hat{\beta} - \beta)$ directly instead of by using the dual norm inequality as is done in Theorem 7.1. We assume the Lipschitz condition (10.2). Write

$$(\dot{R}_n(\beta') - \dot{R}(\beta'))^T(\beta' - \beta) = \frac{1}{n}\sum_{i=1}^{n}[X_i(\beta' - \beta)\dot{\rho}(X_i\beta', Y_i) - \mathbb{E}X_i(\beta' - \beta)\dot{\rho}(X_i\beta', Y_i)].$$

Then note that for $|\xi| \le \eta$ the map $\xi \mapsto \xi\dot{\rho}(\xi)$ is η-Lipschitz. For $\Omega(\beta' - \beta) \le M$ it holds that $\sup_{x\in\mathscr{X}}|x(\hat{\beta} - \beta)| \le K_\Omega M$. Then apply the Symmetrization Theorem (see Theorem 16.1 in Sect. 16.1) combined with the Contraction Theorem (see Theorem 16.2 in Sect. 16.1):

$$\mathbb{E}\left\{\sup_{\beta'\in\mathscr{B}:\ \Omega(\beta'-\beta)\le M}(\dot{R}_n(\beta') - \dot{R}(\beta'))^T(\beta' - \beta)\right\}$$

$$\le 4K_\Omega M\mathbb{E}\left\{\sup_{\beta'\in\mathscr{B}:\ \Omega(\beta'-\beta)\le M}|\tilde{\epsilon}^T X|/n\right\} \le 4K_\Omega M\Omega_*(X^T\tilde{\epsilon})/n.$$

Here, $\tilde{\epsilon}$ is again a Rademacher sequence independent of the observations. Proceed as in Lemma 10.2 to derive a probability inequality, and then use the peeling device (explained in van de Geer (2000) for example).

Chapter 11
The Margin Condition

Abstract The two point margin condition (Condition 7.2.2 in Sect. 7.2) is studied (as well as the one point margin condition which is Condition 7.6.1 in Sect. 7.6). Furthermore, a relation between the two point margin condition and effective sparsity is given.

11.1 Introduction

Let $X_1, \ldots, X_n$ be independent observations with values in $\mathscr{X}$, $\bar{\mathscr{B}}$ a subset of $\mathbb{R}^p$ and let for each $\beta \in \bar{\mathscr{B}}$ be defined a loss function $\rho_\beta : \mathscr{X} \to \mathbb{R}$. The theoretical risk is $R(\beta) := \frac{1}{n} \sum_{i=1}^n \mathbb{E}\rho_\beta(X_i)$. We assume $R(\beta)$ to be differentiable with derivate $\dot{R}(\beta) = \partial R(\beta)/\partial \beta$. We focus in this chapter on the two point margin condition (Condition 7.2.2 in Sect. 7.2). Recall that this condition requires that for some semi-norm τ on $\mathbb{R}^p$ and some increasing strictly convex function G

$$R(\beta) - R(\beta') \geq \dot{R}(\beta')^T(\beta - \beta') + G(\tau(\beta' - \beta)),$$

for all β, β' in a set $\mathscr{B}_{\text{local}} \subset \mathscr{B}$. The set $\mathscr{B}$ is the parameter space used in the estimation. The set $\mathscr{B}_{\text{local}}$ is a "local" set, a set that contains the estimator of under study $\hat{\beta}$ as well as the candidate oracle β. It means that in applications one first needs to localize the problem: show that $\hat{\beta} \in \mathscr{B}_{\text{local}}$.

Of course, if $\mathscr{B}_{\text{local}}$ can be taken as the complete parameter space $\mathscr{B}$ there is no need for localization. This is true when using least squares loss (see Sect. 11.3.3 for the case of fixed design and Sect. 11.5.2 for the case of random design), for linearized least squares loss (see Sect. 11.5.1) and for projection estimators (see Sect. 11.4). Otherwise, one needs to exploit for example the assumed convexity of the loss as is done in Theorem 7.2. Actually, depending on the assumptions, there are at least two ways to proceed (with some similarity to the approaches sketched in Sect. 10.4.2 for dealing with least squares loss with random design). Let β^0 be the target (or reference value) (say the minimizer $\arg \min_{\beta \in \bar{\mathscr{B}}} R(\beta)$).

○ A first approach is funded on "isotropy" conditions on the distribution of the data (see Definition 15.1 in Sect. 15.1). This is outlined in Problem 11.1. The local set $\mathscr{B}_{\text{local}}$ is then a τ-neighbourhood of the target β^0, where τ is the semi-norm

© Springer International Publishing Switzerland 2016
S. van de Geer, *Estimation and Testing Under Sparsity*,
Lecture Notes in Mathematics 2159, DOI 10.1007/978-3-319-32774-7_11

occurring in the margin condition (typically an L_2-norm). We refer moreover to Mendelson (2015) where isotropy conditions are reduced to so-called "small ball" conditions.

o In a second approach $\mathscr{B}_{\text{local}}$ is an Ω-neighbourhood of the target β^0. This is the approach we will take in this chapter. We look at a generalized linear model type of setup, where $\mathscr{X} = \mathbb{R}^p$ and $\rho_\beta(x)$ depends on β only via the linear function $x\beta$ (and possibly also on an integrated version $\int m(x\beta)d\nu(x)$ for some function m and measure ν). The two-point margin condition often holds if at the target value β^0, the function $|x(\beta' - \beta^0)|$ is for all $\beta' \in \mathscr{B}_{\text{local}}$ bounded by a sufficiently small constant, say

$$\sup_{\beta' \in \mathscr{B}_{\text{local}}} \sup_{x \in \mathscr{X}} |x(\beta' - \beta^0)| \leq \eta.$$

Here, the dual norm inequality may be applied:

$$|x(\beta' - \beta^0)| \leq \Omega_*(x^T)\Omega(\beta' - \beta^0).$$

Therefore we will assume $K_\Omega := \sup_x \Omega_*(x^T) < \infty$ (but not in Sect. 11.6 where, depending on the semi-norm τ chosen in the two point margin condition, we arrive at "non-standard" (i.e., non-quadratic) margin behaviour or "non-standard" effective sparsity) . Then for

$$\mathscr{B}_{\text{local}} \subset \{\beta' \in \mathscr{B} : \Omega(\beta' - \beta^0) \leq M\}$$

we get

$$\sup_{\beta' \in \mathscr{B}_{\text{local}}} \sup_{x \in \mathscr{X}} |x(\beta' - \beta^0)| \leq MK_\Omega.$$

The two approaches sketched above rely on different distributional assumptions. The second approach in general leads to more requiring more sparsity. In the ℓ_1-world, the difference is sparsity s_0 of order $n/\log p$ v.s. sparsity s_0 of order $\sqrt{n/\log p}$. In the statistical community, more severe sparsity assumptions are perhaps preferable to isotropy assumptions on the distribution of the data. In the compressed sensing community it may be the other way around, for instance because the design distribution can be chosen by the researcher.

Nevertheless, the two approaches sketched above may both not work. Then a somewhat "unpractical" way out is to take $\mathscr{B} = \mathscr{B}_{\text{local}}$ by assuming for example boundedness conditions on the parameter space. This is done in Sects. 12.5.4 and 12.6.

Throughout this chapter, we assume $\mathscr{B}$ to be convex. Also, Ω is possibly to be replaced throughout this chapter by another norm $\underline{\Omega}$ at which we have control over the $\underline{\Omega}$-estimation error.

11.2 Two Term Taylor Approximations

Since we are dealing with a high-dimensional situation, some caution is in place when doing Taylor approximations. We also need to be explicit about the norms we are using. Of course, in $\mathbb{R}^p$ all norms are equivalent, but this is not of much help as we care about dependence on p for example.

In many examples, the loss function only depends on β via a linear function $x\beta$, such as is the case in generalized linear models. Then one can essentially do one-dimensional Taylor expansions. In this section we do not take such special structure into account.

Let $\mathscr{B}$ be a convex subset of $\mathbb{R}^p$. Fix some reference value $\beta^0 \in \mathbb{R}^p$ and constant M and let the convex set $\mathscr{B}_{\text{local}}$ satisfy

$$\mathscr{B}_{\text{local}} \subset \{\beta' \in \mathscr{B} : \Omega(\beta' - \beta^0) \le M\}.$$

Lemma 11.1 *Suppose that R is twice differentiable at all $\beta' \in \mathscr{B}_{\text{local}}$. Then for all $\beta', \beta \in \mathscr{B}_{\text{local}}$*

$$R(\beta) - R(\beta') = \dot{R}(\beta')^T(\beta - \beta') + \frac{1}{2}(\beta - \beta')^T \ddot{R}(\tilde{\beta})(\beta - \beta'),$$

where also $\tilde{\beta} \in \mathscr{B}_{\text{local}}$.

Proof of Lemma 11.1 Define for $0 \le t \le 1$ the function $g(t) := R((1 - t)\beta' + t\beta)$. Since $\mathscr{B}_{\text{local}}$ is convex and R is twice differentiable on $\mathscr{B}_{\text{local}}$, the function g is also twice differentiable. Its first derivative is

$$\dot{g}(t) = \dot{R}((1 - t)\beta' + t\beta)^T(\beta - \beta')$$

and its second derivative is

$$\ddot{g}(t) = (\beta - \beta')^T \ddot{R}((1 - t)\beta' + t\beta)(\beta - \beta').$$

We may now use the one-dimensional two term Taylor approximation

$$g(1) - g(0) = \dot{g}(0) + \frac{1}{2}\ddot{g}(\tilde{t})$$

for some $0 \le \tilde{t} \le 1$. In other words

$$R(\beta) - R(\beta') = \dot{R}(\beta')^T(\beta - \beta') + \frac{1}{2}(\beta - \beta')^T \ddot{R}(\tilde{\beta})(\beta - \beta')$$

where $\tilde{\beta} := (1 - \tilde{t})\beta' + \tilde{t}\beta$. $\qquad\qquad\square$

In order to be able to discuss the approximation of the quadratic term in the above lemma by one not depending on $\beta, \beta' \in \mathcal{B}_{\text{local}}$ we introduce for a symmetric $p \times p$ matrix A the notation

$$\Lambda_{\Omega_*}(A) := \max_{\Omega(\beta) \leq 1} \beta^T A \beta.$$

Note that when A is positive semi-definite, then $\Lambda_{\|\cdot\|_2}(A) = \Lambda_{\max}(A)$. Furthermore $\Lambda_{\|\cdot\|_\infty}(A) \leq \|A\|_\infty = \|A\|_\infty$ where $\|\cdot\|_{\Omega_*}$ is defined below. Obviously (by the definition of Λ_{Ω_*})

$$(\beta - \beta')^T (\ddot{R}(\tilde{\beta}) - \ddot{R}(\beta^0))(\beta - \beta') \leq 4M^2 \Lambda_{\Omega_*}(\ddot{R}(\tilde{\beta}) - \ddot{R}(\beta^0)), \quad \beta', \beta \in \mathcal{B}_{\text{local}}.$$

It follows that the two point margin condition holds "approximately" provided $M^2 \Lambda_{\Omega_*}^2(\ddot{R}(\tilde{\beta}) - \ddot{R}(\beta^0))$ is "small". In Theorem 7.2 we found bounds for the $\underline{\Omega}$-error where possibly $\underline{\Omega}$ is less strong than the Ω-norm used in the penalty. One can show that the approximate version of the margin condition (Condition 7.6.1) is sufficient for formulating a version of Theorem 7.2 with only minor modifications if, in asymptotic sense, $M \Lambda_{\underline{\Omega}_*}^2 (\ddot{R}(\tilde{\beta}) - \ddot{R}(\beta^0))/\lambda_\epsilon = o(1)$ where $M = M_\beta$ and λ_ϵ are given there. After this localization theorem a similar modification of Theorem 7.1 can be stated.

Closely related with the two term Taylor approximation of $R(\cdot)$ as given in Lemma 11.1 is of course the linear approximation of $\dot{R}(\cdot)$. We present this here but will only need it in Chap. 13.

Lemma 11.2 *Let R be twice differentiable on $\mathcal{B}_{\text{local}}$ with second derivative matrix $\ddot{R}$. Then for all β and β' in $\mathcal{B}_{\text{local}}$ and for $j = 1, \ldots, p$,*

$$\dot{R}_j(\beta) - \dot{R}_j(\beta') = \sum_{k=1}^{p} \ddot{R}_{j,k}(\tilde{\beta}^j)(\beta_k - \beta'_k),$$

where $\tilde{\beta}^j \in \mathcal{B}_{\text{local}}$.

Proof of Lemma 11.2 This follows from the same argument as used in Lemma 11.1: apply the one-dimensional mean value theorem at each coordinate of $\dot{R}$ separately. □

Defining in Lemma 11.2 $\underline{\ddot{R}}_{j,k} = \ddot{R}_{j,k}(\tilde{\beta}^j)$ for $(j,k) \in \{1, \ldots, p\}^2$ and $\ddot{R} := \ddot{R}(\beta')$ (say) we see that if $\|\cdot\|_\infty$ is stronger than Ω_*

$$\Omega_*\left((\underline{\ddot{R}} - \ddot{R})(\beta - \beta')\right) \leq 2M\|\underline{\ddot{R}}^T - \ddot{R}\|_\Omega$$

with for a $p \times q$ matrix $A = (A_1, \ldots, A_q)$

$$\|A\|_\Omega := \max_{1 \leq k \leq q} \Omega(A_k).$$

11.3 Exponential Families

11.3.1 Density Estimation

We consider exponential families where without loss of generality we take the dictionary $\psi(x) = x$, $x \in \mathscr{X} \subset \mathbb{R}^p$. Let $X_1, \ldots, X_n$ be i.i.d. row vectors with distribution P dominated by a sigma-finite measure v and $\mathscr{B} := \{\beta : \int \exp[x\beta] dv(x) < \infty\}$. Define for $\beta \in \mathscr{B}$ the loss function

$$\rho(x) = -x\beta + d(\beta),$$

where $d(\beta) := \log \int \exp[x\beta] dv(x)$. Let $p_\beta(x) := \exp[x\beta - d(\beta)]$. Then

$$\dot{d}(\beta) = \int x^T p_\beta(x) dv(x), \quad \ddot{d}(\beta) = \int x^T x p_\beta(x) dv(x) - \dot{d}(\beta) \dot{d}^T(\beta).$$

We assume that $p^0 := dP/dv$ has the exponential family form

$$p^0(x) = \exp[x\beta^0 - d(\beta^0)], \quad x \in \mathscr{X}.$$

We also assume that $\mathbb{E}X_1 = 0$ (without loss of generality, see Problem 10.1).
   Define now

$$\mathscr{F} := \left\{ f_\beta : \mathscr{X} \to \mathbb{R} : f_\beta(x) = x\beta, \, x \in \mathscr{X}, \, \beta \in \mathscr{B} \right\}.$$

Let $f^0 := x\beta^0$, $x \in \mathscr{X}$. Define for $f \in \mathscr{F}$ the functional

$$d(f) := \log\left(\int \exp[f(x)] dv(x) \right).$$

We now show the two term Taylor expansion of $d(f)$. To avoid technical notation we do not provide explicit constants. We let $\| \cdot \|$ be the $L_2(P)$-norm.

Lemma 11.3 *Suppose*

$$\sup_{f \in \mathscr{F}} \| f - f^0 \|_\infty < \infty.$$

Then for $t \downarrow 0$

$$\sup_{f \in \mathscr{F}} \left| \frac{d((1-t)f^0 + tf) - d(f^0)}{t^2 \| f - f^0 \|^2} - \frac{1}{2} \right| = \mathscr{O}(t).$$

Proof Throughout we let $0 \le \tilde{t} \le t$ be some intermediate point, not the same at each appearance. Define $h := f - f^0$. We have

$$\dot{\mathrm{d}}(f^0 + th) := \frac{\mathrm{dd}(f^0 + th)}{dt} = \frac{\int \exp[f^0 + th]h dv}{\int \exp[f^0 + th]dv}$$

$$= \frac{\int \exp[f^0 - \mathrm{d}(f^0) + th]h dv}{\int \exp[f^0 - \mathrm{d}(f^0) + th]dv} = \frac{P \exp[th]h}{P \exp[th]}.$$

Moreover

$$P \exp[th] = 1 + tPh + t^2 P(\exp[\tilde{t}h]h^2)/2 = 1 + \mathcal{O}(t^2)Ph^2 = 1 + \mathcal{O}(t^2)$$

(here we invoked $Ph = 0$) and

$$P \exp[th]h = Ph + tPh^2 + t^2 P(\exp[\tilde{t}h]h^3)/2$$
$$= tPh^2 + \mathcal{O}(t^2)Ph^2 = tPh^2(1 + \mathcal{O}(t))$$

(again invoking $Ph = 0$). Hence

$$\frac{P \exp[th]h}{P \exp[th]} = \frac{tPh^2(1 + \mathcal{O}(t))}{1 + \mathcal{O}(t^2)} = tPh^2(1 + \mathcal{O}(t)).$$

It follows that

$$\left[\frac{P \exp[th]h}{P \exp[th]}\right]^2 = [tPh^2(1 + \mathcal{O}(t))]^2 = \mathcal{O}(t^2 Ph^2).$$

We moreover have

$$\ddot{\mathrm{d}}(f^0 + th) := \frac{d^2\mathrm{d}(f^0 + th)}{dt^2} = \frac{\int \exp[f^0 + th]h^2 dv}{\int \exp[f^0 + th]dv} - \left[\frac{\int \exp[f^0 + th]h dv}{\int \exp[g^0 + th]dv}\right]^2$$

$$= \frac{P \exp[th]h^2}{P \exp[th]} - \left[\frac{P \exp[th]h}{P \exp[th]}\right]^2.$$

But

$$P \exp[th]h^2 = Ph^2 + tP(\exp[\tilde{t}h]h^3 = Ph^2(1 + \mathcal{O}(t)).$$

So we find

$$\ddot{\mathrm{d}}(tf) = Ph^2(1 + \mathcal{O}(t)) - \mathcal{O}(t^2 Ph^2) = Ph^2(1 + \mathcal{O}(t)).$$

Since $\dot{d}(f^0 + th)|_{t=0} = 0$ it follows that

$$d(f^0 + th) - d(f^0) = \frac{1}{2}t^2\ddot{d}(f^0 + \tilde{t}h) = \frac{1}{2}t^2Ph^2(1 + \mathcal{O}(t)).$$

$\square$

Corollary 11.1 *Let*

$$\mathscr{F}_\infty(\eta) := \{f \in \mathscr{F} : \|f - f^0\|_\infty \le \eta\}.$$

Then for $\eta \downarrow 0$

$$d(f) - d(f^0) = \frac{1}{2}\|f - f^0\|^2(1 + \mathcal{O}(\eta)).$$

The condition of control of sup-norm $\|f - f^0\|_\infty$ may be relaxed at the price of less curvature. We refer to Sect. 11.7 for the line of reasoning.

We impose the following assumptions.

- $\mathscr{B}_{\text{local}} \supset \{\beta' \in \mathscr{B} : \Omega(\beta' - \beta^0) \le M\} \ni M$,
- $K_\Omega := \sup_x \Omega_*(x^T) < \infty$,
- $\eta := K_\Omega M$ is small enough.

Then by Corollary 11.1 the one point margin condition (Condition 7.6.1 in Sect. 7.6) holds with some constant C depending on η, with $\tau^2(\tilde{\beta}) = \tilde{\beta}^T\ddot{R}(\beta^0)\tilde{\beta} = \|X\tilde{\beta}\|^2$, $\tilde{\beta} \in \mathbb{R}^p$, and $G(u) = u^2/(2C^2)$, $u > 0$. The two point margin condition (Condition 7.2.2 in Sect. 7.2) is deferred to Problem 11.2.

11.3.2 Regression with Fixed or Random Design

Let $Y_i \in \mathbb{R}$ and $X_i \in \mathscr{X} \subset \mathbb{R}^p$ $(i = 1, \ldots, n)$. Consider the loss

$$\rho(x, y) = -yx\beta + d(x\beta), \ x \in \mathbb{R}^p, \ y \in \mathbb{R}$$

where $\xi \mapsto d(\xi)$, $\xi \in \mathbb{R}$ is twice differentiable. Write its first derivative as $\dot{d}$ and its second derivative as $\ddot{d}$.

Let us now not insist that the model is well-specified. We let β^0 be some reference value.

We impose the following conditions:

- $\mathscr{B}_{\text{local}} \supset \{\beta' : \Omega(\beta' - \beta^0) \le M\} \ni M$,
- $K_\Omega := \sup_x \Omega_*(x^T) < \infty$,
- $\sup_x |x\beta^0 - \xi^0| \le \eta \ni \eta, \xi^0 \in \mathbb{R}$,
- $\ddot{d}(\xi) \ge 1/C^2, \forall |\xi - \xi^0| \le 2\eta \ni C$,
- $K_\Omega M \le \eta$.

Then for $\beta' \in \mathscr{B}_{\text{local}}$

$$\sup_x |x\beta' - \xi_0| \le K_\Omega M + \eta \le 2\eta.$$

Hence for $\beta, \beta' \in \mathscr{B}_{\text{local}}$

$$R(\beta) - R(\beta') \ge \underbrace{-\frac{1}{n} \sum_{i=1}^n \mathbb{E}((Y_i - \dot{d}(X_i\beta')X_i)(\beta - \beta')}_{\dot{R}(\beta')^T(\beta-\beta')} + (\beta - \beta')^T \Sigma_0 (\beta - \beta')/(2C^2),$$

where $\Sigma_0 = \mathbb{E}X^T X/n$ ($\Sigma_0 = X^T X/n$ for the case of fixed design). It follows that the two point margin condition holds with $G(u) = u^2/(2C^2)$ and $\tau = \|X \cdot \|$ where $\|X\tilde{\beta}\|^2 = \sum_{i=1}^n \mathbb{E}|X_i\tilde{\beta}|^2/n$, $\tilde{\beta} \in \mathbb{R}^p$ ($\| \cdot \| = \| \cdot \|_n$ in the case of fixed design).

11.3.3 Least Squares Regression with Fixed Design

This is a special case of the previous subsection, where localization is not necessary: we may take $\mathscr{B}_{\text{local}} = \mathscr{B}$. Let $Y_i \in \mathbb{R}$ be the response vector and $X_i \in \mathbb{R}^p$ be fixed row vectors ($i = 1, \ldots, n$). Consider the least squares loss

$$\rho(x, y) = (y - x\beta)^2/2.$$

Then the two point margin condition holds with $G(u) = u^2/2$, $u \ge 0$ and with $\tau(\cdot) = \|X \cdot \|_n$.

11.4 Projection Estimators

In this section localization is also unnecessary. We discuss here the margin condition. A sharp oracle inequality is derived in Sect. 12.3. Let $X_1, \ldots, X_n$ be i.i.d. with density p^0 with respect to a sigma-finite measure ν on $\mathscr{X}$. We are given a dictionary $\{\psi_j\}_{j=1}^p \subset L_2(\nu)$ and consider the loss

$$\rho_\beta(x) := -\psi(x)\beta + \frac{1}{2}\|\psi\beta\|_\nu^2$$

where $\psi(x) = (\psi_1(x), \ldots, \psi_p(x))$, $x \in \mathscr{X}$ and where $\| \cdot \|_\nu$ is the $L_2(\nu)$-norm. Then

$$R(\beta) - R(\beta') = -\mathbb{E}\psi(X)(\beta - \beta') + \|\psi\beta\|_\nu^2/2 - \|\psi\beta'\|_\nu^2/2.$$

We have

$$\|\psi\beta\|_v^2/2 - \|\psi\beta'\|_v^2/2 = \|\psi(\beta' - \beta)\|_v^2/2 - \int \psi(\beta' - \beta)\psi\beta'.$$

and hence

$$R(\beta) - R(\beta') = \underbrace{-\int (\psi\beta' - p^0)\psi(\beta - \beta')dv}_{\dot{R}(\beta')^T(\beta-\beta')} + \|\psi(\beta' - \beta)\|_v^2/2.$$

Thus the two point margin condition holds with $G(u) = u^2/2, u \geq 0$ and $\tau^2(\tilde{\beta}) = \|\psi\tilde{\beta}\|_v^2, \tilde{\beta} \in \mathbb{R}^p$.

11.5 The Linear Model with Random Design

Let $Y_i \in \mathbb{R}$ and $X_i \in \mathscr{X} \subset \mathbb{R}^p$ $(i = 1, \ldots, n)$. We consider the model

$$Y = X\beta^0 + \epsilon,$$

where $\mathbb{E}\epsilon = 0$. We define $\hat{\Sigma} := X^T X/n$ and $\Sigma_0 := \mathbb{E}X^T X/n$, and for any $\tilde{\beta} \in \mathbb{R}^p$ $\|X\tilde{\beta}\|_n^2 = \sum_{i=1}^n (X_i\tilde{\beta})^2/n$, $\|X\tilde{\beta}\|^2 = \mathbb{E}\|X\tilde{\beta}\|_n^2$,

11.5.1 Known Co-variance Matrix

Let the loss function be the linearized least squares loss

$$\rho_\beta = -yx\beta + \beta^T \Sigma_0\beta/2.$$

Then

$$R(\beta) - R(\beta') = -\beta^{0T}\Sigma_0(\beta - \beta') + \beta'^T\Sigma_0\beta'/2 - \beta^T\Sigma_0\beta/2$$

$$= \underbrace{-(\beta' - \beta^0)^T\Sigma_0(\beta - \beta')}_{\dot{R}(\beta')^T(\beta-\beta')} + (\beta' - \beta^0)^T\Sigma_0(\beta' - \beta^0)/2.$$

So the two point margin condition holds with $G(u) = u^2/2, u > 0$ and with $\tau(\cdot) = \|X \cdot\|$.

11.5.2 Unknown Co-variance Matrix

We use the least squares loss

$$\rho_\beta(x, y) = (y - x\beta)^2/2.$$

Then

$$R(\beta) - R(\beta') = (\beta - \beta^0)^T \Sigma_0 (\beta - \beta^0)/2 - (\beta' - \beta^0)^T \Sigma_0 (\beta' - \beta^0)/2$$

$$= \underbrace{-(\beta' - \beta^0)^T \Sigma_0 (\beta' - \beta)}_{\dot{R}(\beta')^T(\beta - \beta')} + (\beta' - \beta)^T \Sigma_0 (\beta' - \beta)/2.$$

So the two point margin condition holds with $G(u) = u^2/2$, $u > 0$ and with $\tau(\cdot) = \|X \cdot \|$.

11.6 Generalized Linear Models

Let $Y_i \in \mathbb{R}$ and $X_i \in \mathscr{X} \subset \mathbb{R}^p$ ($i = 1, \ldots, n$). We assume for simplicity that $\{X_i, Y_i\}_{i=1}^n$ are i.i.d.. Consider the loss function

$$\rho_\beta(x, y) = \rho(x\beta, y), \ x \in \mathbb{R}^p, \ y \in \mathbb{R}.$$

Define for $\xi \in \mathbb{R}$

$$r(\xi, x) := \mathbb{E}\rho(\xi, Y_1)|X_1 = x).$$

The target is then typically

$$f^0(x) := \arg\min_{\xi \in \mathbb{R}} r(\xi, x), \ x \in \mathscr{X}.$$

Let us suppose a well-specified model where $f^0(x) = x\beta^0$, $x \in \mathscr{X}$ is indeed linear. If this is not the case we need to assume that f^0 can be approximated by a linear function, as was done in Sect. 11.3.2. Suppose

$$\dot{r}(\xi, x) = \partial r(\xi, x)/\partial \xi, \ \ddot{r}(\xi, x) = \partial^2 r(\xi, x)/\partial \xi^2, \xi \in \mathbb{R}$$

exist.

We impose the following conditions:

- $\mathscr{B}_{local} := \mathscr{B} \cap \{\beta : \Omega(\beta - \beta^0) \le M\} \ni M$,
- for all x and β such that $|x(\beta - \beta^0)| \le \Omega_*(x)M$, we have $\ddot{r}(x\beta, x) \ge 1/C^2(x) \exists C(\cdot)$.

Note that we allow this time the lower bound for the second derivative $\ddot{r}(x\beta, x)$ to depend on x to include more general cases.

For all x and for β and β' in $\mathscr{B}_{\text{local}}$

$$r(x\beta, x) - r(x\beta', x) \geq \dot{r}(x\beta', x)x(\beta' - \beta) + |x(\beta' - \beta)|^2/(2C^2(x)).$$

Then

$$R(\beta) - R(\beta') \geq \underbrace{\mathbb{E}\dot{r}(X_1\beta', X_1)X_1(\beta - \beta')}_{\dot{R}(\beta')^T(\beta-\beta')} + \mathbb{E}|X_1(\beta' - \beta)|^2/(2C^2(X_1)).$$

It follows that the two point margin condition holds with $G(u) = u^2/2$, $u > 0$ together with the squared pseudo-norm

$$\tau^2(\tilde{\beta}) = \mathbb{E}|X_1\tilde{\beta}|^2/C^2(X_1), \quad \tilde{\beta} \in \mathbb{R}^p.$$

11.7   The Two Point Margin Condition and Effective Sparsity

Consider $\mathscr{X} \subset \mathbb{R}^p$ and a positive function $C(\cdot)$ on $\mathbb{R}^p$. Write for $\tilde{\beta} \in \mathbb{R}^p$

$$\|X\tilde{\beta}\|^2 := \frac{1}{n}\sum_{i=1}^{n} \mathbb{E}|X_i\tilde{\beta}|^2$$

and

$$\tau^2(\tilde{\beta}) := \frac{1}{n}\sum_{i=1}^{n} \mathbb{E}|X_i\tilde{\beta}|^2/C^2(X_i).$$

Fix a $\beta \in \mathbb{R}^p$ at which Ω has the triangle property (see Definition 7.4). Write $\Omega^+ := \Omega_\beta^+$, $\Omega^- := \Omega_\beta^-$ and $\underline{\Omega} := \Omega^+ + \Omega_-$. Fix also a stretching factor L and let $\Gamma_\Omega(L, \beta, \|X \cdot \|)$ be the effective sparsity for the pseudo-norm $\|X \cdot \|$ and $\Gamma_\Omega(L, \beta, \tau)$ be the effective sparsity for the pseudo-norm τ (see Definition 7.5).

In Lemma 11.6 we bound $\Gamma_\Omega(L, \beta, \tau)$ in terms of $\Gamma_\Omega(L, \beta, \|X \cdot \|)$. This can be applied for example in the previous section (Sect. 11.6).

First an auxiliary lemma.

Lemma 11.4 *Let U and V be two random variables with $|U| \leq K_U$ for some constant K_U. Let H_1 be a convex function such that*

$$H_1(v) \geq v\mathbb{P}(|V| \leq v), \quad v > 0$$

and let G_1 be the convex conjugate of H_1 (see Definition 7.2). Then

$$\mathbb{E}|UV| \geq K_U G_1(\mathbb{E}|U|/K_U).$$

For example, when taking $V = U^{-r}$ for some $0 < r < 1$ in the above lemma one obtains a type of "reversed" Jensen inequality.

Proof of Lemma 11.4 We have for all $v \geq 0$

$$\begin{aligned}
\mathbb{E}|UV| &= \mathbb{E}|UV|\{|V| > v\} + \mathbb{E}|UV|\{|V| < v\} \\
&\geq v\,\mathbb{E}|U|\{|V| > v\} = v\,\mathbb{E}|U| - v\,\mathbb{E}|U|\{|V| \leq v\} \\
&\geq v\,\mathbb{E}|U| - vK_U\mathbb{P}(|V| \leq v) \geq K_U\left(v\frac{\mathbb{E}|U|}{K_U} - H_1(v)\right).
\end{aligned}$$

Now optimize over v.   □

An application of the previous lemma gives

Lemma 11.5 *Suppose that for a convex function H_1*

$$H_1(v) \geq \frac{v}{n}\sum_{i=1}^{n}\mathbb{P}\left(\underline{\Omega}_*^2(X_i)/C^2(X_i) \leq v\right), \quad v > 0.$$

Let G_1 be the convex conjugate of H_1. Then for all $\tilde{\beta}$

$$\tau^2(\tilde{\beta}) \geq \underline{\Omega}^2(\tilde{\beta})G_1\left(\|X\tilde{\beta}\|^2/\underline{\Omega}^2(\tilde{\beta})\right).$$

Proof of Lemma 11.5 This follows from Lemma 11.4 and from noting that for all $\tilde{\beta}$

$$\tau^2(\tilde{\beta}) = \frac{1}{n}\sum_{i=1}^{n}\mathbb{E}\frac{|X_i\tilde{\beta}|^2}{\underline{\Omega}_*^2(X_i)}\frac{\underline{\Omega}_*^2(X_i)}{C^2(X_i)},$$

and clearly

$$|X_i\tilde{\beta}|^2/\underline{\Omega}_*^2(X_i) \leq \underline{\Omega}^2(\tilde{\beta}).$$

                                                                 □

This leads to the promised bound of one effective sparsity in terms of the other.

Lemma 11.6 *Assume the conditions of Lemma 11.5. Then*

$$\frac{1}{\Gamma_{\underline{\Omega}}^2(L, \beta, \tau)} \geq G_1\left(\frac{1}{\Gamma_{\underline{\Omega}}^2(L, \beta, \|X\cdot\|)(L+1)^2}\right)$$

Proof of Lemma 11.6 Let $\tilde{\beta}$ satisfy $\Omega^-(\tilde{\beta}) \leq L\Omega^+(\tilde{\beta})$. By Lemma 11.5 we have

$$\tau^2(\tilde{\beta}) \geq \underline{\Omega}^2(\tilde{\beta})G_1\left(\|X(\tilde{\beta})\|^2/\underline{\Omega}^2(\tilde{\beta})\right).$$

It follows that

$$\|X(\tilde{\beta})\|^2 \leq \underline{\Omega}^2(\tilde{\beta})G_1^{-1}\left(\frac{\tau^2(\tilde{\beta})}{\underline{\Omega}^2(\tilde{\beta})}\right).$$

By the definition of effective sparsity for $\|X \cdot \|$

$$\underline{\Omega}(\tilde{\beta}) \leq (L+1)\Omega^+(\tilde{\beta}) \leq (L+1)\|X\tilde{\beta}\|\Gamma_\Omega(L, \beta, \|X \cdot \|).$$

Hence

$$\underline{\Omega}^2(\tilde{\beta}) \leq (L+1)^2\underline{\Omega}^2(\tilde{\beta})G_1^{-1}\left(\frac{\tau^2(\tilde{\beta})}{\underline{\Omega}^2(\tilde{\beta})}\right)\Gamma_\Omega^2(L, \beta, \|X \cdot \|).$$

Thus

$$(L+1)^2 G_1^{-1}\left(\frac{\tau^2(\tilde{\beta})}{\underline{\Omega}^2(\tilde{\beta})}\right)\Gamma_\Omega^2(L, \beta, \|X \cdot \|) \geq 1$$

or

$$G_1^{-1}\left(\frac{\tau^2(\tilde{\beta})}{\underline{\Omega}^2(\tilde{\beta})}\right) \geq \left(\Gamma_\Omega^2(L, \beta, \|X \cdot \|)(L+1)^2\right)^{-1}.$$

So

$$\underline{\Omega}^2(\tilde{\beta}) \leq \left[G_1\left(\left(\Gamma_\Omega^2(L, \beta, \|X \cdot \|)(L+1)^2\right)^{-1}\right)\right]^{-1}\tau^2(\tilde{\beta}).$$

$\square$

As an illustration, if for some $\gamma > 0$, $H_1(v) \sim v^{1+\frac{1}{\gamma}}$, $v > 0$, then $G_1(u) \sim u^{1+\gamma}$, $u > 0$. If then for example $\Omega = \| \cdot \|_1$ and $\beta = \beta_S$ where S is a set with cardinality s, we get

$$\|\tilde{\beta}\|_1^2 \leq C_\gamma^2 s^{1+\gamma}\tau^2(\tilde{\beta})/\phi^{2(1+\gamma)}(L, S, \| \cdot \|)$$

where C_γ is a constant depending only on γ and where $\phi^2(L, S, \| \cdot \|)$ is the compatibility constant for the pseudo-metric $\| \cdot \|$. The "standard" case corresponds to $\gamma = 0$.

11.8   Other Models

In Sect. 10.6 we considered briefly the high-dimensional Gaussian mixture model. It is clear that in such models, that is, models where the loss depends on the data x and the parameters β via finitely many linear functions $x\beta^{(k)}$, $k = 1, \ldots, r$, one may establish the two point margin condition assuming appropriate local second order Taylor expansions with non-singular Hessians. Observe that the Gaussian mixture minus log-likelihood is not convex in the parameter so that our theory does not apply. However, one can deal with this by assuming a priori that for all $\beta \in \mathcal{B}$ each $\sup_x |x\beta^{(k)}|$ ($k = 1, \ldots, r$) is bounded (Problem 7.5).

When the loss is non-linear in β and not with the structure described above, localizing high-dimensional problems is a non-trivial task. An example is where the target β^0 is a precision matrix Θ_0 and the Gaussian minus log-likelihood is used as loss function. Then $R_n(\Theta') - R(\Theta')$ is linear in Θ' but $R(\Theta) = -\log\det(\Theta')$ is non-linear. We will study this in Chap. 13.

Problems

11.1 This problem treats the situation of Sect. 11.6. Let $Y_i \in \mathbb{R}$ and $X_i \in \mathcal{X} \subset \mathbb{R}^p$ ($i = 1, \ldots, n$), with $\{X_i, Y_i\}_{i=1}^n$ i.i.d.. Consider the loss function

$$\rho_\beta(x, y) = \rho(x\beta, y), \ x \in \mathbb{R}^p, \ y \in \mathbb{R}.$$

Suppose

$$\dot{\rho}(\xi, x) = \partial\rho(\xi, y)/\partial\xi, \ \ddot{\rho}(\xi, y) = \partial^2\rho(\xi, y)/\partial\xi^2, \xi \in \mathbb{R}$$

exist and let

$$\beta^0 = \arg\min_{\beta \in \mathcal{B}} R(\beta)$$

with $R(\beta) = P\rho_\beta$. Then

$$R(\beta) - R(\beta^0) = \frac{1}{2}\mathbb{E}\left(\ddot{\rho}(X_1\tilde{\beta}(X_1), Y)|X_1(\beta - \beta^0)|^2\right)$$

for some intermediate point $\tilde{\beta}(X_1)$. Suppose

- $\exists\, C > 0$ and $K > 0$ such that $\ddot{\rho}(x\beta, y) \geq 1/C^2$, $\forall\, |x(\beta - \beta^0)| \leq K$,
- for some $m \geq 4$, X_1 is m-th order isotropic with constant C_m (see Definition 15.1 in Sect. 15.1).

Show that for all $0 < \eta < 1$ and for $\|X(\beta - \beta^0)\|^{m-4} \leq K^{m-2}(m-2)\eta/(2C_m^m)$ one has the one point margin condition

$$R(\beta) - R(\beta^0) \geq \frac{1-\eta}{2C^2}\|X(\beta - \beta^0)\|^2.$$

Hint: apply a truncation argument:

$$\mathbb{E}\left(\ddot{\rho}(X_1\tilde{\beta}(X_1), Y)(X_1(\beta - \beta^0))^2\right)$$

$$\geq \frac{1}{C^2}\mathbb{E}(X_1(\beta - \beta^0))^2\{|X_1(\beta - \beta^0)| \leq K\}$$

$$= \frac{1}{C^2}\|X(\beta - \beta^0)\|^2 - \frac{1}{C^2}\mathbb{E}(X_1(\beta - \beta^0))^2\{|X_1(\beta - \beta^0)| > K\}.$$

11.2 Let $X_1, \ldots, X_n$ be i.i.d. row vectors with distribution P with mean zero, ν be a sigma-finite measure on $\mathscr{X}$ and $\mathscr{B} := \{\beta : \int \exp[x\beta]d\nu(x) < \infty\}$. The loss function is

$$\rho_\beta(x) = -x\beta + d(\beta), \ \beta \in \mathscr{B}$$

where $d(\beta) := \log \int \exp[x\beta]d\nu(x)$. Let p^0 be the density $p^0 := dP/d\nu$ and suppose $\log p^0(x) = x\beta^0 - d(\beta^0)$, $x \in \mathbb{R}^p$, i.e., that the model is well-specified. Assume that

- $\mathscr{B}_{\text{local}} := \mathscr{B} \cap \{\beta : \Omega(\beta - \beta^0) \leq M\} \ni M$,
- $K_\Omega := \sup_x \Omega_*(x^T) < \infty$,
- $\eta := 2K_\Omega M$ is small enough.

Verify that the two point margin condition (Condition 7.2.2 in Sect. 7.2) holds with some constant C depending on η: for all $\beta, \beta' \in \mathscr{B}_{\text{local}}$

$$R(\beta) - R(\beta') \geq \dot{R}(\beta')^T(\beta - \beta') + \frac{1}{2C^2}\beta^T\ddot{R}(\beta^0)\beta.$$

Chapter 12
Some Worked-Out Examples

Abstract The sharp oracle inequalities for the Lasso, the square-root Lasso and other structured sparsity estimators are completed with a probability statement saying the result holds with confidence at least $1 - \alpha$ (say), with $\alpha > 0$ a fixed error level. Also a sharp oracle inequality for projection estimators of a density are given. As a "representative" example for the case of non-linear M-estimation with an ℓ_1-penalty, a sharp oracle inequality is presented for logistic regression. For trace regression with nuclear norm penalty, a sharp oracle is given when least squares loss is used, and a non-sharp one when least absolute deviations loss is used. In the latter case, the design is taken as in the matrix completion problem. As final example sparse principal component analysis is addressed.

12.1 The Lasso and Square-Root Lasso Completed

We use the notation of Chaps. 2 and 3. Recall the linear model

$$Y = X\beta^0 + \epsilon$$

with $\epsilon \sim \mathcal{N}_n(0, \sigma_0^2 I)$ and X and a given $n \times p$ matrix. We assume $\operatorname{diag}(X^T X)/n = I$. Define $W := X^T \epsilon / n = (W_1, \ldots, W_p)^T$. Note that $W_j \sim \mathcal{N}(0, \sigma_0^2/n)$ for all j.

Combining Theorem 2.2 with Corollary 8.1 completes the result for the Lasso.

Corollary 12.1 *Consider the Lasso*

$$\hat{\beta} := \arg\min_{\beta \in \mathbb{R}^p} \left\{ \|Y - X\beta\|_n^2 + 2\lambda \|\beta\|_1 \right\}.$$

Let for some $0 < \alpha < 1$

$$\lambda_\epsilon := \sigma_0 \sqrt{\frac{2 \log(2p/\alpha)}{n}}.$$

Let $0 \leq \delta < 1$ *be arbitrary and define for* $\lambda > \lambda_\epsilon$

$$\underline{\lambda} := \lambda - \lambda_\epsilon, \quad \bar{\lambda} := \lambda + \lambda_\epsilon + \delta \underline{\lambda}.$$

© Springer International Publishing Switzerland 2016
S. van de Geer, *Estimation and Testing Under Sparsity*,
Lecture Notes in Mathematics 2159, DOI 10.1007/978-3-319-32774-7_12

and

$$L := \frac{\bar{\lambda}}{(1-\delta)\underline{\lambda}}.$$

Then for all $\beta \in \mathbb{R}^p$ and all S we have with probability at least $1 - \alpha$

$$2\delta\underline{\lambda}\|\hat{\beta} - \beta\|_1 + \|X(\hat{\beta} - \beta^0)\|_n^2 \leq \|X(\beta - \beta^0)\|_n^2 + \frac{\bar{\lambda}^2|S|}{\hat{\phi}^2(L,S)} + 4\lambda\|\beta_{-S}\|_1.$$

We now combine Theorem 3.1 with Lemma 8.2 to complete the result for the square-root Lasso.

Corollary 12.2 *Consider the square-root Lasso*

$$\hat{\beta} := \arg\min_{\beta \in \mathbb{R}^p}\left\{ \|Y - X\beta\|_n + \lambda\|\beta\|_1 \right\}.$$

Define for some positive α and $\underline{\alpha}$ satisfying $\alpha + \underline{\alpha} < 1$ the quantities

$$\lambda_{0,\epsilon} := \sqrt{\frac{2\log(2p/\alpha)}{n-1}}, \quad \underline{\sigma}^2 := \sigma_0^2\left(1 - 2\sqrt{\frac{\log(1/\underline{\alpha})}{n}}\right).$$

Assume for some $\eta > 0$

$$\lambda_0\|\beta^0\|_1 \leq 2\underline{\sigma}\left(\sqrt{1 + (\eta/2)^2} - 1\right), \quad \lambda_0(1-\eta) > \lambda_{0,\epsilon}.$$

For arbitrary $0 \leq \delta < 1$ define

$$\underline{\lambda}_0 := \lambda_0(1-\eta) - \lambda_{0,\epsilon},$$
$$\bar{\lambda}_0 := \lambda_0(1+\eta) + \lambda_{0,\epsilon} + \delta\underline{\lambda}_0$$

and

$$L := \frac{\bar{\lambda}_0}{(1-\delta)\underline{\lambda}_0}.$$

Then for all β and S, with probability at least $1 - \alpha - \underline{\alpha}$ we have

$$2\delta\underline{\lambda}_0\|\hat{\beta} - \beta\|_1\|\epsilon\|_n + \|X(\hat{\beta} - \beta^0)\|_n^2$$
$$\leq \|X(\beta - \beta^0)\|_n^2 + \frac{\bar{\lambda}_0^2|S|\|\epsilon\|_n^2}{\hat{\phi}^2(L,S)} + 4\lambda_0(1+\eta)\|\epsilon\|_n\|\beta_{-S}\|_1.$$

12.2   Least Squares Loss with Ω-Structured Sparsity Completed

We use the notation of Chap. 6. Again the linear model is examined:

$$Y = X\beta^0 + \epsilon$$

with $\epsilon \sim \mathcal{N}_n(0, \sigma_0^2 I)$ and X and a given $n \times p$ matrix with $\mathrm{diag}(X^T X)/n = I$. We set $W := X^T \epsilon / n = (W_1, \ldots, W_p)^T$. As in Sect. 6.9 we let $\mathscr{A}$ be a convex cone in $\mathbb{R}_+^p =: [0, \infty)^p$ and define

$$\Omega(\beta) := \min_{a \in \mathscr{A}} \frac{1}{2} \sum_{j=1}^{p} \left[\frac{|\beta_j|^2}{a_j} + a_j \right], \ \beta \in \mathbb{R}^p.$$

For set S such that $\mathscr{A}_S \subset \mathscr{A}$ (so that S is allowed, see Lemma 6.7) we define $\mathscr{E}_S(\mathscr{A})$ as the set of extreme points of $\mathscr{A}_S \cap \{\|a_S\|_1 \leq 1\}$ and $\mathscr{E}^{-S}(\mathscr{A})$ as the set of extreme points of $\mathscr{A}_{-S} \cap \{\|a_{-S}\|_1 \leq 1\}$. We now assume both $\mathscr{E}_S(\mathscr{A})$ and $\mathscr{E}^{-S}(\mathscr{A})$ are finite and define for positive error levels α_1 and α_2 such that $\alpha_1 + \alpha_2 \leq 1$

$$\frac{n\lambda_S^2}{\sigma_0^2} := \min\left\{ \left(1 + 2\sqrt{\log\left(\frac{|\mathscr{E}_S(\mathscr{A})|}{\alpha_1} \right)} + 2\log\left(\frac{|\mathscr{E}_S(\mathscr{A})|}{\alpha_1} \right) \right), 2\log\left(\frac{2|S|}{\alpha_1} \right) \right\}$$

and

$$\frac{n(\lambda^{-S})^2}{\sigma_0^2}$$

$$:= \min\left\{ \left(1 + 2\sqrt{\log\left(\frac{|\mathscr{E}^{-S}(\mathscr{A})|}{\alpha_2} \right)} + 2\log\left(\frac{|\mathscr{E}^{-S}(\mathscr{A})|}{\alpha_2} \right) \right), 2\log\left(\frac{2(p - |S|)}{\alpha_2} \right) \right\}.$$

We obtain from Lemma 8.5 that with probability at least $1 - \alpha_1 - \alpha_2$,

$$\Omega_*(X_S^T \epsilon)/n \leq \lambda_S, \ \Omega_*^{-S}(X_{-S}^T \epsilon)/n \leq \lambda^{-S}.$$

(Recall that $\Omega_*(X^T \epsilon)/n \leq \max\{\Omega_*(X_S^T \epsilon)/n, \Omega_*^{-S}(X_{-S}^T \epsilon)/n\}$, see Lemma 6.2.)

Theorem 6.1 then leads to the following corollary.

Corollary 12.3 *Let Ω be the norm generated from the convex cone $\mathscr{A}$ and consider the Ω-structured sparsity estimator*

$$\hat{\beta} := \arg\min\left\{ \|Y - X\beta\|_n^2 + 2\lambda\Omega(\beta) \right\}.$$

Assume for all allowed sets S that $\mathscr{E}_S(\mathscr{A})$ and $\mathscr{E}^{-S}(\mathscr{A})$ are finite. Let, for allowed sets S, the constants λ_S and λ^{-S} be defined as above. Let $\delta_1 \geq 0$ and $0 \leq \delta_2 < 1$ be arbitrary. Take $\lambda > \max\{\lambda^{-S} : S \text{ allowed}\}$ and define

$$\underline{\lambda} := \lambda - \lambda^{-S}, \; \bar{\lambda} := \lambda + \lambda_S + \delta_1\underline{\lambda}$$

and

$$L := \frac{\bar{\lambda}}{(1-\delta_2)\underline{\lambda}}.$$

Then for any allowed set S and any β, with probability at least $1 - \alpha_1 - \alpha_2$ it holds that

$$2\delta_1\underline{\lambda}\Omega(\hat{\beta}_S - \beta) + 2\delta_2\underline{\lambda}\Omega^{-S}(\hat{\beta}_{-S}) + \|X(\hat{\beta} - \beta^0)\|_n^2$$

$$\leq \|X(\beta - \beta^0)\|_n^2 + \frac{\bar{\lambda}^2|S|}{\hat{\phi}_\Omega^2(L, S)} + 4\lambda\Omega(\beta_{-S}).$$

For the group Lasso (see Example 6.1) we may improve the lower bound on the tuning parameter. We assume orthogonal design within groups $X_{G_j}^T X_{G_j}/n = I$, $j = 1\ldots,m$. Equivalently, one may define the penalty as

$$\Omega_{\text{group}}(\beta) := \sum_{j=1}^m \sqrt{|G_j|}\|X\beta_{G_j}\|_n, \; \beta \in \mathbb{R}^p.$$

Combining Corollary 8.3 with Theorem 6.1 yields the following.

Corollary 12.4 *Consider the group Lasso as in Example 6.1:*

$$\hat{\beta} := \arg\min\left\{\|Y - X\beta\|_n^2 + 2\lambda\Omega_{\text{group}}(\beta)\right\}.$$

Assume within-group orthogonal design. Let $0 \leq \delta < 1$ be arbitrary. Take

$$\lambda > \lambda_\epsilon := \frac{\sigma_0}{\sqrt{n}}\left(1 + 2\sqrt{\frac{\log(m/\alpha)}{T_{\min}}} + \frac{2\log(m/\alpha)}{T_{\min}}\right)^{1/2},$$

where m is the number of groups and $T_{\min} := \min\{|G_j| : j = 1,\ldots,m\}$ is the minimal group size. Define

$$\underline{\lambda} := \lambda - \lambda_\epsilon, \; \bar{\lambda} := \lambda + \lambda_\epsilon + \delta\underline{\lambda}$$

and

$$L := \frac{\bar{\lambda}}{(1-\delta)\underline{\lambda}}.$$

Then for any allowed set S and any β, with probability at least $1 - \alpha$ it holds that

$$2\delta\underline{\lambda}\Omega_{\text{group}}(\hat{\beta} - \beta) + \|X(\hat{\beta} - \beta^0)\|_n^2 \le \|X(\beta - \beta^0)\|_n^2 + \frac{\bar{\lambda}^2|S|}{\hat{\phi}_{\Omega_{\text{group}}}^2(L, S)}$$

$$+ 4\lambda\Omega_{\text{group}}(\beta_{-S}).$$

The wedge penalty (see Example 6.2) corresponds to taking

$$\Omega_{\text{wedge}}(\beta) = \arg\min_{a_1 \ge \dots \ge a_p \ge 0,\ \|a\|_1 = 1} \sqrt{\sum_{j=1}^{p} \frac{\beta_j^2}{a_j}},\quad \beta \in \mathbb{R}^p.$$

In the case of orthogonal design we have an improved version of the generic Corollary 12.3. For simplicity we take $\delta_1 = \delta_2 =: \delta$ and $\alpha_1 = \alpha_2 =: \alpha$ in this case.

Corollary 12.5 *Consider the wedge estimator from Example 6.2:*

$$\hat{\beta} := \arg\min\left\{ \|Y - X\beta\|_n^2 + 2\lambda\Omega_{\text{wedge}}(\beta) \right\}.$$

Let $0 \le \delta < 1$ be arbitrary. Suppose orthogonal design: $\hat{\Sigma} = I$ (and hence $p = n$). Let $0 < \alpha < 1/2$. Take

$$\lambda > \lambda_\epsilon := \frac{\sigma_0}{\sqrt{n}}\left(1 + 2\sqrt{\log\left(\frac{1+\alpha}{\alpha}\right)} + 2\log\left(\frac{1+\alpha}{\alpha}\right)\right)^{1/2}.$$

Define

$$\underline{\lambda} := \lambda - \lambda_\epsilon,\ \bar{\lambda} := \lambda + \lambda_\epsilon + \delta\underline{\lambda}$$

and

$$L := \frac{\bar{\lambda}}{(1-\delta)\underline{\lambda}}.$$

Apply Lemma 8.7 to find that for any allowed set S and any β, with probability at least $1 - 2\alpha$ it holds that

$$2\delta\underline{\lambda}\,\underline{\Omega}_{\mathrm{wedge}}(\hat{\beta} - \beta) + \|X(\hat{\beta} - \beta^0)\|_n^2 \le \|X(\beta - \beta^0)\|_n^2 + \frac{\bar{\lambda}^2|S|}{\hat{\phi}_{\underline{\Omega}_{\mathrm{wedge}}}^2(L, S)}$$

$$+ 4\lambda\Omega_{\mathrm{wedge}}(\beta_{-S})$$

where $\underline{\Omega}_{\mathrm{wedge}} = \Omega_{\mathrm{wedge}}(\cdot|S) + \Omega_{\mathrm{wedge}}^{-S}$.

12.3 Density Estimation Using Projections

Let $X_1, \ldots, X_n$ be i.i.d. random variables with values in some observation space $\mathscr{X}$ and with distribution P. Write their density with respect to some known sigma-finite measure ν as $p^0 := dP/d\nu$. Given a dictionary $\{\psi_j(\cdot)\}_{j=1}^p \subset L_2(\nu)$, we employ the loss

$$\rho_\beta(x) := -\psi(x)\beta + \frac{1}{2}\|\psi\beta\|_\nu^2, \ x \in \mathscr{X},$$

with $\psi : \mathscr{X} \to \mathbb{R}^p$ the function $\psi(\cdot) = (\psi_1(\cdot), \ldots, \psi_p(\cdot))$ and with $\|\cdot\|_\nu$ the $L_2(\nu)$-norm.

The ℓ_1-penalized density estimator is

$$\hat{\beta} := \arg\min_{\beta\in\mathbb{R}^p}\left\{-\frac{1}{n}\sum_{i=1}^n \psi(X_i)\beta + \frac{1}{2}\|\psi\beta\|_\nu^2 + \lambda\|\beta\|_1\right\}.$$

Combining Sects. 10.3 and 11.4 with Theorem 7.1 in Sect. 7.5 yields the following sharp oracle inequality.

Lemma 12.1 *Suppose that for some positive constants K and σ*

$$\max_{1\le j\le p}\mathbb{E}|\psi_j(X_1) - \mathbb{E}\psi_j(X_1)|^m \le \frac{m!}{2}K^{m-2}\sigma^2, \forall\ m \in \{2, 3, \cdots\}.$$

Take $\lambda \ge \lambda_\epsilon$ where

$$\lambda_\epsilon := \sigma\sqrt{\frac{2\log(2p/\alpha)}{n}} + \frac{K\log(2p/\alpha)}{n}.$$

Define for some $0 \le \delta < 1$

$$\underline{\lambda} := \lambda - \lambda_\epsilon, \ \bar{\lambda} := \lambda + \lambda_\epsilon + \delta\underline{\lambda}$$

and

$$L := \frac{\bar{\lambda}}{(1 - \delta)\underline{\lambda}}.$$

Then for all $\beta \in \mathbb{R}^p$ *and all sets* S, *with probability at least* $1 - \alpha$

$$2\delta\|\hat{\beta} - \beta\|_1 + \|\psi\hat{\beta} - p^0\|_\nu^2 \leq \|\psi\beta - p^0\|_\nu^2 + \frac{\bar{\lambda}^2|S|}{\phi^2(L, S, \|\cdot\|_\nu)} + 4\lambda\|\beta_{-S}\|_1.$$

Here

$$\phi^2(L, S, \|\cdot\|_\nu) := \min\left\{|S|\|\psi\tilde{\beta}\|_\nu^2 : \tilde{\beta} \in \mathbb{R}^p, \ \|\tilde{\beta}_S\|_1 = 1, \ \|\tilde{\beta}_{-S}\|_1 \leq L\right\}$$

is the compatibility constant for the norm $\|\cdot\|_\nu$.

In this situation, the sharpness of the oracle inequality is very useful because typically the model is misspecified: the density p^0 is generally not a linear function. In other words the approximation error $\|\psi\beta - p^0\|_\nu^2$ generally does not vanish for any β. The estimator $\hat{\beta}$ mimics the best possible linear approximation of p^0.

Commonly, one will choose the ψ_j $(j = 1, \ldots, p)$ linearly independent in $L_2(\nu)$. or even orthonormal. In the latter case, $\phi(L, S, \|\cdot\|_\nu) = 1$ for all L and S. More generally, $\phi^2(L, S, \|\cdot\|_\nu)$ is bounded from below by the smallest eigenvalue of Σ_ν where

$$\Sigma_\nu := \int \psi^T \psi d\nu.$$

Proof of Lemma 12.1 Let

$$W := \frac{1}{n}\sum_{i=1}^n (\psi^T(X_i) - \mathbb{E}\psi^T(X_i)).$$

Then as shown in Sect. 10.3, for $R_n(\beta) := -\sum_{i=1}^n \psi(X_i)\beta/n + \|\psi\beta\|_\nu^2/2$,

$$\|\dot{R}_n(\beta) - \dot{R}(\beta)\|_\infty = \|W\|_\infty.$$

From Theorem 9.2 in Sect. 9.1 and the union bound we see that for all $t > 0$

$$\mathbb{P}\left(\|W\|_\infty \geq \sigma\sqrt{\frac{2(t + \log(2p))}{n}} + \frac{K(t + \log(2p))}{n}\right) \leq \exp[-t].$$

Apply this with $t = \log(1/\alpha)$. The two point margin condition was established in Sect. 11.4. The result now follows from Theorem 7.1 in Sect. 7.5 and noting that

$$R(\beta) = -\int \psi\beta p^0 dv + \frac{1}{2}\|\psi\beta\|_v^2$$

$$= \frac{1}{2}\|\psi\beta - p^0\|_v^2 - \frac{1}{2}\|p^0\|_v^2.$$

$\square$

12.4   Logistic Regression

Let $(X_1, Y_1), \ldots, (X_n, Y_n)$ be independent observations, with $Y_i \in \{0, 1\}$ the response variable and $X_i \in \mathscr{X} \subset \mathbb{R}^p$ a co-variable ($i = 1, \ldots, n$). The loss for logistic regression is

$$\rho_\beta(x, y) := -yx\beta + d(x\beta), \ \beta \in \mathbb{R}^p$$

where

$$d(\xi) = \log(1 + e^\xi), \ \xi \in \mathbb{R}.$$

We take the norm Ω in the penalty to be the ℓ_1-norm. Furthermore, we impose no restrictions on β, i.e. $\mathscr{B} := \mathbb{R}^p$. The ℓ_1-regularized logistic regression estimator is then

$$\hat{\beta} := \arg\min_{\beta \in \mathbb{R}^p} \left\{ \frac{1}{n} \sum_{i=1}^n \left[-Y_i X_i\beta + d(X_i\beta) \right] + \lambda\|\beta\|_1 \right\}.$$

Define for $i = 1, \ldots, n$ and $x \in \mathscr{X}$

$$\mu_i^0(x) := \mathbb{E}(Y_i | X_i = x), \ f_i^0(x) = \log\left(\frac{\mu_i^0(x)}{1 - \mu_i^0(x)} \right).$$

We assume the generalized linear model is well-specified: for some β^0

$$f^0(x) = x\beta^0, \forall x \in \mathscr{X}.$$

In the high-dimensional situation with $\text{rank}(X) = n \leq p$, and with fixed design, we can take here $\mathscr{X} = \{X_1, \ldots, X_n\}$ and then there always is a solution β^0 of the equation $f^0(x) = x\beta^0$, $x \in \mathscr{X}$. In what follows we consider fixed and random design, but in both cases we take the risk for fixed design, which we write as

$$R(\beta|X) := \frac{1}{n}\sum_{i=1}^{n}\left[-\dot{d}(X_i\beta^0)X_i\beta + d(X_i\beta)\right], \ \beta \in \mathbb{R}^p.$$

We have

$$\ddot{d}(\xi) = \frac{e^\xi}{(1+e^\xi)^2}.$$

It follows that for $\|\beta - \beta^0\|_1 \leq M$

$$\ddot{d}(x\beta) \geq 1/C_M^2(x),$$

where

$$\frac{1}{C_M^2(x)} = \left(\frac{1}{1+e^{\beta^0 x+\|x^T\|_\infty M}}\right)\left(1 - \frac{1}{1+e^{\beta^0 x-\|x^T\|_\infty M}}\right).$$

12.4.1 Logistic Regression with Fixed, Bounded Design

We assume that X is fixed and that for $\hat{\Sigma} := X^T X/n$, $\text{diag}(\hat{\Sigma}) = I$, i.e., the design is normalized. We write $K_1 := \max_{1\leq i\leq n}|X_i|$ and $K_0 := \max_{1\leq i\leq n}|f^0(X_i)|$ and

$$\frac{1}{C_M^2} := \left(\frac{1}{1+e^{K_0+K_1}}\right)\left(1 - \frac{1}{1+e^{-K_0-K_1}}\right).$$

Theorem 12.1 *Let* $\lambda_\epsilon := \sqrt{2\log(2p/\alpha)}$. *Let further, for some* $0 < \delta < 1$, *the tuning parameter* λ *satisfy* $\lambda \geq 8\lambda_\epsilon/(1-\delta)$. *Define*

$$\underline{\lambda} := \lambda - \lambda_\epsilon, \ \bar{\lambda} := \lambda + \lambda_\epsilon + \delta\underline{\lambda}$$

and

$$L := \frac{\bar{\lambda}}{(1-\delta)\underline{\lambda}}.$$

Furthermore, define for any vector $\beta \in \mathbb{R}^p$ *and set* $S \subset \{1,\dots,p\}$.

$$\delta\lambda M_{\beta,S} := \frac{4C_1^2\lambda^2(1+\delta^2)|S|}{\hat{\phi}^2(L,S)} + 8(R(\beta|X) - R(\beta^0|X)) + 16\lambda\|\beta_{-S}\|_1.$$

For those β and S such that $M_{\beta,S} \leq 1/2$ we have with probability at least $1 - \alpha$

$$\delta\underline{\lambda}\|\hat{\beta} - \beta\|_1 + R(\hat{\beta}|X) \leq R(\beta|X) + \frac{C_1^2\bar{\lambda}^2|S|}{2\hat{\phi}^2(L,S)} + 2\lambda\|\beta_{-S}\|_1.$$

Proof of Theorem 12.1 Let

$$\epsilon_i := Y_i - \mu^0(X_i), \ i = 1, \ldots, n.$$

Then for all i, $|\epsilon_i| \leq 1$. Hence by Hoeffding's inequality (see Corollary 17.1) we have

$$\mathbb{P}\left(\|X^T\epsilon\|_\infty/n \geq \sqrt{2(t + \log p)}\right) \leq 2\exp[-t].$$

For the margin conditions we apply the results of Sect. 11.3.2. We let $\mathscr{B}_M = \{\beta' : \|\beta' - \beta^0\|_1 \leq M\}$. We write $K_1 := \max_{1 \leq i \leq n} |X_i|$ and $K_0 := \max_{1 \leq i \leq n} |f^0(X_i)|$ and

$$\frac{1}{C_M^2} := \left(\frac{1}{1 + e^{K_0+K_1}}\right)\left(1 - \frac{1}{1 + e^{-K_0-K_1}}\right).$$

Then the one point and two point margin condition holds for $\mathscr{B}_{\text{local}} = \mathscr{B}_M$: for all β and β' in $\mathscr{B}_M$

$$R(\beta|X) - R(\beta'|X) \geq \dot{R}(\beta')^T(\beta - \beta')^T + \|X(\beta' - \beta)\|_n^2/(2C_M^2).$$

Now, fix some M, say $M = 1$ with corresponding value of the constant $C_M := C_1$. Let $\mathscr{B}_{\text{local}} := \mathscr{B}_1$. Define

$$\delta\lambda M_\beta := \frac{2C_1^2\lambda^2(1 + \delta^2)|S_\beta|}{\hat{\phi}^2(1/(1 - \delta), S_\beta)} + 8(R(\beta|X) - R(\beta^0|X)) + 16\lambda\|\beta_{-S}\|_1.$$

If we assume $M_\beta \leq 1/2$ we see that if $\beta \in \mathscr{B}_{1/2}$ then $\{\beta' : \|\beta' - \beta\|_1 \leq M_\beta\} \subset \mathscr{B}_1 = \mathscr{B}_{\text{local}}$.

By Theorem 7.2, for $\lambda \geq 8\lambda_\epsilon/\delta$ we find $\|\hat{\beta} - \beta\|_1 \leq M_\beta$ and so $\hat{\beta} \in \mathscr{B}_{\text{local}}$. Now apply Theorem 7.1. $\quad\square$

12.4.2   Logistic Regression with Random, Unbounded Design

In the case of unbounded design, one may apply the approach of Problem 11.1 where isotropy conditions are assumed. Here we take a slightly different route. The final result will be that if only a few input variables are not bounded one gets oracle inequalities. For simplicity (as this is getting quite technical), we take β^0 as candidate oracle in this subsection.

Define for all $\tilde{\beta} \in \mathbb{R}^p$ and $M > 0$

$$\hat{\tau}_M^2(\tilde{\beta}) := \frac{1}{n} \sum_{i=1}^n |X_i \tilde{\beta}|^2 / C_M^2(X_i), \ \tilde{\beta} \in \mathbb{R}^p, \ \tau_M^2(\tilde{\beta}) := \mathbb{E}\hat{\tau}_M^2(\tilde{\beta}).$$

Then the one point margin condition holds with pseudo-norm $\hat{\tau}_M$ for all β' satisfying $\|\beta' - \beta^0\|_1 \le M$. This follows from the same arguments as those used in Sect. 11.6.

Consider the Fisher-information matrix

$$\ddot{R}(\beta^0) = \mathbb{E}\left(X_1^T X_1 \frac{\exp[X_1\beta^0]}{(1 + \exp[X_1\beta^0])^2}\right)$$

We assume it is non-singular and denote its smallest eigenvalue by $\Lambda_{\min}(\ddot{R}(\beta^0))$. Observe that

$$\tau_0^2(\tilde{\beta}) = \tilde{\beta}^T \ddot{R}(\beta^0)\tilde{\beta}, \ \tilde{\beta} \in \mathbb{R}^p.$$

We now show that in the margin condition the pseudo-norm $\hat{\tau}_M$ may be replaced by τ_M. Then we show that the effective sparsity for τ_M may be replaced by the effective sparsity for τ_0 provided M is sufficiently small. However, due to unbounded design, this may have a (slight) increasing effect in terms of the dependency of the cardinality $|S|$ of the active set S involved. To show that $\|\hat{\beta} - \beta^0\|_1 \le M$ with M sufficiently small, we insert Theorem 7.2.

We first show that $\hat{\tau}_M$ can be replaced by its theoretical version τ_M. To this end, we define some positive constants L_0 and C_0

$$\lambda(L_0, C_0) := 2L_0\left(C_0\sqrt{\frac{2\log(2p^2/\alpha_1))}{n}} + \frac{\log(2p^2/\alpha_1)}{n}\right).$$

Lemma 12.2 *Suppose that for some L_0*

$$C_0^2 := \max_{1 \le j \le p} \mathbb{E}\exp[|X_{1,j}|^2/L_0] < \infty.$$

Then for all $M > 0$ with probability at least $1 - \alpha_1$

$$\sup_{\tilde{\beta} \in \mathbb{R}^p} \frac{|\hat{\tau}_M^2(\tilde{\beta}) - \tau_M^2(\tilde{\beta})|}{\|\tilde{\beta}\|_1^2} \le \lambda(L_0, C_0).$$

Proof of Lemma 12.2 Apply Corollary 8.2 to find that for all j and k and for all $t > 0$

$$\mathbb{P}\left(\left|\frac{1}{n}\sum_{i=1}^{n}\frac{X_{i,j}X_{i,k}}{C_M^2(X_i)} - \mathbb{E}\frac{X_{1,j}X_{1,k}}{C_M^2(X_1)}\right| \geq 2L_0\left(C_0\sqrt{\frac{2t}{n}} + \frac{t}{n}\right)\right) \leq 2\exp[-t].$$

The result then follows from the union bound.   □

Because for any set S the ℓ_1-norm is decomposable as $\|\tilde{\beta}\|_1 = \|\tilde{\beta}_S\|_1 + \|\tilde{\beta}_{-S}\|_1$, $\tilde{\beta} \in \mathbb{R}^p$, the effective sparsity at some $\beta = \beta_S$ depends only on the set S: the effective sparsity for $\hat{\tau}_M$ is

$$\Gamma_{\|\cdot\|_1}(L, S, \hat{\tau}_M) = \max\{\hat{\tau}_M^{-1}(\tilde{\beta}) : \|\tilde{\beta}_S\|_1 = 1, \|\tilde{\beta}_{-S}\|_1 \leq L\}$$

and its theoretical variant where $\hat{\tau}_M$ is replaced by τ_M is

$$\Gamma_{\|\cdot\|_1}(L, S, \tau_M) = \max\{\tau_M^{-1}(\tilde{\beta}) : \|\tilde{\beta}_S\|_1 = 1, \|\tilde{\beta}_{-S}\|_1 \leq L\}.$$

Compatibility constants may be defined as

$$\phi^2(L, S, \hat{\tau}_M) := \Gamma_{\|\cdot\|_1}^{-2}(L, S, \hat{\tau}_M)/|S|, \quad \phi^2(L, S, \tau_M) := \Gamma_{\|\cdot\|_1}^{-2}(L, S, \tau_M)/|S|.$$

Corollary 12.6 *Let $M > 0$ be arbitrary. Under the condition of Lemma 12.2 for all sets $S \in \{1,\dots,p\}$, all constants $L > 0$ and all $t > 0$, with probability at least $1 - \alpha_1$ the following inequality is true whenever $(L+1)\lambda(L_0, C_0)\,|S|/\phi^2(L, S, \tau_M) < 1$:*

$$\phi(L, S, \hat{\tau}_M) \geq \phi(L, S, \tau_M)\left(1 - (L+1)\lambda(L_0, C_0)\sqrt{|S|}/\phi(L, S, \tau_M)\right).$$

In the next lemma we assume the existence of the moment generating function (mgf) of the "envelope" $\|X_1^T\|_\infty$. We have in mind the situation where the constants K_0 and D_0 involved do not depend on p, in which case the assumption of the lemma is a rather strong one. It holds for example when essentially all variables are bounded, say up to one or two exceptions. Another case where it holds if each $X_{1,j}$ is the product of a bounded random variable and one which is common to all j ($1 \leq j \leq p$) and which has a mgf.

Lemma 12.3 *Let $0 < \gamma \leq 1$. Suppose that for some constant K_0*

$$D_0 := \mathbb{E}\exp[\|X_1^T\|_\infty/K_0] < \infty.$$

Then for all $v > 0$ and for $K_0 M \leq \gamma$

$$v\mathbb{P}(\exp[-\|X_1^T\|_\infty M] \leq v) \leq D_0 v^{1+\frac{1}{\gamma}}.$$

Moreover for some constant $C(K_0, D_0)$ *and for all* $K_0 M \leq \gamma$

$$\frac{\phi^2(L, S, \tau_M)}{|S|} \geq \left(\frac{\Lambda_{\min}(\ddot{R}(\beta^0))}{|S|} \right)^{1+\gamma} \frac{1}{C(K_0, D_0)}.$$

Proof of Lemma 12.3 We apply Chebyshev's inequality to get

$$\mathbb{P}\left(e^{-\|X_1^T\|_\infty M} \leq v \right) = \mathbb{P}\left(\|X_1^T\|_\infty \geq \frac{1}{M} \log\left(\frac{1}{v}\right) \right)$$

$$\leq D_0 \exp\left[-\frac{\log(\frac{1}{v})}{K_0 M} \right] = D_0 v^{\frac{1}{K_0 M^2}}.$$

The second result follows from Lemma 11.6 and inserting the bound $\phi^2(L, S, \tau_0) \geq \Lambda_{\min}(\ddot{R}(\beta^0))$.   □

Fix some $0 < \gamma \leq 1$ as in Lemma 12.3. Let S^0 be a set with cardinality $|S^0|$ not too large but also not too small:

$$32\|\beta^0_{-S^0}\|_1 \frac{\Lambda_{\min}^{1+\gamma}(\ddot{R}(\beta^0))}{\lambda} \leq |S^0|^{1+\gamma} \leq \frac{\Lambda_{\min}^{1+\gamma}(\ddot{R}(\beta^0))}{\lambda(L_0, C_0)} \frac{1}{6C(K_0, D_0)}. \tag{12.1}$$

Ignoring the constants involved here, the right inequality is true for $|S^0|$ of small order $(n/\log p)^{\frac{1}{2(1+\gamma)}}$. Under the weak sparsity bound

$$\sum_{j=1}^{p} |\beta^0_j|^r \leq \rho_r^r$$

we have the left inequality when S^0 is the set of coefficients which are at least of order $\lambda^{\frac{1}{1-r\gamma}} \rho^{-\frac{r\gamma}{1-r\gamma}}$. We have both right and left inequality if $\rho_r^{r(1-\gamma)}$ is of small order $\lambda^{-\frac{1}{1-r\gamma}}$. In that case $\|\beta^0_{-S^0}\|_1$ is of order $\lambda^{\frac{1-r}{1-r\gamma}} \rho_r^{\frac{r(1-\gamma)}{1-r\gamma}} = o(1)$. For $\gamma \downarrow 0$ this is as in Corollary 2.4.

Theorem 12.2 *Suppose that for some constant* L_0

$$C_0^2 := \max_{1 \leq j \leq p} \mathbb{E} \exp[|X_{1,j}|^2 / L_0] < \infty$$

and for some constant K_0

$$D_0 := \mathbb{E} \exp[\|X_1^T\|_\infty / K_0] < \infty.$$

Take $\lambda_\epsilon := 2\sqrt{\log(2p/\alpha)}$ and $\lambda = 16\lambda_\epsilon$. Take $n \geq N(L_0, C_0)$ in such a way that $\lambda(L_0, C_0) \leq 1$. Fix some $0 < \gamma \leq 1$. Let S^0 be a set satisfying (12.1). Define

$$M_0 := 10\lambda \left(\frac{2|S^0|}{\Lambda_{\min}(\ddot{R}(\beta^0))} \right)^{1+\gamma} C(K_0, D_0).$$

Assume that $M_0 K_0 \leq \gamma$. Then $\|\hat{\beta} - \beta^0\|_1 \leq M_0$ with probability at least $1 - \alpha - \alpha_1$.

Proof of Theorem 12.2 Let $\mathscr{B}_{\text{local}} := \{\beta : \|\beta - \beta^0\|_1 \leq M_0\}$. The one point margin condition holds for the conditional risk $R(\cdot|X)$, with pseudo metric $\hat{\tau}_{M_0}$ and with margin function $G(u) = u^2/2$, $u > 0$ (apply the same arguments as in Sect. 11.6). We take $\delta = 1/2$ and $\beta = \beta^0 := \beta^0_{S^0} + \beta^0_{-S^0}$ in Theorem 7.2, where S^0 is a set satisfying (12.1). Combining Corollary 12.6 with Lemma 12.3 gives that with probability at least $1 - \alpha_1$

$$\frac{\phi^2(L, S^0, \hat{\tau}_{M_0})}{|S^0|} \geq \left(\frac{\Lambda_{\min}(\ddot{R}(\beta^0))}{|S^0|} \right)^{1+\gamma}$$

$$\times \frac{\left(1 - (L+1)\lambda(L_0, C_0)[|S^0|/\phi^2(L, S^0, \tau_0)]^{1+\gamma} C(K_0, D_0) \right)^{1+\gamma}}{C(K_0, D_0)}.$$

With $L = 1/(1 - \delta) = 2$ and by the upper bound on $|S^0|$ in (12.1), we have

$$1 - (L+1)\lambda(L_0, C_0)[|S^0|/\Lambda_{\min}(\ddot{R}(\beta^0))]^{1+\gamma} C(K_0, D_0) \geq 1/2.$$

Hence with probability at least $1 - \alpha_1$

$$\frac{\phi^2(L, S^0, \hat{\tau}_{M_0})}{|S^0|} \geq \left(\frac{\Lambda_{\min}(\ddot{R}(\beta^0))}{2|S^0|} \right)^{1+\gamma} \frac{1}{C(K_0, D_0)}.$$

We have for M_{β^0} defined in Theorem 7.2

$$M_{\beta^0} = \frac{9\lambda|S^0|}{\phi^2(2, S^0, \hat{\tau}_{M_0})} + 32\|\beta^0_{-S^0}\|_1.$$

We obtain that with probability at least $1 - \alpha_1$

$$M_{\beta^0} \leq 9\lambda \left(\frac{2|S^0|}{2\Lambda_{\min}(\ddot{R}(\beta^0))} \right)^{1+\gamma} C(K_0, D_0) + 32\|\beta^0_{-S^0}\|_1$$

$$\leq 10\lambda \left(\frac{2|S^0|}{\Lambda_{\min}(\ddot{R}(\beta^0))} \right)^{1+\gamma} C(K_0, D_0) = M_0$$

where in the last inequality we used the lower bound on $|S^0|$ in (12.1).

We will next apply Theorem 7.2 to show that $\hat{\beta} \in \mathscr{B}_{\text{local}}$. Indeed, it is easy to see that we may take

$$\lambda_\epsilon \leq \sqrt{1 + \lambda(L_0, C_0)} \sqrt{2 \log(2p/\alpha)}$$

as

$$\mathbb{P}\left(\|X^T \epsilon\|_\infty/n \geq \sqrt{1 + \lambda(L_0, C_0)} \sqrt{2 \log(2p/\alpha)} \right)$$

$$\leq \mathbb{P}\left(\|X^T \epsilon\|_\infty/n \geq \sqrt{1 + \lambda(L_0, C_0)} \sqrt{2 \log(2p/\alpha)} \cap \max_j \|X_j\|_n^2 \leq 1 + \lambda(L_0, C_0) \right)$$

$$+ \mathbb{P}\left(\max_j \|X_j\|_n^2 \leq 1 + \lambda(L_0, C_0) \right).$$

The second probability is bounded by α_1 in view of Lemma 12.2. The first probability can be bounded by

$$\mathbb{P}\left(\|X^T \epsilon\|_\infty/n \geq \max_j \|X_j\|_n \sqrt{2 \log(2p/\alpha)} \right)$$

which is at most α in view of Hoeffding's inequality, as in the previous section. Since we assumed $n \geq N(L_0, C_0)$ we see that $1 + \lambda(L_0, C_0) \leq 2$. Hence we may apply Theorem 7.2 with $\lambda_\epsilon = 2\sqrt{\log(2p/\alpha)}$. Finally one may verify that the probability of the set where our results hold is at least $1 - \alpha - \alpha_1$. □

Application of Theorem 7.1 gives a sharp oracle inequality.

Corollary 12.7 *Assume the conditions of Theorem 12.2 and apply the same notation as there. Then for all β satisfying $\|\beta - \beta^0\|_1 \leq M_0$, all S satisfying*

$$|S|^{1+\gamma} \leq \frac{\Lambda_{\min}^{1+\gamma}(\ddot{R}(\beta^0))}{\lambda(L_0, C_0)} \frac{1}{6C(K_0, D_0)}$$

and for $0 < \delta < 1/3$ arbitrary, $\underline{\lambda} := \lambda - \lambda_\epsilon$ and $\bar{\lambda} := \lambda + \lambda_\epsilon + \delta\underline{\lambda}$, with probability at least $1 - \alpha - \alpha_1$ it holds that

$$\delta\underline{\lambda}\|\hat{\beta} - \beta\|_1 + R(\hat{\beta}|X) \leq R(\beta|X) + \bar{\lambda}^2 2^\gamma \left(\frac{|S|}{\Lambda_{\min}(\ddot{R}(\beta^0))} \right)^{1+\gamma} + 2\lambda\Omega(\beta_{-S}).$$

(bounding $L := \bar{\lambda}/((1 - \delta)\underline{\lambda})$ with $\delta \leq 1/3$).

Asymptotics If $M_0 = o(1)$ in Theorem 12.2 and its corollary (Corollary 12.7) one may re-iterate the argument showing that one can choose an arbitrary small value of γ. It means that asymptotically one almost has quadratic margin behaviour.

12.5   Trace Regression with Nuclear Norm Penalization

This section examines the nuclear norm penalty. See Koltchinskii (2011) for an in-depth treatment.
  Suppose

$$Y_i = \text{trace}(X_i B^0) + \epsilon_i, \ i = 1, \dots, n,$$

where B^0 is a $p \times q$ matrix and X_i $(i = 1, \dots, n)$ is a $q \times p$ matrix with $q \leq p$.
  Writing

$$\tilde{X}_i \beta^0 := \text{trace}(X_i B^0),$$

where $\tilde{X}_i^T := \text{vec}(X_i^T)$, $\beta^0 := \text{vec}(B^0)$, we see that this is the linear model:

$$Y_i = \tilde{X}_i \beta^0 + \epsilon_i, \ i = 1, \dots, n.$$

The reason why it is written in trace form is because actually the structure in β^0 is now not assumed to be in the sparsity of the coefficients, but rather in the sparsity of the singular values of B^0. The norm which induces this sparsity structure is the nuclear norm

$$\Omega(\beta) := \|B\|_{\text{nuclear}}, \ B = \text{vec}^{-1}(\beta),$$

where $\| \cdot \|_{\text{nuclear}}$ is the nuclear norm. In what follows, we will identify matrices B with their vectorization $\text{vec}(B)$ and simply write $\Omega(B) = \|B\|_{\text{nuclear}}$. When matrices are concerned the norm $\| \cdot \|_2$ denotes the Frobenius norm. Moreover, for a matrix A we let $\Lambda^2_{\max}(A)$ being the largest eigenvalue of $A^T A$.

12.5.1   Some Useful Matrix Inequalities

Lemma 12.4 *Let A be a $p \times q$ matrix. Then*

$$\|A\|_{\text{nuclear}} \leq \sqrt{\text{rank}(A)}\|A\|_2.$$

Let P be a $p \times s$ matrix with $P^T P = I$ and $s \leq p$. Then

$$\|PP^T A\|_2 \leq \sqrt{s}\Lambda_{\max}(A)$$

and

$$\|PP^T A\|_2 \leq \|A\|_2.$$

Proof of Lemma 12.4 Let $r := \text{rank}(A)$. Write the singular value decomposition of A as

$$A = P_A \Phi_A Q_A^T$$

with $P_A^T P_A = I$, $Q_A^T Q_A = I$ and $\Phi_A = \text{diag}(\phi_{A,1}, \cdots, \phi_{A,r})$ the diagonal matrix of positive singular values of A. Then $\|A\|_{\text{nuclear}} = \sum_{k=1}^{r} \phi_{A,k}$ and $\|A\|_2^2 = \text{trace}(A^T A) = \sum_{k=1}^{r} \phi_{A,k}^2$. The first result thus follows from $\|u\|_1 \leq \sqrt{r}\|u\|_2$ for a vector $u \in \mathbb{R}^r$.

For the second result we introduce the p-dimensional jth unit vector e_j, ($j = 1, \ldots, p$). Then (with $\Lambda_{\max}(A) = \max_k \phi_{A,k}$)

$$e_j^T P P^T A A^T P P^T e_j \leq \Lambda_{\max}^2(A) \|P P^T e_j\|_2^2$$

and hence

$$\|PP^T A\|_2^2 = \text{trace}(PP^T A A^T PP^T) = \sum_{j=1}^{p} e_j^T PP^T A A^T PP^T e_j$$

$$\leq \Lambda_{\max}^2(A) \sum_{j=1}^{P} \|PP^T e_j\|_2^2 = \Lambda_{\max}^2(A)\text{trace}(PP^T)$$

$$= s\Lambda_{\max}^2(A).$$

For the last result we write

$$\|A\|_2^2 = \text{trace}(A^T A)$$

$$= \text{trace}((PP^T A + (I - PP)^T A)^T (PP^T A + (I - PP^T)A))$$

$$= \text{trace}((PP^T A)^T (PP^T A)) + \text{trace}(((I - PP^T)A)^T (I - PP^T)A)$$

$$\geq \text{trace}((PP^T A)^T (PP^T A)) = \|PP^T A\|_2^2.$$

$\square$

12.5.2 *Dual Norm of the Nuclear Norm and Its Triangle Property*

The dual norm of $\Omega = \|\cdot\|_{\text{nuclear}}$ is

$$\Omega_* = \Lambda_{\max}(\cdot).$$

Moreover (see Watson 1992)

$$\partial\|B\|_{\text{nuclear}} = \{Z = PQ^T + (I - PP^T)W(I - QQ^T) : \Lambda_{\max}(W) = 1\}.$$

Let the $p \times q$ matrix B (possibly a candidate oracle) have rank s and singular value decomposition

$$B = P\Phi Q^T,$$

with P a $p \times s$ matrix, Q a $q \times s$ matrix, $P^T P = I$, $Q^T Q = I$, and with $\Phi = \text{diag}(\phi_1, \cdots, \phi_s)$ the diagonal matrix of non-zero singular values, where $\phi_1 \geq \ldots \geq \phi_s > 0$.

Lemma 12.5 *The norm* $\Omega = \| \cdot \|_{\text{nuclear}}$ *has the triangle property at B with, for all $p \times q$ matrices B',*

$$\Omega_B^+(B') = \sqrt{s}(\|PP^T B'\|_2 + \|B'QQ^T\|_2 + \|PP^T B'QQ^T\|_2)$$

and

$$\Omega_B^-(B') = \|(I - PP^T)B'(I - QQ^T)\|_{\text{nuclear}}.$$

Moreover

$$\| \cdot \|_{\text{nuclear}} \leq \Omega_B^+ + \Omega_B^-.$$

Remark 12.1 As for the last result, note the contrast with weakly decomposable norms as defined in Sect. 6.4, which have $\Omega \geq \Omega^+ + \Omega^-$.

Proof of Lemma 12.5 Write for $Z \in \partial\|B\|_{\text{nuclear}}$

$$Z := Z_1 + Z_2, \ Z_1 = PQ^T, \ Z_2 = (I - PP^T)W(I - QQ^T).$$

We have

$$\text{trace}(Z_1^T B') = \text{trace}(QP^T B') = \text{trace}(P^T B'Q)$$
$$= \text{trace}(P^T PP^T B'QQ^T Q) = \text{trace}(QP^T PP^T B'QQ^T)$$
$$\leq \|PP^T B'QQ^T\|_{\text{nuclear}}$$

since $\Lambda_{\max}(PQ^T) = 1$. Moreover

$$\text{trace}(Z_2^T B') = \text{trace}((I - QQ^T)W^T(I - PP^T)B')$$
$$= \text{trace}(W^T(I - PP^T)B'(I - QQ^T)).$$

Hence, there exists a W with $\Lambda_{\max}(W) = 1$ such that

$$\text{trace}(W^T B') = \|(I - PP^T)B'(I - QQ^T)\|_{\text{nuclear}}.$$

We thus see that (replacing B' by $B' - B$)

$$\max_{Z \in \partial \|B\|_{\text{nuclear}}} \text{trace}(Z^T(B' - B))$$

$$= \max_{\Lambda_{\max}(W)=1} \text{trace}((I - QQ^T)W^T(I - PP^T)(B' - B))$$

$$+ \text{trace}(QP^T(B' - B))$$

$$\geq \|(I - PP^T)B'(I - QQ^T)\|_{\text{nuclear}} - \|PP^T(B' - B)QQ^T\|_{\text{nuclear}}.$$

Now use Lemma 12.4 to establish

$$\|PP^T(B' - B)QQ^T\|_{\text{nuclear}} \leq \sqrt{s}\|PP^T(B' - B)QQ^T\|_2 \leq \Omega^+(B' - B).$$

Obtaining the second result of the lemma is almost trivial: for all B'

$$\|B'\|_{\text{nuclear}} = \|PP^T B' + B'QQ^T - PP^T B'QQ^T + (I - PP^T)B'(I - QQ^T)\|_{\text{nuclear}}$$

$$\leq \|PP^T B'\|_{\text{nuclear}} + \|B'QQ^T\|_{\text{nuclear}}$$

$$+ \|PP^T B'QQ^T\|_{\text{nuclear}} + \|(I - PP^T)B'(I - QQ^T)\|_{\text{nuclear}}$$

$$\leq \sqrt{s}(\|PP^T B'\|_2 + \|B'QQ^T\|_2 + \|PP^T B'QQ^T\|_2)$$

$$+ \|(I - PP^T)B'(I - QQ^T)\|_{\text{nuclear}}$$

where we invoked Lemma 12.4. $\qquad\square$

Lemma 12.6 *Let*

$$\underline{\Omega} := \Omega_B^+ + \Omega_B^-$$

with Ω_B^+ and Ω_B^- as in Lemma 12.5 Then

$$\underline{\Omega}_*(\cdot) \leq \Lambda_{\max}(\cdot).$$

Proof of Lemma 12.6 This follows from $\|\cdot\|_{\text{nuclear}} \leq \underline{\Omega}$ (see Lemma 12.5) and the fact that the nuclear norm has dual norm $\Lambda_{\max}$. $\qquad\square$

Notation for the Candidate Oracle We will next provide the notation for the candidate oracle B which we might aim at mimicking. Recall that $q \leq p$. Let

$$B = P\Phi Q^T$$

with P a $p \times q$ matrix, Q a $q \times q$ matrix, $P^T P = I$, $Q^T Q = I$, and $\Phi = \operatorname{diag}(\phi_1, \ldots, \phi_q)$ where $\phi_1 \geq \ldots \geq \phi_q$.

Write

$$B = B^+ + B^-, \quad B^+ = \sum_{k=1}^{s} \phi_k P_k Q_k^T, \quad B^- = \sum_{k=s+1}^{q} \phi_k P_k Q_k^T. \tag{12.2}$$

We see that

$$\|B^-\|_{\text{nuclear}} = \sum_{k=s+1}^{q} \phi_k, \quad \Omega_{B^+}^+(B^-) = 0.$$

Define $\underline{\Omega} := \Omega_{B^+}^+ + \Omega_{B^+}^-$.

12.5.3 An Oracle Result for Trace Regression with Least Squares Loss

We consider the nuclear norm regularized estimator

$$\hat{B} := \arg\min_B \left\{ \frac{1}{n} \sum_{i=1}^{n} (Y_i - \operatorname{trace}(X_i B))^2 + 2\lambda \|B\|_{\text{nuclear}} \right\}.$$

Definition 12.1 Let $L > 0$ be some stretching factor. Suppose B has singular value decomposition $P\Phi Q^T$. Let $s := \operatorname{rank}(B)$. We define the $\| \cdot \|_{\text{nuclear}}$-*compatibility constant* at B as

$$\hat{\phi}_{\text{nuclear}}^2(L, B) := \min \left\{ \frac{s}{n} \sum_{i=1}^{n} \operatorname{trace}^2(X_i B') : \right.$$

$$\sqrt{s}(\|PP^T B'\|_2 + \|B' Q Q^T\|_2 + \|PP^T B' Q Q^T\|_2) = 1,$$

$$\left. \|(I - PP)^T B'(I - QQ)^T\|_{\text{nuclear}} \leq L \right\}.$$

Corollary 12.8 *Application of Theorem 7.1 to the nuclear norm penalty gives the following. Let $B = B^+ + B^-$ where B^+ and B^- are given in (12.2). Let now*

$$\lambda_\epsilon \geq \Lambda_{\max} \left(\frac{1}{n} \sum_{i=1}^{n} \epsilon_i X_i \right).$$

For $\lambda > \lambda_\epsilon$, $\underline{\lambda} := \lambda - \lambda_\epsilon$, $\bar{\lambda} := \lambda + \lambda_\epsilon + \delta\underline{\lambda}$, $L := \bar{\lambda}/((1-\delta)\underline{\lambda})$, we have

$$\delta\underline{\lambda}\Omega(\hat{B} - B)_{\text{nuclear}} + \frac{1}{n}\sum_{i=1}^{n}\text{trace}^2(X_i(\hat{B} - B^0))$$

$$\leq \frac{1}{n}\sum_{i=1}^{n}\text{trace}^2(X_i(B - B^0)) + \frac{s\bar{\lambda}^2}{\hat{\phi}^2_{\text{nuclear}}(L, B^+)} + 4\lambda\|B^-\|_{\text{nuclear}}.$$

We refer to Sect. 9.5 for a probability inequality for the maximal singular value $\Lambda_{\max}(\sum_{i=1}^{n}\epsilon_i X_i)/n$ in the context of matrix completion.

Recall that (see Lemma 12.5) $\|\cdot\|_{\text{nuclear}} \leq \underline{\Omega}$. Hence from Corollary 12.8 one may also establish a bound for the nuclear norm estimation error.

12.5.4 Robust Matrix Completion

Let $\mathscr{B}$ be the collection of $p \times q$ matrices with all entries bounded by some constant $\eta > 0$:

$$\mathscr{B} := \{B : \|B\|_\infty \leq \eta\}.$$

The bounded parameter space $\mathscr{B}$ allows one to take $\mathscr{B}_{\text{local}} = \mathscr{B}$ when applying Theorem 7.2.

Let $\mathscr{X}$ be the space of all $q \times p$ matrices X consisting of zeroes at all entries except for a single entry at which the value equal to one:

$$X = \begin{pmatrix} 0 & \cdots & 0 & \cdots & 0 & 0 \\ \vdots & & \vdots & & \vdots & \vdots \\ 0 & \cdots & 1 & \cdots & 0 & 0 \\ 0 & \cdots & 0 & \cdots & 0 & 0 \\ \vdots & & \vdots & & \vdots & \vdots \\ 0 & \cdots & 0 & \cdots & 0 & 0 \end{pmatrix}.$$

Such matrices—called masks—have also been studied in Sect. 9.5. There are $p \times q$ such matrices. We let $\{X_1, \ldots, X_n\}$ be i.i.d. with values in $\mathscr{X}$.

In Lafond (2015) one can find matrix completion problems with exponential loss. We study the least absolute deviations estimator with nuclear norm penalty

$$\hat{B} := \arg\min_{B \in \mathscr{B}}\left\{\frac{1}{n}\sum_{i=1}^{n}|Y_i - \text{trace}(X_i B)| + \lambda\|B\|_{\text{nuclear}}\right\}.$$

A related estimator for a simpler situation is considered in Problem 12.4.

We will not prove a sharp oracle inequality in this subsection because the loss is not twice differentiable. We conjecture though that lack of differentiability per se is not a reason for impossibility of sharp oracle inequalities.

Theorem 12.3 *Let B be given in (12.2). Suppose that $\epsilon_1, \ldots, \epsilon_n$ are i.i.d. with median zero and with density f_ϵ with respect to Lebesgue measure. Assume that for some positive constant C and some $\eta > 0$.*

$$f_\epsilon(u) \geq 1/C^2 \; \forall \; |u| \leq 2\eta.$$

Define for C_0 a suitable universal constant

$$\lambda_\epsilon := 8C_0 \sqrt{\frac{\log(p+q)}{nq}} + 8C_0 \sqrt{\log(1+q)} \left(\frac{\log(p+q)}{n} \right)$$

$$+ \sqrt{\frac{2}{nq}} \sqrt{\frac{\log(1/\alpha)}{p}} + \frac{16\eta}{3\sqrt{nq}} \frac{\log(1/\alpha)}{p}.$$

Take, for some $0 < \delta < 1$, the tuning parameter $\lambda \geq 8\lambda_\epsilon/\delta$ and define M_B by

$$\delta \lambda M_B = 12C^2 \lambda^2 (1+\delta)^2 pqs + 8 \left(R(B) - R(B^0) \right) + 16\lambda \|B^-\|_{\text{nuclear}}.$$

Then with probability at least $1 - \alpha$ we have $\underline{\Omega}(\hat{B} - B) \leq M_B$ and

$$R(\hat{B}) - R(B) \leq (\lambda_\epsilon + \lambda)M_B + 2\lambda \|B^-\|_{\text{nuclear}}.$$

Asymptotics and Weak Sparsity Suppose that $q \log(1 + q)$ is of small order $n/\log p$. Theorem 12.3 shows that for a suitable value for the tuning parameter λ of order $\lambda \asymp \sqrt{\log p/nq}$ one has

$$R(\hat{B}) - R(B^0) = \mathcal{O}_{\mathbb{P}} \left(\frac{ps \log p}{n} + R(B) - R(B^0) + \sqrt{\frac{\log p}{nq}} \|B^-\|_{\text{nuclear}} \right).$$

This implies

$$\|\hat{B} - B_0\|_2^2 = \mathcal{O}_{\mathbb{P}} \left(\frac{p^2 qs \log p}{n} + pq(R(B) - R(B^0)) + p\sqrt{q}\sqrt{\frac{\log p}{n}} \|B^-\|_{\text{nuclear}} \right).$$

For example, taking $B = B^0$ and letting s_0 be the rank of B^0, we get

$$\|\hat{B} - B_0\|_2^2 = \mathcal{O}_{\mathbb{P}} \left(\frac{p^2 qs_0 \log p}{n} \right).$$

Admittedly, this is a slow rate, but this is as it should be. For each parameter, the rate of estimation is $\sqrt{pq/n}$ because we have only about $n/(pq)$ noisy observations of this parameter. Without penalization, the rate in squares Frobenius norm would thus be

$$pq \times \frac{pq}{n} = \frac{p^2 q^2}{n}.$$

With penalization, the estimator mimics an oracle that only has to estimate ps_0 (instead of pq) parameters, with a $\log p$-prize to be paid.

Instead of assuming B^0 itself is of low rank, one may assume it is only weakly sparse. Let B^0 have singular values $\{\phi_k^0\}_{k=1}^q$. Fix some $0 < r < 1$ and let

$$\rho_r^r := \sum_{k=1}^{q} |\phi_k^0|^r.$$

Then we obtain (Problem 12.2)

$$\|\hat{B} - B^0\|_2^2 = \mathcal{O}_{\mathbb{P}}\left(\frac{p^2 q \log p}{n}\right)^{1-r} \rho_r^{2r}. \tag{12.3}$$

Proof of Theorem 12.3 From Corollary 16.1 in Section 16.2 (a corollary of Bousquet's inequality, see Lemma 16.2) we know that for all $t > 0$

$$\mathbb{P}\left(\underline{Z}_M \geq 8M\mathbb{E}\underline{\Omega}_*(X^T \tilde{\epsilon})/n + M\sqrt{\frac{2t}{npq}} + \frac{16t\eta}{3n}\right) \leq \exp[-t]$$

where $\tilde{\epsilon}_1, \ldots, \tilde{\epsilon}_n$ is a Rademacher sequence independent of $\{(X_i, Y_i)\}_{i=1}^n$ and

$$\underline{Z}_M := \sup_{\tilde{\beta}\in\mathscr{B}:\ \underline{\Omega}(\beta'-\beta)\leq M} \left| \left[R_n(\beta') - R(\beta')\right] - \left[R_n(\beta) - R(\beta)\right]\right|.$$

Here we used that for $\beta' = \text{vec}(B')$ and $\beta = \text{vec}(B)$

$$\mathbb{E}\left[|Y_1 - \text{trace}(X_1 B')| - |Y_1 - \text{trace}(X_1 B)|\right]^2$$

$$\leq \|B' - B\|_2^2/(pq) \leq \|B' - B\|_{\text{nuclear}}^2/(pq) \leq \underline{\Omega}^2(\beta' - \beta)/(pq).$$

Following the arguments of the example in Sect. 9.5 and Problem 9.2 we see that for some constant C_0 (different from the one taken there to simplify the expression)

$$\mathbb{E}\left(\Lambda_{\max}\left(\frac{1}{n}\sum_{i=1}^{n}\tilde{\epsilon}_i X_i\right)\right) \leq C_0\left(\sqrt{\frac{1}{q}}\sqrt{\frac{\log(p+q)}{n}} + \sqrt{\log(1+q)}\left(\frac{\log(p+q)}{n}\right)\right).$$

Putting these results together gives that with probability at least $1 - \exp[-t]$ the inequality $\underline{\mathbf{Z}}_M \leq \lambda_\epsilon M$ is true, where

$$\lambda_\epsilon := 8C_0\left(\sqrt{\frac{\log(p+q)}{nq}} + \sqrt{\log(1+q)}\left(\frac{\log(p+q)}{n}\right)\right) + \sqrt{\frac{2t}{npq}} + \frac{16t\eta}{3p\sqrt{nq}}.$$

We now follow the line of reasoning of Sect. 11.6. Let

$$r(x, B) := \mathbb{E}\left(|Y_1 - \text{trace}(X_1 B)|\,\Big|\, X_1 = x\right).$$

Let x have its 1 at the entry (k, j). Then $xB = B_{j,k}$ and by assumption $|B_{j,k}| \leq \eta$. It follows that

$$r(x, B) - r(x, B^0) \geq |B_{j,k} - B^0_{j,k}|^2/(2C^2).$$

But then

$$R(B) - R(B^0) = \mathbb{E}R(X_1, B) - \mathbb{E}R(X_1, B^0) \geq \|B - B^0\|_2^2/(2C^2 pq).$$

Hence the one point margin condition (Condition 7.6.1) holds with $\tau = \|\cdot\|_2$ and $G(u) = u^2/(2C^2 pq)$.

Next, we need to bound the effective sparsity. By Lemma 12.4 for any $\tilde{B}$

$$\|\tilde{B}\|_2 \geq \max\{\|P^+ P^{+T} \tilde{B}\|_2, \|\tilde{B} Q^+ Q^{+T}\|_2\}.$$

This gives

$$\Gamma^{-1}_{\text{nuclear}}\left(L, B^+, \|\cdot\|_2\right) = \min\{\|\tilde{B}\|_2 :$$

$$\sqrt{s}(\|P^+ P^{+T} \tilde{B}\|_2 + \|\tilde{B} Q^+ Q^{+T}\|_2 + \|P^+ P^{+T} \tilde{B} Q^+ Q^{+T}\|_2) = 1,$$

$$\|(I - P^+ P^{+T})\tilde{B}(I - Q^+ Q^{+T})\|_{\text{nuclear}} \leq L\}$$

$$\geq 1/\sqrt{3s}.$$

Finally apply Theorem 7.2. with convex conjugate function $H(v) = pqC^2 v^2/2$, $v > 0$. $\square$

12.6   Sparse Principal Components

Consider an $n \times p$ matrix X with i.i.d. rows $\{X_i\}_{i=1}^n$. In this section we will assume that $p \log p$ is sufficiently smaller than n. Let $\hat{\Sigma} := X^T X/n$ and $\Sigma_0 := \mathbb{E}\hat{\Sigma}$. We study the estimation of the first principal component $q^0 \in \mathbb{R}^p$ corresponding to the largest eigenvalue $\phi_{max}^2 := \Lambda_{max}(\Sigma_0)$ of Σ_0. The parameter of interest is $\beta^0 := q^0 \phi_{max}$, so that $\|\beta^0\|_2^2 = \phi_{max}^2$ (since the eigenvector q^0 is normalized to have $\|\cdot\|_2$-length one). It is assumed that β^0 is sparse.

Denote the Frobenius norm of a matrix A by $\|A\|_2$:

$$\|A\|_2^2 := \sum_j \sum_k A_{j,k}^2.$$

We use the ℓ_1-penalized estimator

$$\hat{\beta} := \arg\min_{\beta \in \mathscr{B}} \left\{ \frac{1}{4} \|\hat{\Sigma} - \beta\beta^T\|_2^2 + \lambda \|\beta\|_1 \right\},$$

with $\lambda > 0$ a tuning parameter. The estimator is termed a *sparse PCA estimator*.

For the set $\mathscr{B}$ we take an "ℓ_2-local" set:

$$\mathscr{B} := \{\tilde{\beta} \in \mathbb{R}^p : \|\tilde{\beta} - \beta^0\|_2 \le \eta\}$$

with $\eta > 0$ a suitable constant. To get into such a local set, one may have to use another algorithm, with perhaps a slower rate than the one we obtain in Theorem 12.4 below. This caveat is as it should be, see Berthet and Rigollet (2013): the fast rate of Theorem 12.4 cannot be achieved by any polynomial time algorithm unless e.g. one assumes a priori bounds. In an asymptotic setting, the constant η is not required to tend to zero. We will need 3η to be smaller than the gap between the square-root largest and square-root second largest eigenvalue of Σ_0.

In the risk notation: the empirical risk is

$$R_n(\beta) := \|\hat{\Sigma} - \beta\beta^T\|_2^2 = -\frac{1}{2}\beta^T \hat{\Sigma} \beta + \frac{1}{4}\|\beta\|_2^4.$$

Here, it may be useful to note that for a symmetric matrix A

$$\|A\|_2^2 = \text{trace}(A^2).$$

Hence

$$\|\beta\beta^T\|_2^2 = \text{trace}(\beta\beta^T\beta\beta^T) = \|\beta\|_2^2 \text{trace}(\beta\beta^T) = \|\beta\|_2^4.$$

The theoretical risk is

$$R(\beta) = -\frac{1}{2}\beta^T \Sigma_0 \beta + \frac{1}{4}\|\beta\|_2^4.$$

12.6.1   Two Point Margin and Two Point Inequality for Sparse PCA

By straightforward differentiation

$$\dot{R}(\beta) = -\Sigma_0 \beta + \|\beta\|_2^2 \beta.$$

The minimizer β^0 of $R(\beta)$ satisfies $\dot{R}(\beta^0) = 0$, i.e.,

$$\Sigma_0 \beta^0 = \|\beta^0\|_2^2 \beta^0.$$

Indeed, with $\beta^0 = \phi_{\max} q^0$

$$\begin{aligned}
\Sigma_0 \beta^0 &= \phi_{\max} \Sigma_0 q^0 = \phi_{\max}^3 q^0 \\
&= \|\phi_{\max} q_0\|_2^2 \phi_{\max} q^0 = \|\beta^0\|_2^2 \beta^0.
\end{aligned}$$

We moreover have

$$\ddot{R}(\beta) = -\Sigma_0 + \|\beta\|_2^2 I + 2\beta\beta^T,$$

with I denoting the $p \times p$ identity matrix.

Let now the spectral decomposition of Σ_0 be

$$\Sigma_0 := Q\Phi^2 Q^T,$$

with $\Phi = \mathrm{diag}(\phi_1 \cdots \phi_p)$, $\phi_1 \geq \cdots \geq \phi_p \geq 0$, and with $Q = (q_1, \ldots, q_p)$, $QQ^T = Q^T Q = I$. Thus $\phi_{\max} = \phi_1$ and $q^0 = q_1$. We assume the following spikiness condition.

Condition 12.6.1 *For some $\rho > 0$,*

$$\phi_{\max} \geq \phi_j + \rho, \ \forall j \neq 1.$$

Let, for $\tilde{\beta} \in \mathbb{R}^p$, $\Lambda_{\min}(\ddot{R}(\tilde{\beta}))$ be the smallest eigenvalue of the matrix $\ddot{R}(\tilde{\beta})$.

Lemma 12.7 *Assume Condition 12.6.1 and suppose that $3\eta < \rho$. Then for all $\tilde{\beta} \in \mathbb{R}^p$ satisfying $\|\tilde{\beta} - \beta^0\|_2 \leq \eta$ we have*

$$\Lambda_{\min}(\ddot{R}(\tilde{\beta})) \geq 2(\rho - 3\eta).$$

Proof of Lemma 12.7 Let $\tilde{\beta} \in \mathbb{R}^p$ satisfy $\|\tilde{\beta} - \beta^0\|_2 \leq \eta$. The second derivative matrix at $\tilde{\beta}$ is

$$\ddot{R}(\tilde{\beta}) = -\Sigma_0 + \|\tilde{\beta}\|_2^2 I + 2\tilde{\beta}\tilde{\beta}^T$$

$$= \|\tilde{\beta}\|_2^2 \sum_{j=1}^{p} q_j q_j^T - \sum_{j=1}^{p} \phi_j^2 q_j q_j^T + 2\tilde{\beta}\tilde{\beta}^T$$

$$= (\|\tilde{\beta}\|_2^2 - \phi_{max}^2) q_1 q_1^T + \sum_{j=2}^{p} (\|\tilde{\beta}\|_2^2 - \phi_j^2) q_j q_j^T + 2\tilde{\beta}\tilde{\beta}^T.$$

Since by assumption $\|\tilde{\beta} - \beta^0\|_2 \le \eta$, it holds that

$$\|\tilde{\beta}\|_2 \ge \|\beta^0\|_2 - \eta = \phi_{max} - \eta.$$

It follows that

$$\|\tilde{\beta}\|_2^2 \ge \phi_{max}^2 - 2\eta\phi_{max}$$

and hence for all $j \ge 2$

$$\|\tilde{\beta}\|_2^2 - \phi_j^2 \ge 2\rho\phi_{max} - 2\eta\phi_{max} = 2(\rho - \eta)\phi_{max}.$$

Moreover, for all $x \in \mathbb{R}^p$

$$(x^T \tilde{\beta})^2 = (x^T(\tilde{\beta} - \beta^0) + x^T\beta^0)^2$$
$$= (x^T(\tilde{\beta} - \beta^0))^2 + 2(x^T\beta^0)(x^T(\tilde{\beta} - \beta^0)) + (x^T\beta^0)^2$$
$$\ge (x^T\beta^0)^2 - 2\phi_{max}\eta\|x\|_2^2$$

and

$$x^T(\|\tilde{\beta}\|_2^2 - \phi_{max}^2) q_1 q_1^T x \ge -2\eta\phi_{max}\|x\|_2^2.$$

We thus see that

$$x^T \ddot{R}(\tilde{\beta})x \ge 2(x^T\beta^0)^2 - 4\eta\phi_{max}\|x\|_2^2 + 2(\rho - \eta)\phi_{max} \sum_{j=2}^{p}(x^T q_j)^2$$

$$\ge 2(\rho - \eta)\phi_{max} \sum_{j=1}^{p}(x^T q_j)^2 - 4\eta\phi_{max}\|x\|_2^2$$

$$= 2(\rho - 3\eta)\phi_{max}\|x\|_2^2.$$

□

By a two term Taylor expansion (see also Sect. 11.2) we have

$$R(\beta) - R(\beta') = \dot{R}(\beta')^T(\beta - \beta') + \frac{1}{2}(\beta - \beta')^T\ddot{R}(\tilde{\beta})(\beta - \beta')$$

with $\tilde{\beta}$ an intermediate point. Hence the two point margin condition holds with $G(u) = 2(\rho - 3\eta)u^2$, $u > 0$, $\tau = \|\cdot\|_2$, and $\mathscr{B}_{\text{local}} = \mathscr{B} = \{\beta' \in \mathbb{R}^p := \|\beta' - \beta^0\|_2 \leq \eta\}$.

Since $\ddot{R}(\tilde{\beta}) > 0$ for all $\tilde{\beta} \in \mathscr{B}$ we know that $R(\cdot)$ is convex on $\mathscr{B}$. We now will show that also $R_n(\cdot)$ is convex so that the two point inequality holds as well (see also Problem 6.1). This is where we employ the condition that $p \log p$ is sufficiently smaller than n. We will assume that the entries in X_i are bounded ($i = 1, \ldots, n$). Under alternative distributional assumptions (e.g. Gaussianity) one may arrive at less strong requirements on the matrix dimension p. We show convexity of $R_n(\cdot)$ by proving that with high probability $\ddot{R}_n(\cdot)$ is positive definite on $\mathscr{B}$.

Lemma 12.8 *Assume the conditions of Lemma 12.7 and in addition that $K_1 := \|X_1^T\|_\infty < \infty$. Let $t > 0$ be such that $\eta_t < 2(\rho - 3\eta)$ where*

$$\eta_t := K_1\phi_{\max}\sqrt{16p(t + \log(2p))/n} + 4pK_1^2(t + \log(2p))/n.$$

Then with probability at least $1 - \exp[-t]$

$$\Lambda_{\min}(\ddot{R}_n(\tilde{\beta})) \geq 2(\rho - 3\eta) - \eta_t, \ \forall \ \tilde{\beta} \in \mathscr{B}.$$

Proof of Lemma 12.8 Let $\tilde{\beta} \in \mathscr{B}$. We have

$$\ddot{R}(\tilde{\beta}) = \ddot{R}(\tilde{\beta}) - (\hat{\Sigma} - \Sigma_0).$$

But, by Lemma 9.4 in Sect. 9.2, with probability at least $1 - \exp[-t]$ it is true that

$$\Lambda_{\max}(\hat{\Sigma} - \Sigma_0) \leq \eta_t.$$

$\square$

12.6.2   Effective Sparsity and Dual Norm Inequality for Sparse PCA

We have seen in Sect. 12.6.1 that the (two point) margin condition holds with norm $\tau = \|\cdot\|_2$. Clearly for all S

$$\|\tilde{\beta}_S\|_1 \leq \sqrt{s}\|\tilde{\beta}\|_2.$$

The effective sparsity depends only on β via its active set $S := S_\beta$ and does not depend on L:

$$\Gamma^2_{\|\cdot\|_1}(L, \beta, \|\cdot\|_2) = |S|.$$

The empirical process is

$$[R_n(\beta') - R(\beta')] - [R_n(\beta) - R(\beta)] = \frac{1}{2}\beta'^T W\beta' - \frac{1}{2}\beta W\beta,$$

where $W := \hat{\Sigma} - \Sigma_0$. Thus

$$\left|[R_n(\beta') - R(\beta')] - [R_n(\beta) - R(\beta)]\right| \leq 2\left|\beta^T W(\beta' - \beta)\right| + (\beta' - \beta)^T W(\beta' - \beta)$$

$$\leq 2\|\beta' - \beta\|_1 \|W\beta\|_\infty + \|\beta' - \beta\|_1^2 \|W\|_\infty.$$

12.6.3 A Sharp Oracle Inequality for Sparse PCA

Theorem 12.4 *Suppose the spikiness condition (Condition 12.6.1). Let $\mathscr{B} := \{\tilde{\beta} \in \mathbb{R}^p : \|\tilde{\beta} - \beta^0\|_2 \leq \eta\}$ where $3\eta < \rho$. Assume $K_1 := \|X_1^T\|_\infty < \infty$ and that for some $t > 0$ it holds that $\eta_t < 2(\rho - 3\eta)$ where*

$$\eta_t := K_1\phi_{\max}\sqrt{16p(t + \log(2p))/n} + 4pK_1^2(t + \log(2p))/n.$$

Fix some $\beta \in \mathscr{B}$. Let $W := \hat{\Sigma} - \Sigma_0$. and λ_ϵ be a constant (it will be a suitable high probability bound for $2\|W\beta\|_\infty + \|W\|_\infty$, see below). Let $\lambda \geq 8\lambda_\epsilon/\delta$. Define $\underline{\lambda} := \lambda - \lambda_\epsilon$ and $\bar{\lambda} := \lambda + \lambda_\epsilon + \delta\underline{\lambda}$. Furthermore, define for $S \subset \{1, \ldots, p\}$

$$\delta\lambda M_{\beta,S} := \frac{\lambda^2(1 + \delta)^2|S|}{2(\rho - 3\eta)} + 8(R(\beta) - R(\beta^0)) + 16\lambda\|\beta_{-S}\|_1.$$

Assume that $M_\beta \leq 1$. Then, with probability at least $1 - \exp[-t] - \mathbb{P}(2\|W\beta\|_\infty + \|W\|_\infty > \lambda_\epsilon)$,

$$\delta\underline{\lambda}\|\hat{\beta} - \beta\|_1 + R(\hat{\beta}) \leq R(\beta) + \bar{\lambda}^2|S|/8 + 2\lambda\|\beta_{-S}\|_1.$$

Proof of Theorem 12.4 This follows from combining Theorem 7.2 with Theorem 7.1 and implementing the results of Sects. 12.6.1 and 12.6.2. □

Note that we did not provide a high probability bound for $2\|W\beta\|_\infty + \|W\|_\infty$. This can be done invoking $K_1 := \|X_1^T\|_\infty < \infty$. The variable $X_1\beta$, $\beta \in \mathscr{B}$, has a bounded second moment: $\mathbb{E}(X_1\beta)^2 \leq \phi^2_{\max}(\phi_{\max} + \eta)^2$. One can then apply the line

of reasoning suggested in Problem 14.2 namely applying Dümbgen et al. (2010).
One then establishes the following asymptotics.

Asymptotics For simplicity we take $\beta = \beta^0$ and $S = S_0$. Suppose $p \log p/n = o(1)$, $\|X_1\|_\infty = \mathcal{O}(1)$, $\Lambda_{\max} = \mathcal{O}(1)$ and $1/(\rho - 3\eta) = \mathcal{O}(1)$. Then one may take $\lambda \asymp \sqrt{\log p/n}$. Assuming $s_0\sqrt{\log p/n}$ is sufficiently small (to ensure $M_{\beta_0,S_0} \leq 1$) one obtains $\|\hat{\beta} - \beta^0\|_2^2 = \mathcal{O}_{\mathbb{P}}(s_0 \log p/n)$ and $|\hat{\beta} - \beta^0\|_1 = \mathcal{O}_{\mathbb{P}}(s_0\sqrt{\log p/n})$.

Problems

12.1 We revisit the multiple regression model of Problem 6.5. It is

$$Y_{i,t} = X_{i,t}\beta_t^0 + \epsilon_{i,t}, \ i = 1,\ldots,n, \ t = 1,\ldots,T,$$

where $\{\epsilon_{i,t} : i = 1,\ldots,n, \ t = 1,\ldots,T\}$ are i.i.d. zero mean noise variables. Let, as in Problem 6.5

$$\Omega(\beta) = \sum_{j=1}^p \left[\sum_{t=1}^T \beta_{j,t}^2\right]^{1/2},$$

for $\beta^T = (\beta_1^T,\cdots,\beta_t^T) \in \mathbb{R}^{p\times T}$ and let the estimator be

$$\hat{\beta} := \arg\min_{\beta \in \mathbb{R}^{pT}}\left\{\sum_{t=1}^T\sum_{i=1}^n |Y_{i,t} - X_{i,t}\beta_t|^2/(nT) + 2\lambda\Omega(\beta)\right\}.$$

Let $\epsilon \in \mathbb{R}^{p\times T}$ have i.i.d. entries with mean zero and sub-Gaussian tails: for all i and t for a at least some constant a_0 and τ_0

$$\mathbb{P}(|\epsilon_{i,t}| \geq \tau_0\sqrt{a}) \leq 2\exp[-a].$$

Use the result of Problem 8.5 to bound $\Omega_*(X^T\epsilon)/n$ and apply this to complete your finding of Problem 6.5 to a sharp oracle inequality that holds with probability at least $1 - \alpha$.

12.2 Show that (12.3) in Sect. 12.5.4 is true. Hint: take $B = B^0$ in Theorem 12.3 and use similar arguments as in the proof of Lemma 2.3 in Sect. 2.10.

12.3 Can you derive a sharp oracle inequality for the Lasso with random design, with sharpness in terms of the $\|X \cdot \|$ norm? Hint: use the line of reasoning in Sect. 10.4.2. Assume enough (approximate) sparsity so that $\hat{\beta}$ converges in ℓ_1.

12.4 This problem examines a similar situation as in Example 5.1. Let $\{1,\ldots,p\}$ be certain labels and $Z_1,\ldots,Z_n$ be i.i.d. uniformly distributed on $\{1,\ldots,p\}$. Let

$\{(Z_i, Y_i)\}_{i=1}^n$ be i.i.d. with $Y_i \in \mathbb{R}$ a response variable ($i = 1, \ldots, n$). Consider the design

$$X_{i,j} := 1\{Z_i = j\},\ i = 1, \ldots, n,\ j = 1, \ldots, p.$$

Write $X_i := (X_{i,1}, \ldots, X_{i,p})$ and let X be the matrix with rows X_i, $i = 1, \ldots, n$. Assume

$$Y = X\beta^0 + \epsilon,$$

where ϵ has i.i.d. entries with median zero and with density f_ϵ satisfying

$$f_\epsilon(u) \geq 1/C^2,\ |u| \leq 2\eta.$$

We assume $C = \mathcal{O}(1)$ as well as $1/\eta = \mathcal{O}(1)$. Let $\mathscr{B} := \{\beta \in \mathbb{R}^p : \|\beta\|_\infty \leq \eta\}$ and consider the estimator

$$\hat{\beta} := \arg\min_{\beta \in \mathscr{B}} \left\{ \frac{1}{n} \sum_{i=1}^n |Y_i - X\beta| + \lambda \|\beta\|_1 \right\}.$$

Assume $\beta^0 \in \mathscr{B}$. Let $s_0 := |\{j : \beta_j^0 \neq 0\}|$. Show that for a suitable choice of the tuning parameter λ of order $\sqrt{\log p/(np)}$ it holds that

$$\|\hat{\beta} - \beta^0\|_2^2 = \mathcal{O}_{\mathbb{P}}\left(\frac{s_0 p \log p}{n} \right).$$

Chapter 13
Brouwer's Fixed Point Theorem and Sparsity

Abstract In the generalized linear model and its relatives, the loss depends on the parameter via a transformation (the inverse link function) of the linear function (or linear predictor). In this chapter such a structure is not assumed. Moreover, the chapter hints at cases where the effective parameter space is very large and the localization arguments discussed so far cannot be applied. (The graphical Lasso is an example.) With the help of Brouwer's fixed point theorem it is shown that when $\dot{R}(\beta)$ is Ω_*-close to its linear approximation when β is Ω_*-close to the target β^0, then also the Ω-structured sparsity M-estimator $\hat{\beta}$ is Ω_*-close to the target. Here, the second derivative inverse matrix $\ddot{R}^{-1}(\beta^0)$ is assumed to have Ω-small enough rows, where Ω is the dual norm of Ω_*. Next, weakly decomposable norms Ω are considered. A generalized irrepresentable condition on a set S of indices yields that there is a solution $\tilde{\beta}_S$ of the KKT-conditions with zeroes outside the set S. At such a solution $\tilde{\beta}_S$ the problem is under certain conditions localized, so that one can apply the linear approximation of $\dot{R}(\beta)$. This scheme is carried out for exponential families and in particular for the graphical Lasso.

13.1 Brouwer's Fixed Point and Schauder's Extension

Brouwer's fixed point theorem says that a continuous mapping from a compact subset of Euclidian space into itself has a fixed point. This has been generalized to Banach spaces by Schauder (1930). We will apply this generalization, in the sense that we replace the Euclidean (or ℓ_2-)norm on $\mathbb{R}^p$ by an arbitrary norm Ω. The consequence is the following result for Ω-balls. Let $\delta_0 > 0$ let $\mathscr{B}_{\Omega_*}(\delta_0) := \{\beta \in \mathbb{R}^p : \Omega_*(\beta) \leq \delta_0\}$. Let $\mathscr{M}$ be a continuous mapping such that $\mathscr{M}(\mathscr{B}_{\Omega_*}(\delta_0)) \subset \mathscr{B}_{\Omega_*}(\delta_0)$. Then there is a $\Delta_0 \in \mathscr{B}_{\Omega_*}(\delta_0)$ such that $\mathscr{M}(\Delta_0) = \Delta_0$.

The idea to use Brouwer's (or Schauder's) fixed point theorem for high-dimensional problems is from Ravikumar et al. (2011). This approach is useful when the localization argument as given in Sect. 7.6 does not work. Localization arguments are needed when one can only control the behaviour of the theoretical risk in a local neighbourhood of its minimum. This neighbourhood is defined for an appropriate metric. It is not always possible to apply the argument used in Theorem 7.2 of Sect. 7.6 simply because there may be no convergence in the metric used there, which is the metric induced by the regularizing norm Ω or some weaker

© Springer International Publishing Switzerland 2016

199

S. van de Geer, *Estimation and Testing Under Sparsity*,
Lecture Notes in Mathematics 2159, DOI 10.1007/978-3-319-32774-7_13

version Ω thereof. This is for example the case for the graphical Lasso for estimating a $p \times p$ precision matrix: the graphical Lasso is generally not consistent in ℓ_1 when p is large.

13.2 Ω_*-Bound for Ω-Structured Sparsity M-Estimators

Let for $\beta \in \bar{\mathcal{B}} \subset \mathbb{R}^p$ be defined some empirical risk $R_n(\beta)$. The theoretical risk is then $R(\beta) := \mathbb{E}R_n(\beta)$, $\beta \in \bar{\mathcal{B}}$. Let Ω be a norm on $\mathbb{R}^p$ with dual norm Ω_*. Consider a fixed subset $U \subset \{1, \ldots, p\}$. These are the indices of coefficients that are left unpenalized. Let $\mathcal{B} \subset \bar{\mathcal{B}}$ be a convex open set. Consider the Ω-structured sparsity M-estimator

$$\hat{\beta} = \arg \min_{\beta \in \mathcal{B}} \left\{ R_n(\beta) + \lambda \Omega(\beta_{-U}) \right\}$$

with $\lambda > 0$ a given tuning parameter. The target is

$$\beta^0 := \arg \min_{\beta \in \bar{\mathcal{B}}} R(\beta).$$

Let R_n be differentiable at all $\beta \in \bar{\mathcal{B}}$ with derivative $\dot{R}_n$. We also assume throughout that $\Omega(\cdot \mid -U) \leq \Omega$. The KKT-conditions are

$$\dot{R}_n(\hat{\beta}) + \lambda \hat{z}_{-U} = 0, \ \Omega_*(\hat{z}_{-U}) \leq 1, \ \hat{z}_{-U}^T \hat{\beta} = \Omega(\hat{\beta}_{-U}). \qquad (13.1)$$

(See also Sect. 6.12.1.)

Let $\dot{R}(\beta) := \mathbb{E}\dot{R}_n(\beta)$, $\beta \in \bar{\mathcal{B}}$. Suppose throughout that $\dot{R}(\beta^0) = 0$ and that the second derivative

$$\ddot{R}(\beta^0) = \frac{\partial}{\partial \beta^T} \dot{R}(\beta) \bigg|_{\beta = \beta^0}$$

exists and that its inverse $\ddot{R}^{-1}(\beta^0)$ exists.

For a $p \times q$ matrix $A = (A_1, \ldots, A_q)$ we use the notation

$$\|A\|_{\Omega} := \max_{1 \leq k \leq q} \Omega(A_k).$$

The following theorem is a statistical application of Brouwer's (or Schauder's) fixed point theorem. We have in mind here the case where Ω is stronger than the ℓ_1-norm. If instead $\Omega \leq \| \cdot \|_1$ one may want to replace the Ω_* by $\| \cdot \|_\infty$ (in the spirit of Lemma 6.10). More generally, one may want to replace Ω_* by a weaker norm to relax the conditions of Theorem 13.1.

Theorem 13.1 *Suppose that for some $\delta_0 > 0$ the set $\{\beta \in \mathbb{R}^p : \Omega_*(\beta - \beta^0) \leq \delta_0\} \subset \mathscr{B}$ and that for a further constant $\eta > 0$*

$$\sup_{\beta \in \mathbb{R}^p, \; \Omega_*(\beta - \beta^0) \leq \delta_0} \Omega_*\left(\dot{R}(\beta) - \ddot{R}(\beta^0)(\beta - \beta^0)\right) \leq \eta\delta_0.$$

Let moreover

$$\lambda_\epsilon \geq \sup_{\beta \in \mathbb{R}^p, \; \Omega_*(\beta - \beta^0) \leq \delta_0} \Omega_*\left(\dot{R}_n(\beta) - \dot{R}(\beta)\right).$$

Assume

$$|\!|\!|\ddot{R}^{-1}(\beta_0)|\!|\!|_\Omega\left(\lambda + \lambda_\epsilon + \eta\delta_0\right) \leq \delta_0.$$

Then there is a solution $\hat{\beta}$ of the KKT-conditions such that $\Omega_*(\hat{\beta} - \beta^0) \leq \delta_0$.

For bounding

$$\Omega_*\left(\dot{R}(\beta) - \ddot{R}(\beta^0)(\beta - \beta^0)\right)$$

we may think of using the idea following Lemma 11.2 in Sect. 11.2. The condition on $|\!|\!|\ddot{R}^{-1}(\beta^0)|\!|\!|_\Omega$ is in line with the corresponding result for the Lasso of Lemma 4.1 (Sect. 4.4). There $\Omega = \|\cdot\|_1$ and bounds in sup-norm $\|\cdot\|_\infty$ are studied under conditions on $|\!|\!|\Theta_0|\!|\!|_1$. However, Lemma 4.1 cannot be re-derived from the above Theorem 13.1.

Proof of Theorem 13.1 Let for $\beta^0 + \Delta \in \mathscr{B}$

$$\mathscr{M}(\Delta) := \ddot{R}^{-1}(\beta^0)\left(\dot{R}_n(\beta^0 + \Delta) - \dot{R}(\beta^0 + \Delta)\right) - \ddot{R}^{-1}(\beta^0)\lambda z_{-U,\Delta}$$

$$+ \ddot{R}^{-1}(\beta^0)\left(\dot{R}(\beta^0 + \Delta) - \ddot{R}(\beta^0)\Delta\right).$$

Here $\Omega_*(z_{-U,\Delta}) \leq 1$ and $z_{-U,\Delta}^T(\beta^0 + \Delta)_{-U} = \Omega((\beta^0 + \Delta)_{-U})$. Then for $\Omega_*(\Delta) \leq \delta_0$ we have

$$\Omega_*(\mathscr{M}(\Delta)) \leq |\!|\!|\ddot{R}(\beta^0)|\!|\!|_\Omega(\lambda_\epsilon + \lambda + \eta\delta_0) \leq \delta_0.$$

Hence by the fixed point theorem of Schauder (as described in Sect. 13.1) there is a $\hat{\Delta}$ with $\Omega_*(\hat{\Delta}) \leq \delta_0$ such that $\mathscr{M}(\hat{\Delta}) = \hat{\Delta}$. Hence then we have

$$\ddot{R}^{-1}(\beta^0)\left(\dot{R}_n(\beta^0 + \hat{\Delta}) - \dot{R}(\beta^0 + \hat{\Delta})\right) + \ddot{R}^{-1}(\beta^0)\lambda z_{-U,\hat{\Delta}}$$

$$- \ddot{R}^{-1}(\beta^0)\left(\dot{R}(\beta^0 + \hat{\Delta}) + \ddot{R}(\beta^0)\hat{\Delta}\right) = \hat{\Delta}.$$

Rewrite this to

$$\ddot{R}^{-1}(\beta^0)\left(\dot{R}_n(\beta^0 + \hat{\Delta}) - \dot{R}(\beta^0 + \hat{\Delta})\right) + \ddot{R}^{-1}(\beta^0)\lambda z_{-U,\hat{\Delta}}$$

$$+ \ddot{R}^{-1}(\beta^0)\left(\dot{R}(\beta^0 + \hat{\Delta})\right) = 0$$

or

$$\dot{R}_n(\beta^0 + \hat{\Delta}) + \lambda z_{-U,\hat{\Delta}} = 0,$$

i.e. $\beta^0 + \hat{\Delta}$ is a solution of the KKT-conditions (13.1). $\square$

13.3   The Case of No Penalty

The case of no penalty corresponds to $U = \{1, \ldots, p\}$. One may then choose any norm in Theorem 13.1. Choosing Ω equal to the ℓ_2-norm $\|\cdot\|_2$ gives the following standard result as a corollary.

Corollary 13.1 *Suppose that $\{\beta \in \mathbb{R}^p : \|\beta - \beta^0\|_2 \leq \delta_0\} \subset \mathscr{B}$ for some $\delta_0 > 0$, and that for some constant η*

$$\sup_{\beta \in \mathbb{R}^p, \, \|\beta - \beta^0\|_2 \leq \delta_0} \left\| \dot{R}(\beta) - \dot{R}(\beta^0) - \ddot{R}(\beta^0)(\beta - \beta^0) \right\|_2 \leq \eta \delta_0.$$

Let moreover

$$\lambda_\epsilon \geq \sup_{\beta \in \mathbb{R}^p, \, \|\beta - \beta^0\|_2 \leq \delta_0} \left\| \dot{R}_n(\beta) - \dot{R}(\beta) \right\|_2.$$

Assume

$$\Lambda_{\min}^{-1}(\ddot{R}(\beta^0))\left(\lambda + \lambda_\epsilon + \eta \delta_0\right) \leq \delta_0.$$

Then there is a solution $\hat{\beta}$ of the equation $\dot{R}_n(\hat{\beta}) = 0$ such that $\|\hat{\beta} - \beta^0\|_2 \leq \delta_0$.

13.4   The Linear Model with Random Design

Let $\Omega = \|\cdot\|_1$ and $U = \emptyset$. Consider a random $n \times p$ design matrix X and a response variable $Y = X\beta^0 + \epsilon$. Let $\hat{\Sigma} := X^T X/n$. Assume that $\Sigma_0 := \mathbb{E}\hat{\Sigma}$ is known and let the empirical risk be

$$R_n(\beta) = -Y^T X\beta/n + \beta^T \Sigma_0\beta/2, \ \beta \in \mathbb{R}^p$$

(see also Sects. 10.4.1 and 11.5.1). The estimator $\hat{\beta}$ is thus

$$\hat{\beta} = \arg\min_{\beta \in \mathbb{R}^p}\left\{-Y^T X\beta/n + \beta^T \Sigma_0\beta/2 + \lambda\|\beta\|_1\right\}.$$

We will establish the analogue of Lemma 4.1 in Sect. 4.4.

We prove Lemma 13.1 below by applying Theorem 13.1. We can take there $\eta = 0$ because $\dot{R}_n(\beta)$ is linear in β. In other words the result can be easily derived without using any fixed point theorem. Our method of proof rather illustrates that Theorem 13.1 gives the correct answer in a trivial case.

Lemma 13.1 *Suppose* $\Theta_0 := \Sigma_0^{-1}$ *exists. Then*

$$\|\hat{\beta} - \beta^0\|_\infty \leq \|\|\Theta_0\|\|_1\left(\lambda + \|X^T\epsilon\|_\infty/n + \|(\hat{\Sigma} - \Sigma_0)\beta^0\|_\infty\right).$$

Proof of Lemma 13.1 Take $\delta_0 := \|\|\Theta_0\|\|_1\left(\lambda + \|X^T\epsilon\|_\infty/n + \|(\hat{\Sigma} - \Sigma_0)\beta^0\|_\infty\right)$ in Theorem 13.1. We have for all β

$$\dot{R}_n(\beta) - \dot{R}(\beta) = X^T\epsilon/n - (\hat{\Sigma} - \Sigma_0)\beta^0,$$

and

$$\dot{R}(\beta) = \Sigma_0(\beta - \beta^0), \ \ddot{R}(\beta_0) = \Sigma_0.$$

It follows that

$$\dot{R}(\beta) - \ddot{R}(\beta)(\beta - \beta^0) = 0$$

i.e., $\eta = 0$ in Theorem 13.1. Now $\Omega_* = \|\cdot\|_\infty$ and

$$\sup_\beta \|\dot{R}_n(\beta) - \dot{R}(\beta)\|_\infty \leq \|X^T\epsilon\|_\infty/n + \|(\hat{\Sigma} - \Sigma_0)\beta^0\|_\infty =: \lambda_\epsilon.$$

As $\ddot{R}(\beta^0) = \Sigma_0$ we get

$$\|\ddot{R}^{-1}(\beta^0)\|_1(\lambda + \lambda_\epsilon) = \delta_0. \qquad \square$$

Asymptotics Note that $\|(\hat{\Sigma} - \Sigma_0)\beta^0\|_\infty$ is the maximum of p mean-zero random variables. For bounded design and under second moment conditions on $X\beta^0$, one has $\|(\hat{\Sigma} - \Sigma_0)\beta^0\|_\infty = \mathcal{O}_{\mathbb{P}}(\sqrt{\log p/n})$ (apply for example Dümbgen et al. (2010), see Problem 14.2). If also $\|X^T\epsilon\|_\infty/n = \mathcal{O}_{\mathbb{P}}(\sqrt{\log p/n})$ one thus finds

$$\|\hat{\beta} - \beta^0\|_\infty = \|\Theta_0\|_1 \mathcal{O}_{\mathbb{P}}(\sqrt{\log p/n}).$$

Observe that for this result to holds we do not need $\|\hat{\beta} - \beta^0\|_1 = \mathcal{O}_{\mathbb{P}}(1)$ as we did in Sect. 4.4. In particular, if s_0 is the number of non-zero coefficients of β^0—or some weak version thereof—then s_0 is not required to be of small order $\sqrt{n}/\log p$.

For the more general Ω-structured sparsity estimator with $\Omega \geq \|\cdot\|_1$, one obtains analogues of the bounds given in Sect. 6.10.

13.5 De-selection of Variables and the Irrepresentable Condition

To avoid cumbersome notations we take (without loss of generality) $U = \emptyset$ in this section.

Let Ω be a norm on $\mathbb{R}^p$ and $S \subset \{1,\ldots,p\}$ be given. We require S to be an Ω-allowed set:

$$\Omega \geq \Omega(\cdot|S) + \Omega^{-S}$$

(see Definition 6.1 in Sect. 6.4). Recall (see Lemma 6.2) that

$$\Omega_* \leq \max\{\Omega_*(\cdot|S), \Omega_*^{-S}\}.$$

This will be important in the results that follow. It is a reason to insist S to be Ω-allowed.

Recall that for $\beta \in \mathbb{R}^p$ the vector β_S is defined as either $\beta_S = \{\beta_j 1\{j \in S\}\} \in \mathbb{R}^p$ or $\beta_S = \{\beta_j\}_{j \in S} \in \mathbb{R}^{|S|}$. Which version is used should be clear from the context.

We define (assuming for the moment of this definition that $S = \{1,\ldots,s\}$ with $s < p - s$) for a $(p - s) \times s$ matrix A

$$\|A\|_\Omega = \max_{\Omega_*(\beta_S)\leq 1} \Omega_*^{-S}(A\beta_S)$$

and if A is a $(p-s) \times (p-s)$ matrix we let

$$\|A\|_{\Omega} = \max_{\Omega_*^{-S}(\beta_{-S}) \leq 1} \Omega_*^{-S}(A\beta_{-S}).$$

We say that $\tilde{\beta}_S$ satisfies the S-KKT-conditions if

$$\left(\dot{R}_n(\tilde{\beta}_S)\right)_S + \lambda \tilde{z}_S = 0, \ \Omega_*(\tilde{z}_S) \leq 1, \ \tilde{z}_S^T \tilde{\beta}_S = \Omega(\tilde{\beta}_S).$$

Lemma 13.2 *Let $\tilde{\beta}_S \in \bar{\mathscr{B}}$ be a solution of the S-KKT-conditions. Suppose*

$$\Omega_*^{-S}\left(\left(\dot{R}_n(\tilde{\beta}^S)\right)_{-S}\right) \leq \lambda. \tag{13.2}$$

Then $\tilde{\beta}_S$ is also solution of the KKT-conditions

$$\dot{R}_n(\tilde{\beta}_S) + \lambda \hat{z} = 0, \ \Omega_*(\hat{z}) \leq 1, \ \hat{z}^T \tilde{\beta}_S = \Omega(\tilde{\beta}_S).$$

Proof of Lemma 13.2 Define $\hat{z}_S = \tilde{z}_S$ and

$$\hat{z}_{-S} = -\frac{1}{\lambda}\left(\dot{R}_n(\tilde{\beta}^S)\right)_{-S}.$$

Then

$$\dot{R}_n(\tilde{\beta}_S) + \lambda \hat{z} = 0, \ \hat{z}^T \tilde{\beta}_S = \Omega_*(\tilde{\beta}_S)$$

and (see Lemma 6.2 in Sect. 6.4)

$$\Omega_*(\hat{z}) \leq \max\{\Omega_*(\tilde{z}_S), \Omega_*^{-S}(\hat{z}_{-S}) \leq 1. \qquad \square$$

For the case of the Lasso, Lemma 13.2 corresponds to Lemma 4.6 in Sect. 4.8. Condition (13.2) is a generalized version of the irrepresentable condition (see Sect. 4.7).

Let us denote sub-matrices of second derivatives of the theoretical loss as

$$\ddot{R}_{S,S}(\beta) := \left(\frac{\partial^2 R(\beta)}{\partial \beta_j \partial \beta_k}\right)_{j,k \in S}, \ddot{R}_{S,-S}(\beta) := \left(\frac{\partial^2 R(\beta)}{\partial \beta_j \partial \beta_k}\right)_{j \in S, \ k \notin S},$$

and

$$\ddot{R}_{-S,S}(\beta) := \ddot{R}^T_{S,-S}(\beta), \; \ddot{R}_{-S,-S}(\beta) := \left(\frac{\partial^2 R(\beta)}{\partial \beta_j \partial \beta_k}\right)_{j,k \notin S}$$

(whenever they exist).

Lemma 13.3 *Let* $\underline{\Omega} := \Omega(\cdot|S) + \Omega^{-S}$. *Suppose that for some* $\delta_0 > 0$ *we have* $\{\beta \in \mathbb{R}^p : \underline{\Omega}_*(\beta - \beta^0) \leq \delta_0\} \subset \mathscr{B}$ *and for some* $\eta > 0$

$$\sup_{\beta_S \in \mathbb{R}^p, \, \underline{\Omega}_*(\beta_S - \beta^0) \leq \delta_0} \underline{\Omega}_* \left(\left(\dot{R}(\beta_S) \right)_S - \left(\begin{matrix} \ddot{R}_{S,S}(\beta^0) \\ \ddot{R}_{S,-S}(\beta^0) \end{matrix} \right) (\beta_S - \beta^0) \right) \leq \eta \delta_0.$$

Let moreover

$$\lambda_S \geq \sup_{\beta_S \in \mathbb{R}^p, \, \underline{\Omega}_*(\beta_S - \beta^0) \leq \delta_0} \underline{\Omega}_* \left(\left(\dot{R}_n(\beta_S) - \dot{R}(\beta_S) \right)_S \right).$$

Assume

$$\|\ddot{R}^{-1}_{S,S}(\beta_0)\|_{\Omega} \left(\lambda + \lambda_S + \eta \delta_0 \right) \leq \delta_0.$$

Suppose finally that (13.2) it true. Then there is a solution $\hat{\beta} \in \mathscr{B}$ *of the KKT-conditions satisfying* $\underline{\Omega}_*(\hat{\beta} - \beta^0) \leq \delta_0$ *and* $\hat{\beta}_{-S} = 0$.

Proof of Lemma 13.3 By an slight modification of Theorem 13.1 (which adds the term $\ddot{R}_{S,-S}(\beta^0)\beta^0_{-S}$ in the expansion of derivative of the empirical risk) there is a solution $\tilde{\beta}_S$ of the S-KKT-conditions such that $\Omega_*(\tilde{\beta}_S - \beta^0_S) \leq \delta_0$. But then also $\underline{\Omega}_*(\tilde{\beta}_S - \beta^0) = \max\{\Omega_*(\tilde{\beta}_S - \beta^0_S), \Omega^{-S}_*(\beta^0_{-S})\} \leq \delta_0$. Moreover, by Lemma 13.2 $\tilde{\beta}_S$ is a solution of the KKT-conditions. $\qquad\square$

We now further investigate the condition (13.2) used in Lemmas 13.2 and 13.3.

Lemma 13.4 *Let* $\underline{\Omega} := \Omega(\cdot|S) + \Omega^{-S}$. *Suppose that for some* $\delta_0 > 0$ *we have* $\{\beta \in \mathbb{R}^p : \underline{\Omega}_*(\beta - \beta^0) \leq \delta_0\} \subset \mathscr{B}$ *and that for some* $\eta > 0$

$$\sup_{\beta_S \in \mathbb{R}^p : \, \underline{\Omega}_*(\beta_S - \beta^0) \leq \delta_0} \underline{\Omega}_* \left(\dot{R}(\beta_S) - \ddot{R}(\beta^0)(\beta_S - \beta^0) \right) \leq \eta \delta_0.$$

Let moreover λ_S and λ^{-S} satisfy

$$\lambda_S \geq \sup_{\beta_S \in \mathbb{R}^p,\ \underline{\Omega}_*(\beta_S - \beta^0) \leq \delta_0} \Omega_*\left(\left(\dot{R}_n(\beta_S) - \dot{R}(\beta_S)\right)_S\right)$$

$$\lambda^{-S} \geq \sup_{\beta_S \in \mathbb{R}^p,\ \underline{\Omega}_*(\beta_S - \beta^0) \leq \delta_0} \Omega_*^{-S}\left(\left(\dot{R}_n(\beta_S) - \dot{R}(\beta_S)\right)_{-S}\right).$$

Assume

$$\|\ddot{R}_{S,S}^{-1}(\beta_0)\|_{\Omega}\left(\lambda + \lambda_S + \eta\delta_0\right) \leq \delta_0.$$

Then

$$\Omega_*^{-S}\left(\left(\dot{R}_n(\tilde{\beta}_S)\right)_{-S}\right) \leq \lambda^{-S} + \delta_0\eta + \|\ddot{R}_{-S,S}(\beta^0)\ddot{R}_{S,S}^{-1}(\beta^0)\|_{\Omega}(\lambda + \lambda_S + \delta_0\eta)$$

$$+ \|\ddot{R}_{-S,-S}(\beta^0) - \ddot{R}_{-S,S}(\beta^0)\ddot{R}_{S,S}^{-1}(\beta^0)\ddot{R}_{S,-S}(\beta^0)\|_{\Omega}\Omega_*^{-S}(\beta_{-S}^0).$$

Proof of Lemma 13.4 We write $\dot{R}_{n,S}(\cdot) := (\dot{R}_n(\cdot))_S$ and $\dot{R}_S(\cdot) := (\dot{R}(\cdot))_S$. The quantities $\dot{R}_{n,-S}(\cdot)$ and $\dot{R}_{-S}(\cdot)$ are defined similarly with S replaced its complement. By (a slight modification of) Theorem 13.1 there is a solution of the $\tilde{\beta}_S$ of the S-KKT-conditions satisfying $\underline{\Omega}_*(\tilde{\beta}_S - \beta^0) \leq \delta_0$. Hence we have

$$0 = \dot{R}_{n,S}(\tilde{\beta}_S) + \lambda\tilde{z}_S = \left[\dot{R}_{n,S}(\tilde{\beta}_S) - \dot{R}_S(\tilde{\beta}_S)\right] + \dot{R}_S(\tilde{\beta}_S) + \lambda\tilde{z}_S$$

$$= \left[\dot{R}_{n,S}(\tilde{\beta}_S) - \dot{R}_S(\tilde{\beta}_S)\right] + \ddot{R}_{S,S}(\beta^0)(\tilde{\beta}_S - \beta_S^0) - \ddot{R}_{S,-S}(\beta^0)\beta_{-S}^0$$

$$+ \left[\dot{R}_S(\tilde{\beta}_S) - \ddot{R}_{S,S}(\beta^0)(\tilde{\beta}_S - \beta_S^0) - \ddot{R}_{S,-S}(\beta^0)\beta_{-S}^0\right] + \lambda\tilde{z}_S.$$

Hence

$$\tilde{\beta}_S - \beta_S^0 = -\ddot{R}_{S,S}(\beta^0)\left\{\left[\dot{R}_{n,S}(\tilde{\beta}_S) - \dot{R}_S(\tilde{\beta}_S)\right] - \ddot{R}_{S,-S}(\beta^0)\beta_{-S}^0\right.$$

$$\left. + \left[\dot{R}_S(\tilde{\beta}_S) - \ddot{R}_{S,S}(\beta^0)(\tilde{\beta}_S - \beta_S^0) - \ddot{R}_{S,-S}(\beta^0)\beta_{-S}^0\right] + \lambda\tilde{z}_S\right\}.$$

Expanding $\dot{R}_{n,-S}(\tilde{\beta}_S)$ in a similar way and inserting the above expression for $\tilde{\beta}_S - \beta_S^0$ we get

$$
\dot{R}_{n,-S}(\tilde{\beta}_S) = \left[\dot{R}_{n,-S}(\tilde{\beta}_S) - \dot{R}_{-S}(\tilde{\beta}_S) \right] + \ddot{R}_{-S,S}(\beta^0)(\tilde{\beta}_S - \beta_S^0) - \ddot{R}_{-S,-S}(\beta^0)\beta_{-S}^0
$$

$$
+ \left[\dot{R}_{-S}(\tilde{\beta}_S) - \ddot{R}_{-S,S}(\beta^0)(\tilde{\beta}_S - \beta_S^0) - \ddot{R}_{-S,-S}(\beta^0)\beta_{-S}^0 \right]
$$

$$
= \left[\dot{R}_{n,-S}(\tilde{\beta}_S) - \dot{R}_{-S}(\tilde{\beta}_S) \right] - \ddot{R}_{-S,S}(\beta^0)\ddot{R}_{S,S}^{-1}(\beta^0)\left[\dot{R}_{n,S}(\tilde{\beta}_S) - \dot{R}_S(\tilde{\beta}_S) \right]
$$

$$
- \left[\ddot{R}_{-S,-S}(\beta^0) - \ddot{R}_{-S,S}(\beta^0)\ddot{R}_{S,S}^{-1}(\beta^0)\ddot{R}_{S,-S}(\beta^0) \right]\beta_{-S}^0
$$

$$
+ \left[\dot{R}_{-S}(\tilde{\beta}_S) - \ddot{R}_{-S,S}(\beta^0)(\tilde{\beta}_S - \beta_S^0) - \ddot{R}_{-S,-S}(\beta^0)\beta_{-S}^0 \right]
$$

$$
- \ddot{R}_{-S,S}\ddot{R}_{S,S}^{-1}(\beta^0)\left[\dot{R}_S(\tilde{\beta}_S) - \ddot{R}_{S,S}(\beta^0)(\tilde{\beta}_S - \beta_S^0) + \ddot{R}_{S,-S}\beta_{-S}^0 \right]
$$

$$
+ \lambda \ddot{R}_{-S,S}\ddot{R}_{S,S}^{-1}(\beta^0)\tilde{z}_S.
$$

We therefore find

$$
\Omega_*^{-S}(\dot{R}_{n,-S}(\tilde{\beta}_S)) \leq \Omega_*^{-S}\left(\dot{R}_{n,-S}(\tilde{\beta}_S) - \dot{R}_{-S}(\tilde{\beta}_S) \right)
$$

$$
+ \|\ddot{R}_{-S,S}(\beta^0)\ddot{R}_{S,S}^{-1}(\beta^0)\|_\Omega \Omega_*\left(\left(\dot{R}_{n,S}(\tilde{\beta}_S) - \dot{R}_S(\tilde{\beta}_S) \right) \right)
$$

$$
+ \|\ddot{R}_{-S,-S}(\beta^0) - \ddot{R}_{-S,S}(\beta^0)\ddot{R}_{S,S}^{-1}(\beta^0)\ddot{R}_{S,-S}(\beta^0)\|_\Omega \Omega_*^{-S}(\beta_{-S}^0)
$$

$$
+ \|\ddot{R}_{-S,S}R_{S,S}^{-1}(\beta^0)\|_\Omega \Omega_*\left(\dot{R}_S(\tilde{\beta}_S) - \ddot{R}_{S,S}(\beta^0)(\tilde{\beta}_S - \beta_S^0) + \ddot{R}_{S,-S}(\beta^0)\beta_{-S}^0 \right)
$$

$$
+ \Omega_*^{-S}\left(\dot{R}_{-S}(\tilde{\beta}_S) - \ddot{R}_{-S,S}(\beta^0)(\tilde{\beta}_S - \beta_S^0) - \ddot{R}_{-S,-S}(\beta^0)\beta_{-S}^0 \right)
$$

$$
\leq \lambda^{-S} + \delta_0\eta + \|\ddot{R}_{-S,S}(\beta^0)\ddot{R}_{S,S}^{-1}(\beta^0)\|_\Omega(\lambda + \lambda_S + \delta_0\eta)
$$

$$
+ \|\ddot{R}_{-S,-S}(\beta^0) - \ddot{R}_{-S,S}(\beta^0)\ddot{R}_{S,S}^{-1}(\beta^0)\ddot{R}_{S,-S}(\beta^0)\|_\Omega \Omega_*^{-S}(\beta_{-S}^0).
$$

$\square$

Theorem 13.2 *Assume the conditions of Lemma 13.4 and that in addition*

$$
\lambda^{-S} + \delta_0\eta + \|\ddot{R}_{-S,S}(\beta^0)\ddot{R}_{S,S}^{-1}(\beta^0)\|_\Omega(\lambda + \lambda_S + \delta_0\eta)
$$

$$
+ \|\ddot{R}_{-S,-S}(\beta^0) - \ddot{R}_{-S,S}(\beta^0)\ddot{R}_{S,S}^{-1}(\beta^0)\ddot{R}_{S,-S}(\beta^0)\|_\Omega \Omega_*^{-S}(\beta_{-S}^0) \leq \lambda.
$$

Then there is a solution $\hat{\beta}$ of the KKT-conditions satisfying $\hat{\beta}_{-S} = 0$ and $\Omega_*(\hat{\beta} - \beta^0) \leq \delta_0$.

Proof of Theorem 13.2 The conditions of Lemma Lemma 13.4 include those of Lemma 13.3 except condition (13.2). Condition (13.2) follows from Lemma 13.4 by the additional assumption of the theorem.   □

13.6 Brouwer and Exponential Families

We now study the exponential family risk

$$R_n(\beta) = -\sum_{i=1}^{n} \psi(X_i)\beta/n + d(\beta)$$

with $\psi(\cdot) = (\psi_1(\cdot), \ldots, \psi_p(\cdot))$ and $\dot{d}(\beta^0) = \mathbb{E}\psi(X_i)^T$. The KKT-conditions are

$$-\sum_{i=1}^{n} \psi(X_i)^T/n + \dot{d}(\hat{\beta}) + \lambda\hat{z} = 0, \ \Omega_*(z) \leq 1, \ \hat{z}^T\hat{\beta} = \Omega(\hat{\beta}). \tag{13.3}$$

The solution $\hat{\beta}$ is required to be an element of the parameter space $\mathscr{B} \subseteq \{\beta \in \mathbb{R}^p : d(\beta) < \infty\}$ which is generally not the whole Euclidean space $\mathbb{R}^p$. For example, when estimating a log-density with respect to some dominating measure ν, d is the norming constant

$$d(\beta) = \log\left(\int \exp[\psi(x)\beta]d\nu(x)\right)$$

(see also Sects. 10.2.1 and 11.3.1) which is possibly only defined for a strict subset of $\mathbb{R}^p$. Hence when applying Theorem 13.2 one needs to show that $\{\beta \in \mathbb{R}^p : \underline{\Omega}_*(\beta - \beta^0) \leq \delta_0\} \subset \mathscr{B}$. See Lemma 13.5 for an example.

Assume

$$\ddot{d}(\beta^0) := \frac{\partial \dot{d}(\beta)}{\partial \beta^T}\bigg|_{\beta=\beta^0}$$

exists and that its inverse $\ddot{d}^{-1}(\beta^0)$ exists.

Let $W := \sum_{i=1}^{n}(\psi(X_i)^T - \dot{d}(\beta^0))/n$. Clearly

$$\dot{R}_n(\beta) - \dot{R}(\beta) = W.$$

We formulate a corollary of Theorem 13.2 for this case where for simplicity we take a common value $\lambda_S = \lambda^{-S} =: \lambda_\epsilon$.

Corollary 13.2 *Let* $\underline{\Omega} := \Omega(\cdot|S) + \Omega^{-S}$. *Suppose that for some* $\delta_0 > 0$ *and* $\eta > 0$ *and all* $\beta \in \{\beta \in \mathbb{R}^p : \underline{\Omega}_*(\beta_S - \beta^0) \leq \delta_0\} \subset \mathscr{B}$

$$\underline{\Omega}_* \left(\dot{d}(\beta_S) - \dot{d}(\beta^0) - \ddot{d}(\beta^0)(\beta_S - \beta^0) \right) \leq \eta\delta_0.$$

Let moreover

$$\lambda_\epsilon \geq \underline{\Omega}_*(W).$$

Assume

$$\|\ddot{d}_{S,S}^{-1}(\beta_0)\|_\Omega \left(\lambda + \lambda_\epsilon + \eta\delta_0 \right) \leq \delta_0.$$

$$\lambda - \lambda_\epsilon \geq \delta_0\eta + \|\ddot{d}_{-S,S}(\beta^0)\ddot{d}_{S,S}^{-1}(\beta^0)\|_\Omega(\lambda + \lambda_\epsilon + \delta_0\eta)$$
$$+ \|\ddot{d}_{-S,-S}(\beta^0) - \ddot{d}_{-S,S}(\beta^0)\ddot{d}_{S,S}^{-1}(\beta^0)\ddot{d}_{S,-S}(\beta^0)\|_\Omega \Omega_*^{-S}(\beta_{-S}^0).$$

Then there is a solution $\hat{\beta} \in \mathscr{B}$ *of the KKT-conditions*

$$-\sum_{i=1}^{n} \psi(X_i)^T + \dot{d}(\hat{\beta}) + \lambda\hat{z} = 0, \ \Omega_*(z) \leq 1, \ \hat{z}^T\hat{\beta} = \Omega(\hat{\beta})$$

satisfying $\hat{\beta}_{-S} = 0$ *and* $\Omega_*(\hat{\beta} - \beta^0) \leq \delta_0$.

13.7   The Graphical Lasso

Let X be an $n \times p$ data matrix with i.i.d. rows. Let $\hat{\Sigma} := X^T X/n$ and $\Sigma_0 := \mathbb{E}\hat{\Sigma}$ and assume that the inverse $\Theta_0 := \Sigma_0^{-1}$ exists. The matrix Θ_0 is called the precision matrix. The parameter space is the collection of symmetric positive definite (> 0) matrices Θ. As penalty we choose the ℓ_1-norm but we leave the diagonal of the matrix is unpenalized:

$$\|\Theta_{-U}\|_1 := \sum_{j \neq k} |\Theta_{j,k}|.$$

The risk function is log-determinant divergence

$$R_n(\Theta) := \text{trace}(\hat{\Sigma}\Theta) - \log\det(\Theta).$$

This corresponds to exponential family loss with

$$\psi_{j,k}(X_i) = -X_{i,j}X_{i,k}, \; j,k \in \{1,\dots,p\}, \; i = 1,\dots,n,$$

and

$$d(\Theta) = -\log\det(\Theta), \; \Theta > 0.$$

We can write $\dot{d}(\Theta)$ in matrix form as

$$\dot{D}(\Theta) := -\Theta^{-1}$$

and $\ddot{d}(\Theta)$ in kronecker matrix form as

$$\ddot{D}(\Theta) := \Theta^{-1} \otimes \Theta^{-1}.$$

The graphical Lasso (Friedman et al. 2008) is

$$\hat{\Theta} = \arg\min_{\Theta>0}\left\{ \text{trace}(\hat{\Sigma}\Theta) - \log\det(\Theta) + \lambda\|\Theta_U\|_1 \right\}.$$

The KKT-conditions are

$$\hat{\Sigma} - \hat{\Theta}^{-1} + \lambda\hat{Z} = 0,$$

with $\text{diag}(\hat{Z}) = 0$ and $\hat{Z}_{j,k} = \text{sign}(\hat{\Theta}_{j,k})$ whenever $j \neq k$ and $\hat{\Theta}_{j,k} \neq 0$.

We now discuss the results of Ravikumar et al. (2011) and give a slight extension to the approximately sparse case. A different approach can be found in Rothmann et al. (2008) for the case $p = o(n)$.

Write $\Theta_0 =: \{\Theta_{j,k}^0\}$. For a set $S \subset \{1,\dots,p\}^2$ of double indices (j,k) satisfying $[(j,k) \in S \Rightarrow (k,j) \in S]$ and containing $U = \{(j,j) : 1 \leq j \leq p\}$ and for Θ a symmetric matrix, denote by Θ_S the matrix

$$\Theta_S := \{\Theta_{j,k}\mathrm{l}\{j,k \in S\}\}, \Theta_S^0 := \{\Theta_{j,k}^0\mathrm{l}\{j,k \in S\}\}.$$

Let

$$\text{degree}(S) := \max_k |(j,k) : j \in S|.$$

Lemma 13.5 *Let Θ_S be a symmetric matrix satisfying for some constant δ_0*

$$\|\Theta_S - \Theta_0\|_\infty \leq \delta_0, \; \text{degree}(S)\delta_0 + \|\|\Theta_{-S}^0\|\|_1 \leq \Lambda_{\min}(\Theta_0).$$

Then $\Theta_S > 0$.

Proof of Lemma 13.5 For a symmetric matrix A, it holds that $\Lambda_{\max}(A) \leq \|\|A\|\|_1$. We will apply this with $A = \Theta_S - \Theta_0$. We have

$$\Lambda_{\max}(\Theta_S - \Theta_0) \leq \|\|\Theta_S - \Theta_0\|\|_1 \leq \mathrm{degree}(S)\delta_0 + \|\|\Theta^0_{-S}\|\|_1.$$

Let $v \in \mathbb{R}^p$ be an arbitrary vector. Then

$$|v^T(\Theta_S - \Theta^0)v| \leq \|v\|_2^2 \Lambda_{\max}(\Theta_S - \Theta_0)$$

$$\leq \|v\|_2^2 \Big(\mathrm{degree}(S)\delta_0 + \|\|\Theta^0_{-S}\|\|_1 \Big) < \|v\|_2^2 \Lambda_{\min}(\Theta_0).$$

But then

$$v^T \Theta_S v \geq v^T \Theta_0 v - |v^T(\Theta_S - \Theta^0)v| > 0. \qquad \square$$

Next we deal with the remainder term in the linear approximation of $\dot{D}(\Theta_S) = -\Theta_S^{-1}$.

Lemma 13.6 *Suppose that for some δ_0 and η_0*

$$\|\|\Sigma_0\|\|_1 \Big(\delta_0 \mathrm{degree}(S) + \|\|\Theta^0_{-S}\|\|_1 \Big) \leq \eta_0 < 1.$$

Then we have for all Θ such that $\Theta_S \geq 0$ and $\|\Theta_S - \Theta_0\|_\infty \leq \delta_0$ the inequality

$$\left\| -\Big(\Theta_S^{-1} - \Theta_0^{-1}\Big) - \Theta_0^{-1}\Big(\Theta_S - \Theta_0\Big)\Theta_0^{-1} \right\|_\infty \leq \eta_0 \delta_0^2 \|\|\Sigma_0\|\|_1^2/(1 - \eta_0).$$

Proof of Lemma 13.6 We first note that $\|\|\Sigma_0\|\|_1 \geq \Lambda_{\max}(\Sigma_0) = 1/\Lambda_{\min}(\Theta_0)$ so the first condition of the lemma implies

$$\delta_0 \mathrm{degree}(S) + \|\|\Theta^0_{-S}\|\|_1 \leq \Lambda_{\min}(\Theta_0).$$

It follows from Lemma 13.5 that $\Theta_S > 0$.
Let $\Delta := \Theta_S - \Theta_0$ and

$$\mathrm{rem}(\Delta) := -\Big(\Theta_S^{-1} - \Theta_0^{-1}\Big) - \Theta_0^{-1}\Delta\Theta_0^{-1}.$$

It holds that

$$\mathrm{rem}(\Delta) = (\Theta_0^{-1}\Delta)^2(I + \Theta_0^{-1}\Delta)^{-1}\Theta_0^{-1}.$$

But

$$\|\|(I + \Theta_0^{-1}\Delta)^{-1}\|\|_1 \le \sum_{m=0}^{\infty} \|\|(\Theta_0^{-1}\Delta)^m\|\|_1$$

and

$$\|\|(\Theta_0^{-1}\Delta)^m\|\|_1 \le \|\|\Theta_0^{-1}\Delta\|\|_1^m \le (\|\|\Theta_0^{-1}\|\|_1\|\|\Delta\|\|_1)^m.$$

We have

$$\|\|\Theta_S - \Theta_0\|\|_1 \le \delta_0 \text{degree}(S) + \|\|\Theta_{-S}^0\|\|_1.$$

It follows that

$$\|\|\Theta_0^{-1}\|\|_1\|\|\Theta_S - \Theta_0\|\|_1 \le \|\|\Theta_0^{-1}\|\|_1 \left(\delta_0 \text{degree}(S) + \|\|\Theta_{-S}^0\|\|_1 \right) \le \eta_0$$

and so

$$\|\|(I - \Theta_0^{-1}\Delta)\|\|_1 \le 1/(1 - \eta_0).$$

We moreover have

$$\|e_j^T(\Theta_0^{-1}(\Theta_S - \Theta_0))^2\|\|_1 \le \|e_j^T\Theta_0^{-1}(\Theta_S - \Theta_0)\Theta_0^{-1}\|\|_1\delta_0 \le \|\|\Theta_0^{-1}(\Theta_S - \Theta_0)\Theta_0^{-1}\|\|_1\delta_0$$

$$\le (\delta_0\text{degree}(S) + \|\|\Theta_{-S}^0\|\|_1)\|\|\Theta_0^{-1}\|\|_1^2\delta_0 \le \eta_0\delta_0\|\|\Theta_0^{-1}\|\|_1.$$

Also

$$\|(I + \Theta_0^{-1}(\Theta_S - \Theta_0))^{-1}\Theta_0^{-1}e_k\|_\infty \le \|\|(I + \Theta_0^{-1}(\Theta_S - \Theta_0))^{-1}\|\|_1\|\|\Theta_0^{-1}\|\|_1$$

$$\le \|\|\Theta_0^{-1}\|\|_1/(1 - \eta_0).$$

So we find

$$\|\text{rem}(\Delta)\|_\infty \le \eta_0\delta_0\|\|\Theta_0^{-1}\|\|_1^2/(1 - \eta_0). \qquad \square$$

Once we established control of the remainder term, we can apply Corollary 13.2 to obtain a bound for $\|\hat{\Theta} - \Theta_0\|_\infty$. The result reads as follows.

Theorem 13.3 *Let*

$$\lambda_\epsilon \geq \|\hat{\Sigma} - \Sigma_0\|_\infty.$$

Assume that for some constant δ_0

$$\|\Sigma_0\|_1 \left(\delta_0 \text{degree}(S) + \|\Theta^0_{-S}\|_1 \right) \leq \eta_0 < 1.$$

and for $\eta = \eta_0 \delta_0 \|\Sigma_0\|_1^2 / (1 - \eta_0)$

$$\|\ddot{D}^{-1}_{S,S}(\Theta_0)\|_1 \left(\lambda + \lambda_\epsilon + \eta\delta_0 \right) \leq \delta_0.$$

and

$$\lambda - \lambda_\epsilon \geq \delta_0 \eta + \|\ddot{D}_{-S,S}(\Theta_0)\ddot{D}^{-1}_{S,S}(\Theta_0)\|_1 (\lambda + \lambda_\epsilon + \delta_0 \eta)$$

$$+ \|\ddot{D}_{-S,-S}(\Theta_0) - \ddot{D}_{-S,S}(\Theta_0)\ddot{D}^{-1}_{S,S}(\Theta_0)\ddot{D}_{S,-S}(\Theta_0)\|_1 \|\Theta^0_{-S}\|_\infty.$$

Then there is a solution $\hat{\Theta} > 0$ *of the KKT-conditions*

$$\hat{\Sigma} - \hat{\Theta}^{-1} + \lambda\hat{Z} = 0, \ \hat{Z}_U = 0, \ \hat{Z}_{-U} \in \partial\|\hat{\Theta}_{-U}\|_1,$$

satisfying $\hat{\Theta}_{-S} = 0$ *and* $\|\hat{\Theta} - \Theta^0\|_\infty \leq \delta_0.$

Proof of Theorem 13.3 This follows from combining Lemmas 13.5 and 13.6 with Corollary 13.2.   □

Asymptotics For simplicity we assume that $\Theta_0 = \Theta^0_{S_0}$ for some S_0 and we take $S = S_0$ (the strongly sparse case). We let $d_0 := \text{degree}(S_0)$ and define $\kappa := \|\Sigma_0\|_1$ and $\kappa_{S_0} := \|\ddot{D}^{-1}_{S_0,S_0}\|_1$. We impose the irrepresentable condition which requires that $\|\ddot{D}_{-S_0,S_0}(\Theta_0)\ddot{D}^{-1}_{S_0,S_0}(\Theta_0)\|_1$ is sufficiently smaller than one. We moreover assume $\|\ddot{D}^{-1}_{S_0,S_0}\|_\infty = \mathcal{O}(1)$. Then if $\kappa_{S_0}\kappa^2 d_0$ is sufficiently smaller than $1/\lambda$ one may take $\delta_0 \asymp \lambda\kappa_{S_0}$.

Problems

13.1 Verify the KKT conditions (13.1).

13.2 What structured sparsity norm would you use if you believe Θ_0 is approximately a banded matrix? What if Θ_0 corresponds to a graph with possibly disconnected components?

Chapter 14
Asymptotically Linear Estimators of the Precision Matrix

Abstract This chapter looks at two approaches towards establishing confidence intervals for the entries in high-dimensional precision matrix. The first approach is based on the graphical Lasso, whereas the second one invokes the square-root node-wise Lasso as initial estimator. In both cases the one-step adjustment or "de-sparsifying step" is numerically very simple. Under distributional and sparsity assumptions, the de-sparsified estimator of the precision matrix is asymptotically linear. Here, the conditions are stronger when using the graphical Lasso than when using the square-root node-wise Lasso

14.1 Introduction

In this chapter, X is again an $n \times p$ matrix with rows $\{X_i\}_{i=1}^n$ being i.i.d. copies of a random row-vector $X_0 \in \mathbb{R}^p$. Write $\hat{\Sigma} := X^T X / n$ and $\Sigma_0 := E X_0^T X_0 = \mathbb{E} \hat{\Sigma}$ and define $W := \hat{\Sigma} - \Sigma_0$. When X_0 is centered, Σ_0 is the population co-variance matrix, and $\hat{\Sigma}$ the sample co-variance matrix. In actual practice one will usually center the observations. We assume the precision matrix $\Theta_0 := \Sigma_0^{-1}$ exists. We will construct estimators $\tilde{\Theta}$ such that

$$\tilde{\Theta} - \Theta_0 = -\Theta_0 W \Theta_0 + \text{rem}, \quad \|\text{rem}\|_\infty = o_{\mathbb{P}}(n^{-1/2}).$$

Such an estimator is called *asymptotically linear*. Using Lindeberg conditions, asymptotic linearity can be invoked to establish confidence intervals for $\Theta_{j,k}^0$ for a fixed pair of indices (j, k). This step is omitted here. We refer to Janková and van de Geer (2015a) and Janková and van de Geer (2015b). See also Ren et al. (2015) for another asymptotically linear and asymptotically normal estimator of the entries of Θ_0, not treated here.

© Springer International Publishing Switzerland 2016

S. van de Geer, *Estimation and Testing Under Sparsity*,
Lecture Notes in Mathematics 2159, DOI 10.1007/978-3-319-32774-7_14

215

14.2 The Case p Small

Before going to the more complicated high-dimensional case, let us first examine the case where p is small so that $\hat{\Sigma}$ is invertible for n large enough. Then one can estimate the precision matrix using the estimator

$$\hat{\Theta} := \hat{\Sigma}^{-1}.$$

Lemma 14.1 *We have the decomposition*

$$\hat{\Theta} - \Theta_0 = \underbrace{-\Theta_0 W \Theta_0}_{\text{linear term}} - \text{rem}_1,$$

where

$$\|\text{rem}_1\|_\infty \le \|\Theta_0 W\|_\infty \|\!|\hat{\Theta} - \Theta_0|\!\|_1.$$

Proof of Lemma 14.1 We may write

$$\hat{\Theta} - \Theta_0 = \hat{\Sigma}^{-1} - \Sigma_0^{-1} = \Sigma^{-1} \underbrace{(\Sigma_0 - \hat{\Sigma})}_{=-W} \hat{\Sigma}^{-1}$$

$$= -\Theta_0 W \hat{\Theta} = -\Theta_0 W \Theta_0 - \Theta_0 W (\hat{\Theta} - \Theta_0).$$

Thus $\text{rem}_1 = \Theta_0 W (\hat{\Theta} - \Theta_0)$ so that $\|\text{rem}_1\|_\infty \le \|\Theta_0 W\|_\infty \|\!|\hat{\Theta} - \Theta_0|\!\|_1$. $\square$

Asymptotics Suppose p is fixed and in fact that X_0 has a fixed distribution P with finite fourth moments. Then $\|\Theta_0 \bar{W}\|_\infty = \mathcal{O}_{\mathbb{P}}(1/\sqrt{n})$ and $\|\!|\hat{\Theta} - \Theta_0|\!\|_1 = o_{\mathbb{P}}(1)$ and hence $\|\text{rem}_1\|_\infty = o_{\mathbb{P}}(1/\sqrt{n})$. Moreover by the multivariate central limit theorem (assuming $\{X_i\}_{i=1}^n$ are the first n of an infinite i.i.d. sequence) $\sqrt{n}(\hat{\Theta} - \Theta_0)$ is asymptotically normally distributed.

14.3 The Graphical Lasso

The graphical Lasso (Friedman et al. 2008, see also Sect. 13.7) is defined as

$$\hat{\Theta} = \arg\min_{\Theta > 0} \left\{ \text{trace}(\hat{\Sigma} \Theta) - \log \det(\Theta) + \lambda \|\Theta_{-U}\|_1 \right\},$$

where $\|\Theta_{-U}\|_1 := \sum_{k \ne j} |\Theta_{j,k}|$ and $\lambda > 0$ is a tuning parameter. The minimization is carried out over all positive definite (> 0) matrices. The graphical Lasso is examined in Rothmann et al. (2008) for the case where $p = o(n)$. In Ravikumar et al. (2011) and Sect. 13.7, the dimension p is allowed to be larger than n

but otherwise rather severe restrictions are imposed (such as the irrepresentable condition). We will need these restrictions here as well. One may wonder therefore why we base our asymptotically linear estimator on the graphical Lasso and not for example on the (square-root) node-wise Lasso (which needs much less restrictions) discussed in the next section. The reason is that in simulations both approaches are comparable (see Janková and van de Geer 2015a and Janková and van de Geer 2015b) so that the theory is perhaps somewhat too careful. Note moreover that the graphical Lasso is per definition positive definite (which is not true for the node-wise Lasso which is not even symmetric). Finally, it is theoretically of interest to compare the de-sparsifying step for graphical Lasso and node-wise Lasso.

The KKT-conditions are now

$$\hat{\Sigma} - \hat{\Theta}^{-1} + \lambda \hat{Z} = 0$$

where

$$\hat{Z}_{j,k} = \text{sign}(\hat{\Theta}_{j,k}), \ \hat{\Theta}_{j,k} \neq 0, \ j \neq k, \ \hat{Z}_{j,j} = 0, \ \|\hat{Z}\|_\infty \leq 1.$$

We define the de-sparsified graphical Lasso (Janková and van de Geer 2015a) as

$$\hat{\Theta}_{\text{de-sparse}} := \hat{\Theta} + \lambda \hat{\Theta} \hat{Z} \hat{\Theta}$$

One can show that (Problem 14.1)

$$\hat{\Theta}_{\text{de-sparse}} = 2\hat{\Theta} - \hat{\Theta} \hat{\Sigma} \hat{\Theta}. \tag{14.1}$$

Lemma 14.2 *We have*

$$\hat{\Theta}_{\text{de-sparse}} - \Theta_0 = - \underbrace{\Theta_0 W \Theta_0}_{\text{linear term}} - \text{rem}_1 - \text{rem}_2$$

where

$$\|\text{rem}_1\|_\infty \leq \|\Theta_0 W\|_\infty \|\hat{\Theta} - \Theta_0\|_1, \ \|\text{rem}_2\|_\infty \leq \lambda \|\hat{\Theta}\|_1 \|\hat{\Theta} - \Theta_0\|_1.$$

Proof of Lemma 14.2 Write

$$\hat{\Theta}_{\text{de-sparse}} - \Theta_0 = \Theta_0 W \Theta_0 - \underbrace{\Theta_0 W (\hat{\Theta} - \Theta_0)}_{:=\text{rem}_1} - \underbrace{(\hat{\Theta} - \Theta_0)(\hat{\Theta}^{-1} - \hat{\Sigma})\hat{\Theta}}_{\text{rem}_2}.$$

Using the KKT-conditions one obtains

$$\|\text{rem}_2\|_\infty = \lambda \|(\hat{\Theta} - \Theta_0)\hat{Z}\hat{\Theta}\|_\infty \leq \lambda \|\hat{\Theta} - \Theta_0\|_1 \|\hat{Z}\hat{\Theta}\|_\infty$$

$$\leq \lambda \|\hat{\Theta} - \Theta_0\|_1 \|\hat{Z}\|_\infty \|\hat{\Theta}\|_1 \leq \lambda \|\hat{\Theta} - \Theta_0\|_1 \|\hat{\Theta}\|_1.$$

$\square$

Asymptotics Suppose $\max_j \Theta^0_{j,j} = \mathcal{O}(1)$ (note that $\Theta^0_{j,j} \leq 1/\Lambda_{\min}(\Sigma_0)$ for all j, see Problem 14.3). If the vector X_0 is sub-Gaussian (see Definition 15.2) with constants not depending on n, then

$$\|\Theta_0 \bar{W}\|_\infty = \mathcal{O}_{\mathbb{P}}(\sqrt{\log p/n})$$

(see Problem 14.2). Let $S_0 := \{(j,k) : \Theta^0_{j,k} \neq 0\}$ be the active set of Θ_0. Using the notation of Sect. 13.7, suppose $\Theta_0 = \Theta^0_{S_0}$, Let $d_0 = \text{degree}(S_0)$ (the maximal number of non-zeros in a row of Θ_0) and $\kappa := \|\Sigma_0\|_1$, $\kappa_{S_0} := \||\ddot{D}^{-1}_{S_0,S_0}\||_1$ and impose the irrepresentable condition which requires that $\||\ddot{D}_{-S_0,S_0}(\Theta_0)\ddot{D}^{-1}_{S_0,S_0}(\Theta_0)\||_1$ is sufficiently smaller than one. Assume moreover that $\|\ddot{D}^{-1}_{S_0,S_0}\|_\infty = \mathcal{O}(1)$ and that $\kappa_{S_0}\kappa^2 d_0$ is sufficiently smaller than $1/\lambda$. We have seen in Sect. 13.7 (see the asymptotics paragraph following Theorem 13.3) that then

$$\|\hat{\Theta} - \Theta_0\|_\infty = \mathcal{O}_{\mathbb{P}}(\lambda\kappa_{S_0}).$$

Hence if in addition $\kappa_{S_0}\|\Theta_0\|_1 d_0 = o(\sqrt{n}/\log p)$ the de-sparsified estimator is asymptotically linear:

$$\hat{\Theta}_{\text{de-sparse}} - \Theta_0 = \Theta_0 W \Theta_0 + \text{rem}, \quad \|\text{rem}\|_\infty = o_{\mathbb{P}}(n^{-1/2}).$$

See Jankova and van de Geer (2015a) for results on asymptotic normality. Note that the imposed irrepresentable condition implies $\hat{\Theta} = \hat{\Theta}_{S_0}$ (i.e., the graphical Lasso has no false positives) with high probability. However, confidence sets for any entry in Θ_0 are still of interest because we have no control on the false negatives (unless one assumes strong enough signal, which is in contradiction with the idea of confidence intervals). Recall moreover that in Sect. 13.7 we have extended the situation to approximately sparse Θ_0.

14.4   The Square-Root Node-Wise Lasso

We recall the square-root node-wise Lasso. For $j = 1, \ldots, p$ we consider the square-root Lasso for the regression of the jth node on the other nodes with tuning parameter $\lambda_\sharp$:

$$\hat{\gamma}_j := \arg\min_{\gamma_j \in \mathbb{R}^p} \left\{ \|X_j - X_{-j}\gamma_j\|_n + \lambda_\sharp \|\gamma_j\|_1 \right\}.$$

Write

$$\hat{\tau}_j := \|X_j - X_{-j}\hat{\gamma}_j\|_n, \quad \tilde{\tau}_j^2 = \hat{\tau}_j^2 + \lambda_\sharp \hat{\tau}_j \|\hat{\gamma}_j\|_1$$

and

$$\hat{C}_{k,j} := \begin{cases} 1 & k = j \\ -\hat{\gamma}_{k,j} & k \neq j \end{cases}.$$

Then

$$X_j - X_{-j}\hat{\gamma}_j = X\hat{C}_j.$$

The KKT-conditions read

$$\hat{\Sigma}\hat{C} - \begin{pmatrix} \tilde{\tau}_1^2 & \cdots & 0 \\ \vdots & \ddots & \vdots \\ 0 & \cdots & \tilde{\tau}_p^2 \end{pmatrix} = \lambda_\sharp \hat{Z} \begin{pmatrix} \hat{\tau}_1 & \cdots & 0 \\ \vdots & \ddots & \vdots \\ 0 & \cdots & \hat{\tau}_p \end{pmatrix}$$

where $\|\hat{Z}\|_\infty \leq 1$ and

$$\hat{Z}_{k,j} = \begin{cases} 0 & k = j \\ \text{sign}(\hat{\gamma}_{k,j}) & k \neq j, \ \hat{\gamma}_{k,j} \neq 0 \end{cases}.$$

Let

$$\hat{\tau} := \text{diag}(\hat{\tau}_1, \dots, \hat{\tau}_p), \ \tilde{\tau} := \text{diag}(\tilde{\tau}_1, \dots, \tilde{\tau}_p)$$

and

$$\hat{\Theta} := \hat{C}\tilde{\tau}^{-2}.$$

Thus, the matrix $\hat{\Theta}$ is equal to $\hat{\Theta} = \hat{\Theta}_\sharp$ given in Sect. 5.2. The KKT-conditions can be rewritten as

$$\hat{\Sigma}\hat{\Theta} - I = \lambda_\sharp \hat{Z}\hat{\tau}\tilde{\tau}^{-2}.$$

We invert the KKT-conditions for the node-wise Lasso to get the *de-sparsified node-wise Lasso*:

$$\hat{\Theta}_{\text{de-sparse}} := \hat{\Theta} + \hat{\Theta}^T - \hat{\Theta}^T \hat{\Sigma}\hat{\Theta}.$$

Lemma 14.3 *We have*

$$\hat{\Theta}_{\text{de-sparse}} - \Theta_0 = -\underbrace{\Theta_0 W \Theta_0}_{\text{linear term}} - \text{rem}_1 - \text{rem}_2$$

where

$$\|\text{rem}_1\|_\infty \leq \|\Theta_0 W\|_\infty \||\hat{\Theta} - \Theta_0\||_1, \ \|\text{rem}_2\|_\infty \leq \lambda_\sharp \|\hat{\tau}^{-1}\|_\infty \||\hat{\Theta} - \Theta_0\||_1.$$

Proof of Lemma 14.3 Write

$$\hat{\Theta}_{\text{de-sparse}} - \Theta_0 = -\Theta_0 W \Theta_0 \underbrace{- \Theta_0 W(\hat{\Theta} - \Theta_0)}_{:=\text{rem}_1} \underbrace{- (\hat{\Theta} - \Theta_0)^T (\hat{\Sigma}\hat{\Theta} - I)}_{:=\text{rem}_2}$$

and note that

$$\text{rem}_2 := (\hat{\Theta} - \Theta_0)^T \underbrace{(\hat{\Sigma}\hat{\Theta} - I)}_{=\lambda_\sharp \hat{Z} \hat{\tau} \hat{\tau}^{-2}}$$

Then use that $\|\hat{Z}\|_\infty \leq 1$ and $\|\hat{\tau}\hat{\tau}^{-1}\|_\infty \leq 1$. □

Asymptotics Let us for simplicity assume the data are Gaussian.[1] Let $\Lambda_{\min}(\Sigma_0)$ be the smallest eigenvalue of Σ_0. Assume that $1/\Lambda_{\min}(\Sigma_0) = \mathcal{O}(1)$ and $\|\Sigma_0\|_\infty = \mathcal{O}(1)$. Choose $\lambda_\sharp \asymp \sqrt{\log p/n}$ large enough to ensure the result $\||\hat{\Theta} - \Theta_0\||_1 = \mathcal{O}_\mathbb{P}(d_0\sqrt{\log p/n})$ where d_0 is the maximum degree of the nodes. This can be shown using Theorem 3.1 for all node-wise regressions and Theorem 15.2 for dealing with (empirical) compatibility constants. One can moreover check that the results are uniform in j. To bound $\|\hat{\tau}^{-1}\|_\infty$ we may apply the arguments of Lemma 3.1. The final conclusion is asymptotic linearity when $d_0 = o(\sqrt{n}/\log p)$.

Problems

14.1 Show that the representation (14.1) in Sect. 14.3 holds true.

14.2 Let $W := \hat{\Sigma} - \Sigma_0$. Suppose $\max_j \Theta_{j,j}^0 = \mathcal{O}(1)$.

(i) Show that when the vector in X_0 is sub-Gaussian (see Definition 15.2) with constants not depending on n, then

$$\|\Theta_0 W\|_\infty = \mathcal{O}_\mathbb{P}(\sqrt{\log p/n}).$$

---

[1]This can be generalized to sub-Gaussian data.

(ii) Let $\beta \in \mathbb{R}^p$ be a fixed vector with $\mathbb{E}(X_0\beta)^2 = \mathcal{O}(1)$. Show that when the entries in X_0 are bounded by a constant not depending on n then $\|W\beta\|_\infty = \mathcal{O}_\mathbb{P}(\sqrt{\log p/n})$ (apply for example Dümbgen et al. 2010).

(iii) Let $\Theta^0 := (\Theta_1^0, \ldots, \Theta_p^0)$. Show that when the entries in X_0 are bounded by a constant not depending on n, then for each j

$$\|W\Theta_j^0\|_\infty = \mathcal{O}_\mathbb{P}(\sqrt{\log p/n}).$$

14.3 Let Θ be a symmetric positive definite matrix with largest eigenvalue $\Lambda_{\max}(\Theta)$. Show that $0 \leq \Theta_{j,j} \leq \Lambda_{\max}(\Theta)$.

Chapter 15
Lower Bounds for Sparse Quadratic Forms

Abstract Lower bounds for sparse quadratic forms are studied. This has its implications for effective sparsity (or compatibility constants): the effective sparsity with empirical semi-norm $\|X \cdot \|_n$ is bounded in terms of the effective sparsity with theoretical semi-norm $\|X \cdot \|$. The results are an extension of van de Geer and Muro (Electron. J. Stat. 8:3031–3061, 2014) to more general sparsity inducing norms Ω.

15.1 Introduction

Let X be an $n \times p$ matrix with i.i.d. rows $\{X_i\}_{i=1}^n$ Recall for $\beta \in \mathbb{R}^p$ the notation

$$\|X\beta\|_n^2 := \frac{1}{n} \sum_{i=1}^n |(X_i\beta)|^2, \ \|X\beta\|^2 := \mathbb{E}\|X\beta\|_n^2$$

and the Gram matrices $\hat{\Sigma} := X^T X / n$ and $\Sigma_0 := \mathbb{E}\hat{\Sigma}$. Hence $\|X\beta\|_n^2 = \beta^T \hat{\Sigma} \beta$ and $\|X\beta\|^2 := \beta^T \Sigma_0 \beta$.

Let Ω be some norm on $\mathbb{R}^p$. Aim is to study the behaviour of

$$\inf_{\|X\beta\|=1, \ \Omega(\beta) \leq M} \|X\beta\|_n^2 - 1,$$

where M is some given constant.

Obviously,

$$\sup_{\|X\beta\|=1, \ \Omega(\beta) \leq M} \left| \|X\beta\|_n^2 - 1 \right| \leq \Lambda_{\Omega_*}(\hat{\Sigma} - \Sigma_0) M^2,$$

where for a symmetric $p \times p$ matrix A

$$\Lambda_{\Omega_*}(A) := \max_{\Omega(\beta) \leq 1} \beta^T A \beta$$

© Springer International Publishing Switzerland 2016

S. van de Geer, *Estimation and Testing Under Sparsity*,
Lecture Notes in Mathematics 2159, DOI 10.1007/978-3-319-32774-7_15

223

(see also Sect. 11.2). For example, for $\Omega = \|\cdot\|_1$,

$$\sup_{\|X\beta\|=1, \|\beta\|\leq M} \left| \|X\beta\|_n^2 - 1 \right| \leq \Lambda_{\|\cdot\|_\infty} (\hat{\Sigma} - \Sigma_0) M^2 \leq \|\hat{\Sigma} - \Sigma_0\|_\infty M^2. \tag{15.1}$$

Under certain distributional assumptions (sub-Gaussian entries of the vectors $\{X_i\}_{i=1}^n$, or bounded design) one has

$$\|\hat{\Sigma} - \Sigma_0\|_\infty = \mathcal{O}_{\mathbb{P}}(\sqrt{\log p/n}).$$

Then the left-hand side in (15.1) is $o_{\mathbb{P}}(1)$ when

$$M = o((n/\log p)^{1/4}).$$

This chapter requires alternative distributional assumptions termed *(weak) isotropy conditions*, which say roughly that in any direction β, the projection $X_i\beta$ has certain higher order moments proportional to its second moment or looks " similar" in the sense of its tail behaviour ($i = 1, \ldots, n$). Under such assumptions, and with $\Omega \geq \|\cdot\|_1$ for example, one can get "good" results with $M = o(\sqrt{n/\log p})$ as opposed to $M = o((n/\log p)^{1/4})$.

Definition 15.1 Let $m \geq 2$ and let $X_0 \in \mathbb{R}^p$ be a random row vector with $\Sigma_0 := \mathbb{E}X_0^T X_0 < \infty$. The vector X_0 is called *m*th order isotropic with constant C_m if for all $\beta \in \mathbb{R}^p$ with $\beta^T \Sigma_0 \beta = 1$ it holds that

$$\mathbb{E}|X_0\beta|^m \leq C_m^m.$$

It is called *weakly mth order isotropic* with constant C_m if for all $\beta \in \mathbb{R}^p$ with $\beta^T \Sigma_0 \beta = 1$ one has

$$\mathbb{P}(|X_0\beta| > t) \leq (C_m/t)^m \ \forall \ t > 0.$$

A sub-Gaussian vector X_0 is a special case: it is (weakly) *m*th order isotropic with suitable constant C_m for all m. Sub-Gaussianity of a vector is defined as follows.

Definition 15.2 Let $X_0 \in \mathbb{R}^p$ be a random row vector with $\mathbb{E}X_0^T X_0 = \Sigma_0$. Then X_0 is sub-Gaussian with constant t_0 and τ_0 if for all $\beta \in \mathbb{R}^p$ with $\beta^T \Sigma_0 \beta = 1$ it holds that

$$\mathbb{P}(|X_0\beta| > \tau_0\sqrt{t}) \leq \exp[-t] \ \forall \ t \geq t_0.$$

A Gaussian vector X_0 is a sub-Gaussian vector. Also, if $X_0 = U_0\Sigma_0^{1/2}$ where U_0 is a row vector with independent bounded entries, then X_0 is sub-Gaussian (Problem 15.2).

Other examples of sub-Gaussianity and isotropy are given in van de Geer and Muro (2014).

15.2   Lower Bound

We let $\epsilon_1, \ldots, \epsilon_n$ be a Rademacher sequence (that is, i.i.d. random variables each taking values ± 1 with probability $1/2$) independent of X.

Theorem 15.1 *Suppose that for some $m > 2$ the random vectors X_i ($i = 1, \ldots, n$) are weakly mth order isotropic with constant C_m and define*

$$D_m := [2C_m]^{\frac{m}{m-1}} (m-1)/(m-2). \tag{15.2}$$

Then for all $t > 0$ with probability at least $1 - \exp[-t]$

$$\inf_{\|X\beta\|=1, \, \Omega(\beta)\leq M} \|X\beta\|_n^2 - 1 \geq -\Delta_n(M, t) \tag{15.3}$$

where

$$\Delta_n(M, t) := D_m \left(16M\delta_n + \sqrt{\frac{2t}{n}} \right)^{\frac{m-2}{m-1}} + \frac{8D_m^2}{3} \left(\frac{t}{n} \right)^{\frac{m-2}{m-1}} \tag{15.4}$$

with $\delta_n := \mathbb{E}\Omega_*(X^T\epsilon)/n$.

For norms at least as strong as the ℓ_1-norm, and with sub-exponential design with uniform constants, we have $\Omega_*(X^T\epsilon)/n \asymp \sqrt{\log p / n}$ (Problem 15.3). So then, for C_m fixed and for a given t, the term $\Delta_n(M, t) = o(1)$ as soon as $M = o(\sqrt{n/\log p})$. This is the improvement promised in Sect. 15.1 over $M = o((n/\log p)^{1/4})$.

To prove this result we need the following elementary lemma.

Lemma 15.1 *Let Z be a real-valued random variable satisfying for some constant C_m and $m > 2$*

$$\mathbb{P}(|Z| > t) \leq (C_m/t)^m, \; \forall \, t > 0.$$

Then for any $K > 0$

$$\mathbb{E}(Z^2 - K^2)1_{|Z|>K} \leq 2C_m^m K^{-(m-2)}/(m-2).$$

Proof of Lemma 15.1 This is a straightforward calculation:

$$\mathbb{E}(Z^2 - K^2)1_{\{|Z|>K\}} = \int_0^\infty \mathbb{P}(Z^2 - K^2 > t)dt$$

$$= \int_0^\infty \mathbb{P}(|Z| > \sqrt{K^2 + t})dt$$

$$\leq C_m \int_0^\infty \frac{1}{(K^2 + t)^{m/2}}dt$$

$$= C_m^m \frac{2K^{-(m-2)}}{m - 2}.$$

□

Proof of Theorem 15.1 For $Z \in \mathbb{R}$, and $K > 0$, we introduce the truncated version

$$[Z]_K := \begin{cases} -K, & Z < -K \\ Z, & |Z| \leq K \\ K, & Z > K \end{cases}.$$

We obviously have for any $K > 0$ and $\beta \in \mathbb{R}^p$

$$\|X\beta\|_n^2 \geq \|[X\beta]_K\|_n^2 \tag{15.5}$$

where $[X\beta]_K$ is the vector $\{[X_i\beta]_K : i = 1, \ldots, n\}$. Moreover, whenever $\|X\beta\| = 1$ by the weak isotropy and Lemma 15.1

$$1 - \|[X_i\beta]_K\|^2 \leq 2C_m^m K^{-(m-2)}/(m-2), \ i = 1, \ldots, n.$$

Let for $\beta \in \mathbb{R}_p$, $\|[X\beta]_K\|^2 := \mathbb{E}\|[X\beta]_K\|_n^2$. We note that

$$\sup_{\|X\beta\|=1, \ \Omega(\beta)\leq M} \left| \|[X\beta]_K\|_n^2 - \|[X\beta]_K\|^2 \right|$$

$$= K^2 \sup_{\|X\beta\|=1/K, \ \Omega(\beta)\leq M/K} \left| \|[X\beta]_1\|_n^2 - \|[X\beta]_1\|^2 \right|.$$

Denote the right-hand side without the multiplication factor K^2 by

$$\mathbf{Z} := \sup_{\|X\beta\|=1/K, \ \Omega(\beta)\leq M/K} \left| \|[X\beta]_1\|_n^2 - \|[X\beta]_1\|^2 \right|.$$

By symmetrization (see Theorem 16.1) and contraction (see Theorem 16.2),

$$\mathbb{E}\mathbf{Z} \le 2\mathbb{E} \sup_{\|X\beta\|=1/K,\ \Omega(\beta)\le M/K} \left| \frac{1}{n}\sum_{i=1}^{n} \epsilon_i ([X_i\beta]_1)^2 \right|$$

$$\le 8\mathbb{E} \sup_{\|X\beta\|=1/K,\ \Omega(\beta)\le M/K} \left| \frac{1}{n}\sum_{i=1}^{n} \epsilon_i X_i\beta \right|$$

since the mapping $Z \mapsto [Z]_1^2$ is 2-Lipschitz. Continuing with the last bound, we invoke

$$\mathbb{E} \sup_{\Omega(\beta)\le M/K} \left| \frac{1}{n}\sum_{i=1}^{n} \epsilon_i X_i\beta \right| = \left(\frac{M}{K}\right) \mathbb{E}\Omega_*(X^T\epsilon)/n \le \frac{M}{K}\delta_n.$$

In other words

$$\mathbb{E}\mathbf{Z} \le \frac{8M\delta_n}{K}.$$

Next, apply the result of Bousquet (2002), in particular Corollary 16.1, to $\mathbf{Z}$. One obtains that for all $t > 0$

$$\mathbb{P}\left(\mathbf{Z} \ge 2\mathbb{E}\mathbf{Z} + \frac{\sqrt{2t/n}}{K} + \frac{8t}{3n} \right) \le \exp[-t]$$

and hence

$$\mathbb{P}\left(\mathbf{Z} \ge \frac{16}{K}M\delta_n + \frac{\sqrt{2t/n}}{K} + \frac{8t}{3n} \right) \le \exp[-t].$$

So with probability at least $1 - \exp[-t]$ and using that $\|X\beta\|_n^2 \ge \|[X\beta]_K\|_n^2$

$$\inf_{\|X\beta\|=1,\ \Omega(\beta)\le M} \|X\beta\|_n^2 - 1 \ge -\frac{2C_m^m}{(m-2)K^{m-2}} - 16KM\delta_n - K\sqrt{\frac{2t}{n}} - \frac{8K^2t}{3n}.$$

We now let

$$K := [2C_m^m]^{\frac{1}{m-1}} b^{-\frac{1}{m-1}},$$

where

$$b := 16M\delta_n + \sqrt{\frac{2t}{n}}.$$

Then

$$\frac{2C_m^m}{(m-2)K^{m-2}} + Kb = D_m b^{\frac{m-2}{m-1}},$$

and

$$\frac{8K^2t}{3n} \le \frac{8D_m^2}{3}\left(\frac{t}{n}\right)^{\frac{m-2}{m-1}}.$$

It follows that with probability at least $1 - \exp[-t]$

$$\inf_{\|X\beta\|=1,\ \Omega(\beta)\le M} \|X\beta\|_n^2 - 1 \ge -D_m\left(16M\delta_n + \sqrt{\frac{2t}{n}}\right)^{\frac{m-2}{m-1}} - \frac{8D_m^2}{3}\left(\frac{t}{n}\right)^{\frac{m-2}{m-1}}.$$

□

Example 15.1 As an example, take in Theorem 15.1 the norm

$$\Omega(\beta) := \|X\beta\| = (\beta^T \Sigma_0 \beta)^{1/2}, \ \beta \in \mathbb{R}^p.$$

Then $\delta_n = \sqrt{p/n}$ (Problem 15.4) and so (taking $M = 1$)

$$\inf_{\|X\beta\|=1} \|X\beta\|_n^2 - 1 \ge -\Delta_n(t)$$

where

$$\Delta_n(t) := D_m\left(16\sqrt{\frac{p}{n}} + \sqrt{\frac{2t}{n}}\right)^{\frac{m-2}{m-1}} + \frac{8D_m^2}{3}\left(\frac{t}{n}\right)^{\frac{m-2}{m-1}}.$$

If Σ_0 is the identity matrix and C_m is fixed, this says asymptotically that when $p = o(n)$ the smallest eigenvalue of $\hat{\Sigma}$ converges to that of Σ_0 (which is 1).

Asymptotics We are primarily interested in showing that

$$\inf_{\|X\beta\|=1,\ \Omega(\beta)\le M} \|X\beta\|_n^2 \ge 1 - o_{\mathbb{P}}(1).$$

If the vectors X_i have sub-exponential entries ($i = 1, \ldots, n$) and $\Omega = \|\cdot\|_1$ we then need to require $M = o(\sqrt{n/\log p})$ (Problem 15.3). If $\Omega = \|\cdot\|_{\text{nuclear}}$ and (replacing p by pq) for $i = 1, \ldots n$ the vectors X_i are vectorized versions of the $q \times p$ matrices X_i of the matrix completion example (see Sect. 9.5 or Sect. 12.5) then we need to choose $M = o(\sqrt{nq/\log p})$.

15.3   Application to Effective Sparsity

Let Ω^+ and Ω^- be two semi-norms on $\mathbb{R}^p$ such that $\underline{\Omega} := \Omega^+ + \Omega^-$ is a norm. Define for $L > 0$ a constant the effective sparsity

$$\hat{\Gamma}_{\underline{\Omega}}^2(L) := \left(\min \left\{ \|X\beta\|_n^2 : \Omega^+(\beta) = 1, \ \Omega^-(\beta) \leq L \right\} \right)^{-1},$$

$$\Gamma_{\underline{\Omega}}^2(L) := \left(\min \left\{ \|X\beta\|^2 : \Omega^+(\beta) = 1, \ \Omega^-(\beta) \leq L \right\} \right)^{-1}.$$

Theorem 15.2 *Under the conditions of Theorem 15.1 and using its notation with* $\Omega := \underline{\Omega}$ *(i.e.* $\delta_n := \mathbb{E}\underline{\Omega}_*(X^T\epsilon)/n$*), for all* $t > 0$ *and* $L > 0$ *such that*

$$\Delta_n\left((L+1)\Gamma_\Omega(L), t \right) < 1,$$

with probability at least $1 - \exp[-t]$ *the following inequality holds true:*

$$\hat{\Gamma}_{\underline{\Omega}}^2(L) \leq \frac{\Gamma_{\underline{\Omega}}^2(L)}{1 - \Delta_n\left((L+1)\Gamma_\Omega(L), t \right)}.$$

Proof of Theorem 15.2 By Theorem 15.1 we know that uniformly in β with $\underline{\Omega}(\beta) \leq M\|X\beta\|$ with probability at least $1 - \exp[-t]$

$$\|X\beta\|_n^2 \geq (1 - \Delta_n(M, t))\|X\beta\|^2.$$

Now by the definition of $\Gamma_\Omega(L)$ we have for all β satisfying $\Omega^-(\beta) \leq L\Omega^+(\beta)$ the inequality

$$\Omega^+(\beta) \leq \|X\beta\|\Gamma_\Omega(L)$$

and hence for such β

$$\underline{\Omega}(\beta) \leq (L+1)\|X\beta\|\Gamma_\Omega(L).$$

So we find that for all such β with probability at least $1 - \exp[-t]$

$$\|X\beta\|_n^2 \geq \left[1 - \Delta_n\left((L+1)\Gamma_\Omega, t \right) \right] \|X\beta\|^2.$$

$\square$

Example 15.2 In this example the result of Theorem 15.2 is applied to compatibility constants. Let $\Omega := \|\cdot\|_1$, and for any $\beta \in \mathbb{R}^p$, $\Omega^+(\beta) := \|\beta_S\|_1$ and $\Omega^-(\beta) := \|\beta_{-S}\|_1$ where $S \subset \{1, \ldots, p\}$ is a given set with cardinality $s := |S|$. Then

$$\hat{\Gamma}^2_{\|\cdot\|_1}(L) = s/\hat{\phi}^2(L, S),$$

where $\hat{\phi}^2(L, S)$ is the (empirical) compatibility constant (see Definition 2.1). In the same way, one may define the theoretical compatibility constant

$$\phi^2(L, S) := s/\Gamma^2_{\|\cdot\|_1}(L).$$

The result of Theorem 15.2 then reads

$$\hat{\phi}^2(L, S) \geq \phi^2(L, S)\left[1 - \Delta_n\left(\frac{(L+1)\sqrt{s}}{\phi^2(L, S)}, t\right)\right].$$

Clearly

$$\phi^2(L, S) \geq \Lambda_{\min}(\Sigma_0)$$

where $\Lambda_{\min}(\Sigma_0)$ is the smallest eigenvalue of Σ_0. Assuming $1/\Lambda_{\min} = \mathcal{O}(1)$, $C_m = \mathcal{O}(1)$ and $L = \mathcal{O}(1)$, one obtains for $s = o(n/\log p)$ that also $1/\hat{\phi}^2(L, S) = \mathcal{O}(1)$. The paper of Lecué and Mendelson (2014) has results assuming the so-called "small ball property". This approach can be extended to more general norms as well.

Problems

15.1 Suppose that $X_1, \ldots, X_n$ are i.i.d. copies of a sub-Gaussian vector X_0 with constants t_0 and τ_0. Let $\epsilon_1, \ldots, \epsilon_n$ be a Rademacher sequence independent of $X_1, \ldots, X_n$. Define $\lambda_\epsilon := \mathbb{E}\Omega_*(\epsilon^T X)/n$ where X is the $n \times p$ matrix with rows X_i, $i = 1, \ldots, n$. Use similar arguments as in Theorem 15.1 to show that for $\lambda_\epsilon M \leq 1$ and a suitable constant C depending on t_0 and τ_0 one has for all t, with probability at least $1 - \exp[-t]$,

$$\inf_{\|X\beta\|=1, \, \Omega(\beta)\leq M} \|X\beta\|_n^2 - 1 \geq -C(M\lambda_\epsilon + \sqrt{t/n} + t\log n/n).$$

15.2 Let $X_0 = U_0\Sigma_0^{1/2}$ where U_0 has independent bounded entries. Show that X_0 is sub-Gaussian. Hint: apply Hoeffding's inequality (Corollary 17.1 at the end of Sect. 17.3).

15.3 Assume sub-exponential design: for some constant L_0

$$C_0^2 := \mathbb{E}\exp[|X_1|/L_0] < \infty.$$

Let $\epsilon_1, \ldots, \epsilon_n$ be a Rademacher sequence independent of X. Show that for some constant C depending on L_0 and C_0

$$\mathbb{E}\|X^T\epsilon\|_\infty/n \le C\left(\sqrt{\frac{\log p}{n}} + \frac{\log p}{n}\right).$$

15.4 Show that for a Rademacher sequence $\epsilon_1, \ldots, \epsilon_n$ independent of X

$$\mathbb{E}\sup_{\|X\beta\|=1} |\epsilon^T X\beta|/n = \sqrt{p/n}.$$

Chapter 16
Symmetrization, Contraction and Concentration

Abstract This chapter reviews symmetrization results, presents the contraction theorem of Ledoux and Talagrand (1991) and a multivariate extension and also the concentration (or actually the "deviation" part of) theorems of Bousquet (2002) and Massart (2000). The chapter gathers a collection of tools which have been used in Chap. 10. No proofs are given except for Theorem 16.3 on the multivariate contraction inequality (as it is not a standard formulation) and Lemma 16.2 (as this proof connects well with results from Chap. 8).[1]

16.1 Symmetrization and Contraction

Let $X_1, \ldots, X_n$ be independent random variables with values in some space $\mathscr{X}$ and let $\mathscr{F}$ be a class of real-valued functions on $\mathscr{X}$.

A *Rademacher sequence* is a sequence $\varepsilon_1, \ldots, \varepsilon_n$ of i.i.d. random variables taking values in $\{\pm 1\}$, with $\mathbb{P}(\varepsilon_i = +1) = \mathbb{P}(\varepsilon_i = -1) = 1/2$ $(i = 1, \ldots, n)$.

Theorem 16.1 (Symmetrization of Expectations (van der Vaart and Wellner 1996)) *Let $\varepsilon_1, \ldots, \varepsilon_n$ be a Rademacher sequence independent of $X_1, \ldots, X_n$. Then for any $m \geq 1$,*

$$\mathbb{E}\left(\sup_{f \in \mathscr{F}} \left| \sum_{i=1}^n (f(X_i) - \mathbb{E}f(X_i)) \right|^m \right) \leq 2^m \mathbb{E}\left(\sup_{f \in \mathscr{F}} \left| \sum_{i=1}^n \varepsilon_i f(X_i) \right|^m \right). \tag{16.1}$$

The case $m = 1$ is much called upon in applications. See Problem 16.1 for a proof for that case.

Symmetrization is also very useful in combination with the contraction theorem (see Theorem 16.2). It can moreover be invoked in combination with concentration inequalities which say that the supremum $\sup_{f \in \mathscr{F}} |\sum_{i=1}^n (f(X_i) - \mathbb{E}f(X_i))|$ concentrates (under certain conditions) around its mean. Probability inequalities thus follow fairly easy from inequalities for the mean. In some cases however one

---

[1] All statements are true "modulo measurability": the quantities involved may not be measurable. Easiest way out is to replace the supremum over an uncountable class by one over a countable class.

© Springer International Publishing Switzerland 2016
S. van de Geer, *Estimation and Testing Under Sparsity*,
Lecture Notes in Mathematics 2159, DOI 10.1007/978-3-319-32774-7_16

may want to study the probability inequalities directly. Symmetrization also works for probabilities.

Lemma 16.1 (Symmetrization of Probabilities (Pollard 1984; van de Geer 2000)) *Let $\varepsilon_1, \ldots, \varepsilon_n$ be a Rademacher sequence independent of $X_1, \ldots, X_n$. Then for $R^2 := \sup_{f \in \mathscr{F}} \sum_{i=1}^n \mathbb{E} f^2(X_i)/n$ and for all $t \geq 4$,*

$$\mathbb{P}\left(\sup_{f \in \mathscr{F}} \left| \frac{1}{n} \sum_{i=1}^n (f(X_i) - \mathbb{E} f(X_i)) \right| \geq 4R \sqrt{\frac{2t}{n}} \right) \leq 4\mathbb{P}\left(\sup_{f \in \mathscr{F}} \left| \sum_{i=1}^n \varepsilon_i f(X_i) \right| \geq R \sqrt{\frac{2t}{n}} \right).$$

Observe that $f \mapsto \sum_{i=1}^n \epsilon_i f(X_i)$ is a signed measure. Say when f is bounded, with the bound $|\sum_{i=1}^n \epsilon_i f(X_i)| \leq \|f\|_\infty \sum |\epsilon_i|$ one looses the signs. The contraction theorem is in the spirit of saying that one can apply this type of bound without loosing the signs.

Theorem 16.2 (Contraction Theorem (Ledoux and Talagrand 1991)) *Let $x_1, \ldots, x_n$ be non-random elements of $\mathscr{X}$ and let $\mathscr{F}$ be a class of real-valued functions on $\mathscr{X}$. Consider Lipschitz functions $\rho_i : \mathbb{R} \to \mathbb{R}$, i.e.*

$$|\rho_i(\xi) - \rho_i(\tilde{\xi})| \leq |\xi - \tilde{\xi}|, \ \forall \ \xi, \tilde{\xi} \in \mathbb{R}.$$

Let $\varepsilon_1, \ldots, \varepsilon_n$ be a Rademacher sequence. Then for any function $f^* : \mathscr{X} \to \mathbb{R}$, we have

$$\mathbb{E}\left(\sup_{f \in \mathscr{F}} \left| \sum_{i=1}^n \varepsilon_i \{\rho_i(f(x_i)) - \rho_i(f^*(x_i))\} \right| \right) \leq 2\mathbb{E}\left(\sup_{f \in \mathscr{F}} \left| \sum_{i=1}^n \varepsilon_i (f(x_i) - f^*(x_i)) \right| \right).$$

Theorem 16.3 (Multivariate Contraction Theorem) *Let $x_1, \ldots, x_n$ be non-random elements of $\mathscr{X}$ and let ,for some integer r, $\mathscr{F}$ be a class of $\mathbb{R}^r$-valued functions on $\mathscr{X}$. Consider ℓ_1-Lipschitz functions $\rho_i : \mathbb{R}^r \to \mathbb{R}$, i.e.*

$$|\rho_i(\xi) - \rho_i(\tilde{\xi})| \leq \|\xi - \tilde{\xi}\|_1, \ \forall \ \xi, \tilde{\xi} \in \mathbb{R}^r.$$

Let $\epsilon_1, \ldots, \epsilon_n$ be a Rademacher sequence and $\{V_{i,k} : 1 \leq i \leq n, \ 1 \leq k \leq n\}$ be i.i.d. $\mathcal{N}(0, 1)$ random variables independent of $\{\epsilon_i : 1 \leq i \leq n\}$. Then there exists a universal constant $C > 0$ such that for any function $f^* : \mathscr{X} \to \mathbb{R}^r$, we have

$$\mathbb{E}\left(\sup_{f=(f_1, \ldots, f_r) \in \mathscr{F}} \left| \sum_{i=1}^n \varepsilon_i \{\rho_i(f(x_i)) - \rho_i(f^*(x_i))\} \right| \right)$$
$$\leq C 2^{r-1} \mathbb{E}\left(\sup_{f \in \mathscr{F}} \left| \sum_{k=1}^r \sum_{i=1}^n V_{i,k}(f_k(x_i) - f_k^*(x_i)) \right| \right).$$

Proof The result follows from Theorem 2.1.5 in Talagrand's book (Talagrand 2005). All we need to do is to check that the tail-behaviour of the original process is like the one of the Gaussian process. Note first that for all $\mathbb{R}^r$-valued $f, \tilde{f}$,

$$\mathbb{E}\left|\sum_{k=1}^{r}\sum_{i=1}^{n} V_{i,k}(f_k(x_i) - f_k^*(x_i)) - \sum_{k=1}^{r}\sum_{i=1}^{n} V_{i,k}(\tilde{f}_k(x_i) - f_k^*(x_i))\right|^2 = n\sum_{k=1}^{r}\|f_k - \tilde{f}_k\|_n^2.$$

Moreover

$$n\|\rho(f) - \rho(\tilde{f})\|_n^2 \le \sum_{i=1}^{n}\left|\sum_{k=1}^{r}|f_k(x_i) - \tilde{f}_k(x_i)|\right|^2$$

$$\le n2^{r-1}\sum_{k=1}^{r}\|f_k - \tilde{f}_k\|_n^2.$$

By Hoeffding's inequality (see Theorem 9.1 in Sect. 9.1 or Corollary 17.1 in Sect. 17.3)

$$\mathbb{P}\left(\left|\sum_{i=1}^{n}\varepsilon_i\{\rho_i(f(x_i)) - \rho_i(\tilde{f}(x_i))\}\right| \ge \|\rho(f) - \rho(\tilde{f})\|_n\sqrt{2t/n}\right) \le 2\exp[-t].$$

$\square$

16.2 Concentration Inequalities

Concentration inequalities have been derived by Talagrand (1995) and further studied by Ledoux (1996), Ledoux (2005), Massart (2000) Bousquet (2002) and others. We refer to Boucheron et al. (2013) for a good overview and recent results.

Let $X_1, \ldots, X_n$ be independent random variables with values in some space $\mathscr{X}$ and let $\mathscr{F}$ be a class of real-valued functions on $\mathscr{X}$. Define

$$\mathbf{Z} := \sup_{f \in \mathscr{F}}\left|\frac{1}{n}\sum_{i=1}^{n}(f(X_i) - \mathbb{E}f(X_i))\right|.$$

The class of functions $\mathscr{F}$ can be very rich. The mean $\mathbb{E}\mathbf{Z}$ depends on this richness, on the complexity of the class $\mathscr{F}$ (see Chap. 17 for a quantification of complexity). Concentration inequalities say that the deviation form the mean however does not depend on the complexity of the class

We will show two inequalities concerning the right-sided deviation of $\mathbf{Z}$ from its mean. Left-sided versions are also available but these are not exploited in this monograph. The combination of left- and right-sided deviation inequalities are genuine concentration inequalities. We nevertheless still refer to the one-sided versions as concentration inequalities as this term points more clearly to the many results in literature and to concentration of measure. We remark here that left-sided versions are also important in statistics, for example for establishing lower bounds.

For Gaussian random variables the picture is rather complete and results go back to Borell (1975) for example.

16.2.1 *Massart's Inequality*

Theorem 16.4 (Massart's Concentration Theorem (Massart 2000))   *Suppose* $\mathbb{E}f(X_i) = 0$ *and* $|f(X_i)| \leq c_{i,f}$ *for some real numbers* $c_{i,f}$ *and for all* $1 \leq i \leq n$ *and* $f \in \mathscr{F}$. *Write*

$$R^2 := \sup_{f \in \mathscr{F}} \sum_{i=1}^{n} c_{i,f}^2/n.$$

For any positive t

$$\mathbf{P}\left(\mathbf{Z} \geq \mathbb{E}\mathbf{Z} + R\sqrt{\frac{8t}{n}}\right) \leq \exp[-t].$$

16.2.2 *Bousquet's Inequality*

Theorem 16.5 (Bousquet's Concentration Theorem for the Laplace Transform (Bousquet 2002))   *Suppose*

$$\mathbb{E}f(X_i) = 0, \forall\, i, \; \forall f \in \mathscr{F}, \; \frac{1}{n}\sum_{i=1}^{n} \sup_{f \in \mathscr{F}} \mathbb{E}f^2(X_i) \leq R^2$$

and moreover, for some positive constant K,

$$\|f\|_\infty \leq K, \; \forall f \in \mathscr{F}.$$

Then for all L > 0,

$$\log \mathbb{E}\exp[n(\mathbf{Z} - \mathbb{E}\mathbf{Z})/L] \leq n\left[e^{K/L} - 1 - K/L\right]\left[R^2/K^2 + 2\mathbb{E}\mathbf{Z}/K\right].$$

Lemma 16.2 (Bousquet's Concentration Theorem (Bousquet 2002)) *Suppose the conditions of Theorem 16.5 hold. We have for all $t > 0$,*

$$\mathbb{P}\left(\mathbf{Z} \geq \mathbb{E}\mathbf{Z} + R\sqrt{\frac{2t}{n}}\sqrt{1 + \frac{2K\mathbb{E}\mathbf{Z}}{R^2}} + \frac{tK}{3n}\right) \leq \exp[-t].$$

Proof of Lemma 16.2 It holds that for $L > K/3$

$$e^{K/L} - 1 - K/L = \sum_{m=2}^{\infty} \frac{1}{m!}\left(\frac{K}{L}\right)^m \leq \frac{K^2}{2(L^2 - KL/3)}.$$

Now apply Lemma 8.3 of Sect. 8.4. $\qquad\square$

A somewhat more convenient version (with less good constants) is given in the following corollary. In this corollary, the (right-sided) concentration of $\mathbf{Z}$ at its mean is lost, but one still sees that $\mathbf{Z}$ is not far from its mean.

Corollary 16.1 *Suppose the conditions of Theorem 16.5. Then for all $t > 0$*

$$\mathbb{P}\left(\mathbf{Z} \geq 2\mathbb{E}\mathbf{Z} + R\sqrt{\frac{2t}{n}} + \frac{4tK}{3n}\right) \leq \exp[-t].$$

Problems

16.1 Let $X_1, \ldots, X_n$ be independent random variables and $\mathbf{X}' := (X_1', \ldots, X_n')$ be an independent copy of $\mathbf{X} := (X_1, \ldots, X_n)$.

(i) Show that

$$\mathbb{E}\left(\sup_{f \in \mathscr{F}}\left|\sum_{i=1}^{n}(f(X_i) - \mathbb{E}f(X_i))\right|\right) \leq \mathbb{E}\left(\sup_{f \in \mathscr{F}}\left|\sum_{i=1}^{n}(f(X_i) - f(X_i'))\right|\right).$$

Hint: write $P_n f := \sum_{i=1}^{n} f(X_i)/n$, $P_n'f := \sum_{i=1}^{n} f(X_i')/n$ and $Pf := \mathbb{E}P_n f(= \mathbb{E}P_n'f)$. Let $\mathbb{E}(\cdot|\mathbf{X})$ be the conditional expectation given $\mathbf{X}$. Then

$$(P_n - P)f = \mathbb{E}((P_n - P_n')f|\mathbf{X}).$$

(ii) Let $\epsilon_1, \ldots, \epsilon_n$ be a Rademacher sequence independent of $\mathbf{X}$ and $\mathbf{X}'$. Show that

$$\mathbb{E}\left(\sup_{f \in \mathscr{F}} \left| \sum_{i=1}^{n} (f(X_i) - f(X_i')) \right| \right) = \mathbb{E}\left(\sup_{f \in \mathscr{F}} \left| \sum_{i=1}^{n} \epsilon_i (f(X_i) - f(X_i')) \right| \right)$$

$$\leq \mathbb{E}\left(\sup_{f \in \mathscr{F}} \left| \sum_{i=1}^{n} \epsilon_i f(X_i) \right| \right).$$

Chapter 17
Chaining Including Concentration

Abstract The chaining method is discussed, as well as the more general generic chaining method developed by Talagrand, see e.g. Talagrand (2005). This allows one to bound suprema of random processes. Concentration inequalities are refined probability inequalities, for instance for suprema of random processes, see e.g. Ledoux (2005), Boucheron et al. (2013) and Sect. 16.2 in this monograph. This chapter combines the two. A deviation inequality is obtained using (generic) chaining.

17.1 Introduction

In Chap. 2 we encountered the empirical process $\{\epsilon^T X \beta / n : \beta \in \mathbb{R}^p, \|\beta\|_1 = 1\}$ with $\epsilon \in \mathbb{R}^n$ a noise vector. This chapter is about suprema of empirical processes. For example for the empirical process of Chap. 2 this is

$$\sup_{\|\beta\|_1 = 1} |\epsilon^T X \beta| / n.$$

We then applied the dual norm (in)equality

$$\sup_{\|\beta\|_1 = 1} |\epsilon^T X \beta| / n = \|X^T \epsilon\|_\infty / n.$$

Thus, the problem of understanding the behaviour of the supremum of a infinite collection of random variables is reduced to understanding the behaviour of the maximum of finitely many, namely p, random variables.

There are now several reasons to study (generic) chaining of which we mention two. The first reason is of course that in general a dual norm inequality type of argument may not be applicable. The second reason is the fact that the bound obtained by generic chaining is sharp (up to constants) in the Gaussian case. Hence the dual norm argument (when applicable) must give the same result as generic

© Springer International Publishing Switzerland 2016

S. van de Geer, *Estimation and Testing Under Sparsity*,
Lecture Notes in Mathematics 2159, DOI 10.1007/978-3-319-32774-7_17

chaining. In other words, can we prove a bound of the form $\mathbb{E}\|X^T\epsilon\|_\infty/n \asymp \sqrt{\log p/n}$ for the expectation of the supremum of the empirical process $\{|\epsilon^T X\beta|/n : \|\beta\|_1 = 1\}$ by the geometric argument of generic chaining (instead of by the dual norm inequality)? We touch upon this question in the next chapter without coming to a final answer. See also Talagrand (2005) who first formulated this question. In fact, the dual norm argument also applies to the supremum of the empirical process $\{|\epsilon^T X\beta|/n : \Omega(\beta) = 1\}$. But there, bounding the supremum invoking the geometric argument of generic chaining is as far as we know an open problem. One may argue that since both the dual norm inequality and generic chaining yield sharp bounds, these bounds must be equal and there is nothing to prove. Yet, from a mathematical point of view it is of interest to see a direct (geometric) proof of this fact.

The motivation for this chapter is partly also that a reader interested in the topic of this book may need the tools presented in this chapter when studying other problems not treated in this book. Moreover, this chapter may help to understand a rough version of the concentration phenomenon, without going into detail about the fundamentals of the results cited in Sect. 16.2.

17.2 Notation

For a vector $v \in \mathbb{R}^n$ we use the notation $\|v\|_n^2 := v^T v/n$.

Let $\mathscr{F}$ be a (countable[1]) space and, for $i = 1,\ldots,n$ and $f \in \mathscr{F}$, let $Z_i(f)$ be a real-valued random variable. We write $Z_i := Z_i(\cdot) := \{Z_i(f) : f \in \mathscr{F}\}$, $i = 1,\ldots,n$, and assume that $Z_1,\ldots,Z_n$ are independent.

Example 17.1 (Empirical Processes) Let $X_1,\ldots,X_n$ be i.i.d. $\mathscr{X}$-valued random variables and $\mathscr{F}$ a class of real-valued functions on $\mathscr{X}$. Take $Z_i(f) = f(X_i)$, $i = 1,\ldots,n, f \in \mathscr{F}$.

Example 17.2 (Empirical Risk Minimization) Let $\rho : \mathbb{R}^2 \to \mathbb{R}$ be a loss function and let $(X_i, Y_i) \in \mathbb{R}^{p+1}$ be input and output variables $i = 1,\ldots,n$. Take $\mathscr{F}$ to be a class of real-valued functions on $\mathbb{R}^p$ and take $Z_i(f) := \rho(f(X_i), Y_i), i = 1,\ldots,n$. In generalized linear models the space $\mathscr{F}$ is the linear space $\{f_\beta(x) = x\beta : \beta \in \mathbb{R}^p\}$ or a subset thereof.

Example 17.3 (Finite Class $\mathscr{F}$) Let X be an $n \times p$ matrix with rows X_i. Let $\mathscr{F} := \{1,\ldots,p\}$. Then (for $i = 1,\ldots,n$) $Z_i(f)$ is for example the fth coordinate of X_i.

---

[1]See the footnote on the first page of Chap. 16.

17.3   Hoeffding's Inequality

This section establishes Hoeffding's inequality (Hoeffding 1963) in a number of elementary steps.

Lemma 17.1 *Consider a real-valued random variable Z with*

$$\mathbb{E}Z = 0, \ |Z| \leq c,$$

where $c > 0$ is some constant. Then for any convex function g the mean of $g(Z)$ can be bounded by the average of the extremes in $\pm c$:

$$\mathbb{E}g(Z) \leq \frac{1}{2}(g(c) + g(-c)).$$

Proof of Lemma 17.1 For all $0 \leq \alpha \leq 1$ and all u and v we have

$$g(\alpha u + (1-\alpha)v) \leq \alpha g(u) + (1-\alpha)g(v).$$

Apply this with $\alpha := (c - Z)/(2c)$, $u := -c$ and $v := c$. Then $Z = \alpha u + (1-\alpha)v$, so

$$g(Z) \leq \alpha g(-c) + (1-\alpha)g(c)$$

and since $\mathbb{E}\alpha = 1/2$

$$\mathbb{E}g(Z) \leq \frac{1}{2}g(-c) + \frac{1}{2}g(c).$$

$\square$

In the rest of the section we examine independent real-valued random variables $Z_1, \ldots, Z_n$ (i.e., the space $\mathscr{F}$ is a singleton). Moreover, $c := (c_1, \ldots, c_n)^T$ is a vector of positive constants and $\varepsilon = (\varepsilon_1, \ldots, \varepsilon_n)^T$ is a Rademacher sequence independent of $Z_1, \ldots, Z_n$. Recall that a Rademacher sequence $\varepsilon_1, \ldots, \varepsilon_n$ consists of independent random variables with $\mathbb{P}(\varepsilon_i = 1) = \mathbb{P}(\varepsilon_i = -1) = \frac{1}{2}$, $i = 1, \ldots, n$.

Lemma 17.2 *Let $Z_1, \ldots, Z_n$ be independent real-valued random variables satisfying $\mathbb{E}Z_i = 0$ and $|Z_i| \leq c_i$ for all $i = 1, \ldots, n$. Then for all $\lambda > 0$*

$$\mathbb{E}\exp\left[\lambda \sum_{i=1}^{n} Z_i\right] \leq \mathbb{E}\exp\left[\lambda \varepsilon^T c\right].$$

Proof of Lemma 17.2 Let $\lambda > 0$. The map $u \mapsto \exp[\lambda u]$ is convex. The result thus follows from Lemma 17.1 and the independence assumptions. $\square$

Lemma 17.3 *For any $z \in \mathbb{R}$,*

$$\frac{1}{2}\exp[-z] + \frac{1}{2}\exp[z] \leq \exp[z^2/2].$$

Proof of Lemma 17.3 For any $z \in \mathbb{R}$

$$\frac{1}{2}\exp[-z] + \frac{1}{2}\exp[z] = \sum_{k \text{ even}} \frac{1}{k!}z^k = \sum_{k=0}^{\infty} \frac{1}{(2k)!}z^{2k}.$$

But for $k \in \{0, 1, 2, \ldots\}$ we have

$$\frac{1}{(2k)!} \leq \frac{1}{k!2^k}$$

so that

$$\frac{1}{2}\exp[-z] + \frac{1}{2}\exp[z] \leq \sum_{k=0}^{\infty} \frac{1}{k!}\frac{z^{2k}}{2^k} = \exp[z^2/2].$$

$\square$

Lemma 17.4 *For all $\lambda \in \mathbb{R}$*

$$\mathbb{E}\exp\left[\lambda \varepsilon^T c\right] \leq \exp\left[\frac{n\lambda^2\|c\|_n^2}{2}\right].$$

Proof of Lemma 17.4 In view of Lemma 17.3 we have

$$\mathbb{E}\exp\left[\lambda \varepsilon^T c\right] = \prod_{i=1}^{n}\left(\frac{1}{2}\exp[-\lambda c_i] + \frac{1}{2}\exp[\lambda c_i]\right)$$

$$\leq \prod_{i=1}^{n}\exp\left[\frac{\lambda^2 c_i^2}{2}\right] = \exp\left[\frac{n\lambda^2\|c\|_n^2}{2}\right].$$

$\square$

Theorem 17.1 *Let $Z_1, \ldots, Z_n$ be independent real-valued random variables satisfying $\mathbb{E}Z_i = 0$ and $|Z_i| \leq c_i$ for all $i = 1, \ldots, n$. Then for all $\lambda > 0$*

$$\mathbb{E}\exp\left[\lambda \sum_{i=1}^{n} Z_i\right] \leq \exp\left[n\lambda^2\|c\|_n^2/2\right].$$

Proof of Theorem 17.1 This follows from combining Lemmas 17.2 and 17.4. $\square$

Corollary 17.1 (Hoeffding's Inequality) *Suppose that for all $1 \le i \le n$*

$$\mathbb{E} Z_i = 0, \ |Z_i| \le c_i.$$

Then for all $t > 0$

$$\mathbb{P}\left(\frac{1}{n} \sum_{i=1}^{n} Z_i > \|c\|_n \sqrt{\frac{2t}{n}}\right) \le \exp[-t]$$

(see Problem 17.1).

17.4 The Maximum of p Random Variables

Lemma 17.5 *Let $V_1, \ldots, V_p$ be real-valued random variables. Assume that for all $j \in \{1, \ldots, p\}$ and all $\lambda > 0$*

$$\mathbb{E} \exp[\lambda |V_j|] \le 2 \exp[\lambda^2/2].$$

Then

$$\mathbb{E} \max_{1 \le j \le p} |V_j| \le \sqrt{2 \log(2p)},$$

and for all $t > 0$

$$\mathbb{P}\left(\max_{1 \le j \le p} |V_j| \ge \sqrt{2 \log(2p) + 2t}\right) \le \exp[-t]. \tag{17.1}$$

Example 17.4 Consider independent random vectors $Z_1, \ldots, Z_n$ with values in $\mathbb{R}^p$. Suppose that $\mathbb{E} Z_i = 0$ and $|Z_{i,j}| \le c_{i,j}$ for all i and j where $c_j := (c_{1,j}, \ldots, c_{n,j})^T$, $j = 1, \ldots, p$, are n-dimensional vectors of constants. Let $R := \max_{1 \le j \le p} \|c_j\|_n$. In view of Theorem 17.1, one can apply Lemma 17.5 to $V_j := \sum_{i=1}^{n} Z_{i,j}/\sqrt{nR^2}$, $j = 1, \ldots, p$. A special case is $Z_{i,j} = \epsilon_i X_{i,j}$ for all i and j, with $\epsilon = (\epsilon_1, \ldots, \epsilon_n)^T$ a Rademacher sequence, and $X = (X_{i,j})$ a fixed (nonrandom) $n \times p$ matrix. Then $c_{i,j}$ can be taken as $|X_{i,j}|$ for all i and j (so that $R = \max_{1 \le j \le p} \|X_j\|_n$). Then $V_j = X_j^T \epsilon/\sqrt{nR^2}$, $j = 1, \ldots, p$.

Example 17.5 Let $\epsilon \sim \mathcal{N}_n(0, I)$ and let $X = (X_1, \ldots, X_p)$ be a fixed $n \times p$ matrix. Let $R = \max_{1 \le j \le p} \|X_j\|_n$. Lemma 17.5 applies to $V_j := X_j^T \epsilon/\sqrt{nR^2}$, $j = 1, \ldots, p$.

The inequality 17.1 can be seen as a rudimentary form of a concentration inequality. The logarithmic term $\sqrt{2\log(2p)}$ is a bound for the expectation $\mathbb{E}\max_{1\leq j\leq p}|V_j|$ and the deviation from this bound does not depend on p. The logarithmic term $\log(2p)$ quantifies (up to constants) as the "complexity" of the set $\{1,\ldots,p\}$. Say p is some power of 2. To describe a member $j\in\{1,\ldots,p\}$ in binary code one needs $\log_2(p)$ digits.

Proof of Lemma 17.5 Let $\lambda>0$ be arbitrary. We have

$$\mathbb{E}\max_{1\leq j\leq p}|V_j| = \frac{1}{\lambda}\mathbb{E}\log\exp\left[\lambda\max_{1\leq j\leq p}|V_j|\right]$$
$$\leq \frac{1}{\lambda}\log\mathbb{E}\exp\left[\lambda\max_{1\leq j\leq p}|V_j|\right]$$
$$\leq \frac{1}{\lambda}\log\left((2p)\exp[\lambda^2/2]\right)$$
$$= \frac{\log(2p)}{\lambda}+\frac{\lambda}{2}.$$

Now choose $\lambda=\sqrt{2\log(2p)}$ to minimize the last term.

For the second result one may use Chebyshev's inequality as in Problem 17.1 to establish that for all $t>0$ and all j

$$\mathbb{P}\left(|V_j|\geq\sqrt{2t}\right)\leq 2\exp[-t].$$

Then the inequality for the maximum over j follows from the union bound. □

17.5 Expectations of Positive Parts

We let $[x]_+ := x\vee 0$ denote the positive part of $x\in\mathbb{R}$.

Lemma 17.6 *Let for some $S\in\mathbb{N}$ and for $s=1,\ldots,S$, $\mathbf{Z}_s$ be non-negative random variables that satisfy for certain positive constants $\{H_s\}$ and for all $t>0$*

$$\mathbb{P}(\mathbf{Z}_s\geq\sqrt{H_s+t})\leq\exp[-t].$$

Then

$$\mathbb{E}\max_{1\leq s\leq S}\left[\mathbf{Z}_s-\sqrt{H_s+s\log 2}\right]_+^2\leq 1.$$

This lemma will be applied in the next section with, for each $s \in \{1, \ldots, S\}$, $\mathbf{Z}_s$ being the maximum of a finite number, say N_s, of random variables, and with $H_s := \log(2N_s)$.

Proof of Lemma 17.6 Clearly

$$\mathbb{E}\left[\mathbf{Z}_s - \sqrt{H_s + s\log 2}\right]_+^2 = \int_0^\infty \mathbb{P}\left(\mathbf{Z}_s - \sqrt{H_s + s\log 2} \ge \sqrt{t}\right)dt$$

$$\le \int_0^\infty \mathbb{P}\left(\mathbf{Z}_s \ge \sqrt{H_s + s\log 2 + t}\right)dt$$

$$\le \int_0^\infty \exp[-s\log 2 - t]dt = 2^{-s}.$$

Hence

$$\mathbb{E}\max_{1 \le s \le S}\left[\mathbf{Z}_s - \sqrt{H_s + s\log 2}\right]_+^2 \le \sum_{s=1}^S \mathbb{E}\left[\mathbf{Z}_s - \sqrt{H_s + s\log 2}\right]_+^2 \le \sum_{s=1}^S 2^{-s} \le 1.$$

$\square$

17.6 Chaining Using Covering Sets

Consider a subset $\mathscr{F}$ of a metric space $(\bar{\mathscr{F}}, d)$.

Definition 17.1 Let $\delta > 0$ be some constant. A *δ-covering set* of $\mathscr{F}$ is a set of points $\{f_1, \ldots f_N\} \subset \mathscr{F}$ such that for all $f \in \mathscr{F}$

$$\min_{1 \le j \le N} d(f, f_j) \le \delta.$$

The *δ-covering number* $N(\delta)$ is the size of a minimal covering set (with $N(\delta) = \infty$ if no such finite covering exists). The *δ-entropy* is the logarithm $H(\delta) := \log N(\delta)$.

The idea is now to approximate $\mathscr{F}$ by finite coverings. Fix some $f_0 \in \mathscr{F}$ and denote the *radius* of $\mathscr{F}$ by

$$R := \sup_{f \in \mathscr{F}} d(f, f_0).$$

Fix also some $S \in \mathbb{N}$. For each $s = 1, \ldots, S$, let $\mathscr{G}_s \subset \mathscr{F}$ be a $2^{-s}R$ covering set of $\mathscr{F}$. Take $\mathscr{F}_0 := \{f_0\}$. The finest approximation considered is $\mathscr{G}_S$. The value S is to be chosen in such a way that $\mathscr{G}_S$ is "fine enough".

For a given $f \in \mathscr{F}$ we let its "parent" in the finest covering set $\mathscr{G}_S$ be

$$\mathrm{pa}_S(f) := \arg\min_{j \in \mathscr{G}_S} d(f,j)$$

(with an arbitrary choice if $d(f,j)$ is minimized by several $j \in \mathscr{G}_S$). We also call $\mathrm{pa}_S(f) =: \mathrm{an}_S(f)$ the "ancestor" of f in generation S. We let for $s = 1,\ldots,S$, $\mathrm{an}_{s-1}(f)$ be the parent of $\mathrm{an}_s(f)$:

$$\mathrm{an}_{s-1}(f) := \arg\min_{j \in \mathscr{G}_{s-1}} d(\mathrm{an}_s(f),j).$$

Hence, $\mathrm{an}_0(f) = f_0$ for all $f \in \mathscr{F}$. In other words, $\mathrm{an}_s(f)$ is the ancestor of f in generation $s \in \{0,1,\ldots,S\}$. A family tree is built, and the covering sets $\mathscr{G}_s$, $s = 0,\ldots,S$, form the generations ($\mathscr{F}$ itself being the "current" generation). The "parent" mapping is defined for each $s = 1,\ldots,S$ as:

$$\mathrm{pa}_{s-1}(\cdot) : \mathscr{G}_s \to \mathscr{G}_{s-1}, \ \mathrm{pa}_{s-1}(k) = \arg\min_{j \in \mathscr{G}_{s-1}} d(k,j), \ k \in \mathscr{G}_s.$$

17.7   Generic Chaining

Generic chaining (Talagrand 2005) is a more general way to built family trees: not necessarily through covering sets. Let again $\mathscr{F}$ be a subset of a metric space $(\bar{\mathscr{F}}, d)$. Let $\{\mathscr{G}_s\}_{s=1}^S$ be a sequence of finite non-empty subsets of $\mathscr{F}$ and let $\mathscr{G}_0 := \{f_0\}$. We call $\{\mathscr{G}_s\}_{s=0}^S$ the *generations*, with f_0 being everybody's ancestor. Consider maps $f \mapsto \mathrm{pa}_S(f) \in \mathscr{G}_S$. That is, each element $f \in \mathscr{F}$ is assigned a parent in generation S. We take $\mathrm{pa}_S(f) = f$ if $f \in \mathscr{G}_S$. Moreover, for $s = 1,\ldots,S$, each $j \in \mathscr{G}_s$ is assigned a parent $\mathrm{pa}_{s-1}(j)$ in generation $s-1$ with $\mathrm{pa}_{s-1}(j) = j$ if $j \in \mathscr{G}_{s-1}$. We then define the ancestors

$$\mathrm{an}_{S-k}(f) = \underbrace{\mathrm{pa}_{S-k} \circ \cdots \circ \mathrm{pa}_S}_{k+1 \times}(f), \ f \in \mathscr{F}, \ k = 0,\ldots,S.$$

Define $d_s(j) := d(\mathrm{an}_s(j), \mathrm{an}_{s-1}(j)), \ j \in \mathscr{G}_s, \ s = 1,\ldots,S$. Thus, $d_s(j)$ is the length of that piece of the branch of the tree that goes from generation s to generation $s-1$ and that ends at family member j in generation S ($j \in \mathscr{G}_S, \ s = 1,\ldots,S$). Note that

$$d_s(j) = d(\mathrm{an}_s(j), \mathrm{pa}_{s-1}(\mathrm{an}_s(j))), \ j \in \mathscr{G}_S, \ s = 1,\ldots,S.$$

The total length of the branch ending at family member $j \in \mathscr{G}_S$ is $\sum_{s=1}^{S} d_s(j)$. The maximal branch length over all members in generation S is

$$R_S := \max_{j \in \mathscr{G}_S} \sum_{s=1}^{S} d_s(j).$$

Set $H_s := \log(2|\mathscr{G}_s|)$, $s = 1, \ldots, S$.

17.8 Bounding the Expectation

We use the notation of the previous section concerning $(\mathscr{F}, d)$.

Let $V = \{V(f) : f \in \mathscr{F}\}$ be a random element of $L_\infty(\mathscr{F})$.

Example 17.6 A special case for which the theory to come is useful is $V = \sum_{i=1}^{n} Z_i / \sqrt{n}$, where $Z_1, \ldots, Z_n$ are independent mean zero random elements of $L_\infty(\mathscr{F})$ satisfying the required tail conditions.

Theorem 17.2 *Assume that for each $f, \tilde{f} \in \mathscr{F}$ and all $\lambda > 0$,*

$$\mathbb{E} \exp\left[\lambda \left|V(f) - V(\tilde{f})\right|\right] \le 2 \exp\left[\frac{\lambda^2 d^2(f, \tilde{f})}{2}\right].$$

Then

$$\mathbb{E}\|V - V(f_0)\|_\infty \le \gamma_S + R_S \sqrt{2} + \mathbb{E}\|\mathscr{V}_S\|_\infty$$

where

$$\gamma_S := \max_{j \in \mathscr{G}_S} \sum_{s=1}^{S} d_s(j) \sqrt{2(H_s + s \log 2)}$$

and

$$\mathscr{V}_S(f) := V(f) - V(\mathrm{pa}_S(f)), \quad f \in \mathscr{F}.$$

Observe that if $H_s \ge 1$ for all s (which will be true in typical cases) one can bound $R_S \sqrt{2}$ by γ_S giving

$$\mathbb{E}\|V - V(f_0)\|_\infty \le 2\gamma_S + \mathbb{E}\|\mathscr{V}_S\|_\infty.$$

The term $\mathbb{E}\|\mathscr{V}_S\|_\infty$ is the "remainder". The idea is to choose S by trading off γ_S and $\mathbb{E}\|\mathscr{V}_S\|_\infty$. See also Remark 17.1 formulated after the proof of the theorem.

Proof of Theorem 17.2 Let $j \in \mathscr{G}_S$. We may write for all $f \in \mathscr{F}$ with $\mathrm{pa}_S(f) = j$

$$V(f) - V(f_0) = V(j) - V(f_0) + \mathscr{V}_S(f)$$

$$= \sum_{s=1}^{S} \Big(V(\mathrm{an}_s(j)) - V(\mathrm{an}_{s-1}(j)) \Big) + \mathscr{V}_S(f).$$

Therefore

$$\sup_{f \in \mathscr{F}} |V(f) - V(f_0)| \leq \max_{j \in \mathscr{G}_S} \sum_{s=1}^{S} \Big| V(\mathrm{an}_s(j)) - V(\mathrm{an}_{s-1}(j)) \Big| + \|\mathscr{V}_S\|_\infty.$$

Write shorthand

$$V_{j,s} := V(\mathrm{an}_s(j)) - V(\mathrm{an}_{s-1}(j)), \ j \in \mathscr{G}_S, \ s = 1, \ldots, S.$$

We have

$$\max_{j \in \mathscr{G}_S} \sum_{s=1}^{S} |V_{j,s}| - \gamma_S = \max_{j \in \mathscr{G}_S} \sum_{s=1}^{S} \Big(\frac{|V_{j,s}|}{d_s(j)} - \sqrt{2(H_s + s \log 2)} \Big) d_s(j)$$

$$\leq \max_{j \in \mathscr{G}_S} \sum_{s=1}^{S} \Big[\frac{|V_{j,s})|}{d_s(j)} - \sqrt{2(H_s + s \log 2)} \Big]_+ d_s(j)$$

$$\leq \max_{j \in \mathscr{G}_S} \max_{s \in \{1,\ldots,S\}} \Big[\frac{|V_{j,s}|}{d_s(j)} - \sqrt{2(H_s + s \log 2)} \Big]_+ \sum_{s=1}^{S} d_s(j)$$

$$\leq \max_{j \in \mathscr{G}_S} \max_{s \in \{1,\ldots,S\}} \Big[\frac{|V_{j,s}|}{d_s(j)} - \sqrt{2(H_s + s \log 2)} \Big]_+ R_S.$$

$$= \max_{s \in \{1,\ldots,S\}} \Big[\max_{j \in \mathscr{G}_S} \frac{|V_{j,s}|}{d_s(j)} - \sqrt{2(H_s + s \log 2)} \Big]_+ R_S.$$

Since the ancestor in generation $s - 1$ of $\mathrm{an}_s(j) \in \mathscr{G}_s$ is its parent $\mathrm{pa}(\mathrm{an}_s(j), s - 1)$, the pair $[\mathrm{an}_s(j), \mathrm{an}_{s-1}(j)]$ is determined by its first member, and so there are at most N_s pairs $[\mathrm{an}_s(j), \mathrm{an}_{s-1}(j)]$, $s = 1, \ldots, S, j \in \mathscr{G}_S$. Therefore by Lemma 17.5 for all $t > 0$ and $s \in \{1, \ldots, S\}$,

$$\mathbb{P} \Big(\max_{j \in \mathscr{G}_S} \frac{|V_{j,s}|}{d_s(j)} \geq \sqrt{\frac{2(H_s + t)}{n}} \Big) \leq \exp[-t].$$

Combine this with Lemma 17.6 to find that

$$\mathbb{E} \max_{s \in \{1,\ldots,S\}} \Big[\max_{j \in \mathscr{G}_S} \frac{|V_{j,s}|}{d_s(j)} - \sqrt{2(H_s + s \log 2)} \Big]_+ \leq \sqrt{2}$$

It follows that

$$\mathbb{E}\left(\max_{j\in\mathscr{G}_S}\sum_{s=1}^{S}|V_{j,s}|\right) \le \gamma_S + R_S\sqrt{2}$$

and hence

$$\mathbb{E}\sup_{f\in\mathscr{F}}\left|V(f)-V(f_0)\right| \le \gamma_S + R_S\sqrt{2} + \mathbb{E}\|\mathscr{V}_S\|_\infty.$$

□

Remark 17.1 Because we only use finitely many generations there appears a term $\mathscr{V}_S$ expressing the approximation error using $\mathscr{G}_S$ instead of $\mathscr{F}$. In many situations, one can indeed do with a finite S because the remainder $\mathbb{E}\|\mathscr{V}_S\|_\infty$ is small. Because ancestors are only defined for a population with finitely many generations, it is not possible use a tree with infinitely many (as the family tree is built up starting with the last generation). With infinitely many generations it is less clear how to construct, for all s, pairs $[j,k]$ with $j\in\mathscr{G}_s$ and $k\in\mathscr{G}_{s-1}$ and a map $j\mapsto k$. However, the number of pairs $[j,k]$ with $j\in\mathscr{G}_s$ and $k\in\mathscr{G}_{s-1}$ can be bounded by $|\mathscr{G}_s|\times|\mathscr{G}_{s-1}|$ which is up to constants as good as the bound $|\mathscr{G}_s|$ we applied in the proof of Theorem 17.2. In other words, the reason for considering finitely many generations is mainly to avoid possibly diverging infinite sums.

For fixed (sufficiently large) S the set $\mathscr{G}_S$ may be chosen to be a $2^{-S}R$-covering set of $\mathscr{F}$. This is useful when $V(f)$ is Lipschitz with (random) Lipschitz constant $\mathbf{L}$:

$$|V(f)-V(\tilde{f})| \le \mathbf{L}d(f,\tilde{f}),\ f,\tilde{f}\in\mathscr{F}.$$

It leads to choosing S by a trade off between γ_S and $2^{-S}\mathbb{E}\mathbf{L}$.

The case $S\to\infty$ could formally be described as follows. Suppose that for each $s\in\{0,1,2,\ldots\}$ generations $\mathscr{G}_s\subset\mathscr{F}$ are given, with $\mathscr{G}_0=\{f_0\}$. For each S, mappings $f\mapsto \mathrm{an}_s(f,S)$ are given too as in Sect. 17.7, possibly depending on the depth S of the family tree. The lengths $d_s(j,S)$ now possibly also depend on S. Suppose

$$\lim_{S\to\infty}\mathbb{E}\left(\sup_{f\in\mathscr{F}}|V(f)-V(\mathrm{an}_S(f,S))|\right)=0.$$

Then

$$\mathbb{E}\|V-V(f_0)\|_\infty \le \gamma_\infty + R_\infty\sqrt{2},$$

where

$$\gamma_\infty := \lim_{S\to\infty} \max_{j\in\mathscr{G}_S} \sum_{s=1}^{S} d_s(j,S)\sqrt{2(H_s + s\log 2)}$$

and

$$R_\infty := \lim_{S\to\infty} \sum_{s=1}^{S} d_s(j,S)$$

(both assumed to exist).

17.9 Concentration

We use the same notation as in the previous section. We show a probability inequality for $\|V-V(f_0)\|_\infty$ where the deviation from the bound for the expectation, established in the previous section, does not depend on the sizes of the generations $\{\mathscr{G}_s\}_{s=0}^{S}$.

Theorem 17.3 *Assume that for each* $f,\tilde{f} \in \mathscr{F}$ *and all* $\lambda > 0$,

$$\mathbb{E}\exp\left[\lambda\left|V(f) - V(\tilde{f})\right|\right] \le 2\exp\left[\frac{\lambda^2 d^2(f,\tilde{f})}{2}\right].$$

Then for all $t > 0$

$$\mathbb{P}\left(\|V - V(f_0)\|_\infty \ge \gamma_S(t) + \|\mathscr{V}_S\|_\infty\right) \le \exp[-t]$$

where

$$\gamma_S(t) := \max_{j\in\mathscr{G}_S} \sum_{s=1}^{S} d_s(j)\sqrt{2H_s + 2(1+s)(1+t)}$$

and

$$\mathscr{V}(f) := V(f) - V(\mathrm{pa}_S(f)),\ f \in \mathscr{F}.$$

Proof of Theorem 17.3 Let $t > 0$ be arbitrary and define for $s = 1,\ldots,S$

$$\alpha_s := \sqrt{2H_s + 2(1+s)(1+t)}.$$

Using similar arguments as in the proof of Theorem 17.2 we find

$$\sup_{f \in \mathscr{F}} |V(f) - V(f_0)| \le \max_{j \in \mathscr{G}_S} \sum_{s=1}^{S} |V_{j,s}| + \|\mathscr{V}_S\|_\infty,$$

where

$$V_{j,s} := V(\mathrm{an}_s(j)) - V(\mathrm{an}_{s-1}(j)), \ j \in \mathscr{G}_S, \ s = 1, \ldots, S.$$

Also in parallel with the proof of Theorem 17.2

$$\max_{j \in \mathscr{G}_S} \sum_{s=1}^{S} |V_{j,s}| = \max_{j \in \mathscr{G}_S} \sum_{s=1}^{S} \frac{|V_{j,s}|}{d_s(j)\alpha_s} d_s(j)\alpha_s$$

$$\le \max_{1 \le s \le S} \max_{j \in \mathscr{G}_S} \frac{|V_{j,s}|}{d_s(j)\alpha_s} \sum_{s=1}^{S} d_s(j)\alpha_s$$

$$\le \max_{1 \le s \le S} \max_{j \in \mathscr{G}_S} \frac{|V_{j,s}|}{d_s(j)\alpha(s)} \max_{j \in \mathscr{G}_S} \sum_{s=1}^{S} d_s(j)\alpha_s$$

$$= \max_{1 \le s \le S} \max_{j \in \mathscr{G}_S} \frac{|V_{j,s}|}{d_s(j)\alpha(s)} \gamma_S(t).$$

In view of Lemma 17.5

$$\mathbb{P}\left(\max_{j \in \mathscr{G}_S} \frac{|V_{j,s}|}{d_s(j)\alpha_s} \ge 1 \right) \le \exp[-(1+s)(1+t)], \ s = 1, \ldots, S.$$

Hence

$$\mathbb{P}\left(\max_{1 \le s \le S} \max_{j \in \mathscr{G}_S} \frac{|V_{j,s}|}{d_s(j)\alpha_s} \ge 1 \right) \le \sum_{s=1}^{S} \exp[-(1+s)(1+t)] \le \exp[-t].$$

$\square$

Remark 17.2 Note that $\gamma_S(t)$ (and γ_S) depend on the choice of the generations $\{\mathscr{G}_s\}_{s=0}^{S}$. Fix $\mathscr{G}_S$ for the moment. Choose the other generations in such a way that γ_S is minimized. Then the bound γ_S for the expectation of $\max_{j \in \mathscr{G}_S} |V(j) - V(f_0)|$ as given in Theorem 17.2 (with $\mathscr{F} := \mathscr{G}_S$) is up to universal constants sharp in the Gaussian case (Talagrand 2005). Note further that

$$\gamma_S(t) \le \gamma_S(0) + R_S \sqrt{2t}.$$

The first term does not depend on t, is up to constants equal to γ_S and a bound for the mean of $\max_{j \in \mathscr{G}_S} |V(j) - V(f_0)|$. The second term describes the deviation from this bound. This term does not depend on the sizes of the generations but only on the maximal length of the branches of the tree. See Problem 17.1 for an example.

Problems

17.1 To verify Hoeffding's inequality in Corollary 17.1 (Sect. 17.3), invoke Chebyshev's inequality: for all $\lambda > 0$

$$\mathbb{P}\left(\frac{1}{n}\sum_{i=1}^{n} Z_i \geq \|c\|_n \sqrt{\frac{2t}{n}}\right) \leq \exp\left[\lambda^2/2 - \lambda\sqrt{2t}\right].$$

Now choose $\lambda = \sqrt{2t}$.

17.2 Let $\mathscr{G}_s$ be a minimal 2^{-s}-covering set of $\mathscr{F}$, $s = 1, \ldots, S$.

(i) Show that

$$\gamma_S = R \sum_{s=1}^{S} 2^{-s+1} \sqrt{2(\log(2|\mathscr{G}_s|) + s\log 2)}.$$

This is called (modulo some constants) the Dudley entropy integral, after Dudley (1967).

(ii) Show that

$$\gamma_S(t) \leq R \sum_{s=1}^{S} 2^{-s+1} \sqrt{2(\log(2|\mathscr{G}_s|) + 2(1+s))} + R\sqrt{2t} \sum_{s=1}^{S} 2^{-s+1} \sqrt{1+s}.$$

The second term does not depend on the complexity of the problem, where complexity is in terms of the entropy integral. Note that

$$\sum_{s=1}^{S} 2^{-s+1} \sqrt{(1+s)} \leq 4.$$

17.3 Here is an example dealing with the remainder term $\|\mathscr{V}\|_\infty$ where $\mathscr{V}(f) := V(f) - V(\mathrm{pa}_S(f)), f \in \mathscr{F}$. Let

$$V(f) = \sum_{i=1}^{n} \epsilon_i f_i/\sqrt{n}, f \in \mathscr{F},$$

where $\mathscr{F} \subset \mathbb{R}^n$ and ϵ is a Rademacher sequence.

(i) Show that

$$\|\mathcal{V}\|_\infty \le \sup_{f\in\mathscr{F}} \sqrt{n}\|f - \mathrm{pa}_S(f)\|_n.$$

(ii) Let $\mathscr{G}_s$ be a minimal 2^{-s}-covering set of $(\mathscr{F}, \|\cdot\|_n)$, $s = 1,\dots,S$. Show that

$$\mathbb{E}\|V\|_\infty \le \gamma_S + 2R\sqrt{2} + 2^{-S}R\sqrt{n}.$$

(iii) Suppose that for $s = 1,2,\dots$ the covering numbers $N_s := N(2^{-s}R)$ of $(\mathscr{F}, \|\cdot\|_n)$ satisfy for some constants A and a

$$H_s := \log(2N_s) \le A2^{2s} + as$$

Show that

$$\gamma_S \le 2RS\sqrt{2A} + 2R\sum_{s=1}^{S} 2^{-s}\sqrt{2(\log 2 + a)s\log 2}.$$

(iv) Choose $\frac{1}{2}\log_2 n \le S \le \frac{1}{2}\log_2 n + 1$ to find

$$\|V - V(f_0)\|_\infty \le R\sqrt{2A}\log_2 n + 2R\sqrt{2A} + 2R\sum_{s=1}^{S} 2^{-s}\sqrt{2(\log 2 + a)s\log 2}.$$

Chapter 18
Metric Structure of Convex Hulls

Abstract This chapter investigates generic chaining for the supremum of a random process when this process is the convex hull of a simpler one. By geometric arguments a generic chaining result is obtained under eigenvalues conditions. The case where the coefficients of the convex combination have finite entropy is also considered. Moreover, sparse approximations of convex hulls are studied. The problem of deriving the dual norm inequality via generic chaining remains open.

18.1 Introduction

Let $\epsilon_1, \ldots, \epsilon_n$ be an i.i.d. sequence of standard Gaussians, and let X be a given $n \times p$ matrix with $\mathrm{diag}(\hat{\Sigma}) = I$ where $\hat{\Sigma} = X^T X / n$. Consider the supremum $\mathbf{Z} := \sup_{\beta \in \mathbb{R}^p : \|\beta\|_1 = 1} |\epsilon^T X \beta| / n$.

We apply Theorem 17.2. In the notation used there, the space $\mathscr{F}$ is

$$\mathscr{F} := \{f = X\beta : \beta \in \mathbb{R}^p, \|\beta\|_1 = 1\} \subset \mathbb{R}^n, \ f_0 \equiv 0,$$

Endow $\mathscr{F}$ with the normalized Euclidean metric $d(f, \tilde{f}) := \|f - \tilde{f}\|_n$, $f \in \mathscr{F}$. The radius R of $\mathscr{F}$ is bounded by 1:

$$R := \sup_{f \in \mathscr{F}} \|f\|_n^2 = \sup_{\beta : \in \mathbb{R}^p : \|\beta\|_1 = 1} \|X\beta\|_n$$

$$\leq \sup_{\beta : \in \mathbb{R}^p : \|\beta\|_1 = 1} \sum_{j=1}^{p} |\beta_j| \|X_j\|_n$$

$$= 1.$$

Theorem 17.2 with $V(f) := \epsilon^T f / \sqrt{n}$, $f \in \mathscr{F}$, yields a bound for $\mathbf{E}\mathbf{Z}$ as follows. Let $\{\mathscr{G}_s\}_{s=0}^{S}$ be sequence of subsets of $\mathscr{F}$, with $\mathscr{G}_0 := \{0\}$. We let $\mathscr{G}_S$ be a 2^{-S}-covering

© Springer International Publishing Switzerland 2016
S. van de Geer, *Estimation and Testing Under Sparsity*,
Lecture Notes in Mathematics 2159, DOI 10.1007/978-3-319-32774-7_18

set of $\mathscr{F}$. Moreover, we take S in such a way that

$$\frac{1}{2\sqrt{n}} \leq 2^{-S} \leq \frac{1}{\sqrt{n}}.$$

We use the notation of Sect. 17.7 assuming the typical case $H_s \geq 1$ for all $s \geq 1$.

Lemma 18.1 *Application of the generic chaining method of Theorem 17.2 yields*

$$\mathbb{E}\mathbf{Z} \leq (2\gamma_S + 1)/\sqrt{n}.$$

where

$$\gamma_S = \max_{j \in \mathscr{G}_S} \sum_{s=1}^{S} d_s(j)\sqrt{2(H_s + s\log 2)}.$$

Proof of Lemma 18.1 We have a special case of Lipschitz functions as described in Remark 17.1. Indeed, by the Cauchy-Schwarz inequality

$$|V(f) - V(\tilde{f}))| = |\epsilon^T(f - \tilde{f})/\sqrt{n}| \leq \sqrt{n}\|\epsilon\|_n\|f - \tilde{f}\|_n.$$

Moreover

$$\mathbb{E}\|\epsilon\|_n \leq (\mathbb{E}\|\epsilon\|_n^2)^{1/2} = 1.$$

Therefore, with $\mathscr{V}_S(f) := |V(f) - V(\mathrm{pa}_S(f))|, f \in \mathscr{F}$,

$$\mathbb{E}\|\mathscr{V}_S\|_\infty \leq 1.$$

$\square$

From Talagrand (2005) we can learn that up to universal constants γ_S is also a lower bound for $\mathbb{E}\mathbf{Z}$ when the generations $\{\mathscr{G}_s\}_{s=1}^{S-1}$ are suitably chosen. On the other hand by Lemma 17.5 and the dual norm inequality

$$\mathbb{E}\mathbf{Z} \leq \sqrt{\frac{2\log(2p)}{n}}.$$

Hence we must have for a universal constant C

$$\gamma_S \leq C\sqrt{2\log(2p)}.$$

The question is: can one prove this directly using geometric arguments? This chapter discusses the issue but the question remains open. We refer to van Handel (2015) for important new geometric insights.

18.2 Possible Generations

Recall that for a vector $\beta \in \mathbb{R}^p$, the set $S_\beta{}^1$ denotes its active set, i.e., $S_\beta = \{j : \beta_j \neq 0\}$. Fix some $s \in \{1, \ldots, S\}$. Let $\mathscr{F}_s := \{f = X\beta \in \mathscr{F} : |S_\beta| \leq 2^s\}$. We say that $\mathscr{F}_s$ is the space of functions with strong sparsity 2^s. The number of subspaces of $\mathscr{F}$ with dimension 2^s is

$$\binom{p}{2^s} \leq p^{2^s}.$$

For each $0 < \delta \leq 1$ a δ-covering number of a ball with unit radius in Euclidean space $\mathbb{R}^{2^s}$ needs at most $(3/\delta)^{2^s}$ elements (see e.g. Bühlmann and van de Geer 2011, Lemma 14.27). Let $\mathscr{G}_s$ be a minimal δ-covering of $\mathscr{F}_s$. Then

$$|\mathscr{G}_s| \leq p^{2^s}(3/\delta)^{2^s}.$$

It follows that

$$H_s = \log(2|\mathscr{G}_s|) \leq \log 2 + 2^s \log p + 2^s \log(3/\delta).$$

Now we choose $\delta = 2^{-S}$. Then $1/\delta = 2^S \leq 2\sqrt{n} \leq 2n$. It follows that

$$H_s \leq \log 2 + 2^s \log(3np).$$

In other words, assuming $p \geq n$ and ignoring constants, it holds that $H_s \asymp 2^s \log p$.

We will now show that for all s it is the approximation of $\mathscr{F}$ by $\mathscr{F}_s$ which might cause difficulties, and not the approximation of $\mathscr{F}_s$ by the finite $\mathscr{G}_s$. First, it is clear that each $j \in \mathscr{G}_s$ is an approximation of an element of $\mathscr{F}_s$, say $\mathrm{pre}_s(j) \in \mathscr{F}_s$, such that

$$\|j - \mathrm{pre}_s(j)\|_n \leq 2^{-S}.$$

Hence by the triangle inequality

$$\begin{aligned}
d_s(j) &= \|\mathrm{an}_s(j) - \mathrm{an}_{s-1}(j)\|_n \\
&\leq \|\mathrm{pre}_s(\mathrm{an}_s(j)) - \mathrm{pre}_{s-1}(\mathrm{an}_{s-1}(j))\|_n \\
&\quad + \|\mathrm{pre}_s(\mathrm{an}_s(j)) - \mathrm{an}_s(j)\|_n + \|\mathrm{pre}_{s-1}(\mathrm{an}_{s-1}(j)) - \mathrm{an}_{s-1}(j)\|_n \\
&\leq \|\mathrm{pre}_s(\mathrm{an}_s(j)) - \mathrm{pre}_{s-1}(\mathrm{an}_{s-1}(j))\|_n + 2^{-S+1}.
\end{aligned}$$

---

[1] There is a small clash of notation. In this section S is throughout the index of the last generation, and is not to be confused with an active set S_β, $\beta \in \mathbb{R}^p$.

Since (omitting some constants)

$$\sum_{s=1}^{S} 2^{-S+1}\sqrt{2(H_s + s\log 2)} \asymp \sum_{s=1}^{S} 2^{-S} 2^{s/2}\sqrt{\log p}$$

$$\leq \sum_{s=0}^{\infty} 2^{-s/2}\sqrt{\log p} \leq \frac{\sqrt{2}}{\sqrt{2}-1}\sqrt{\log p},$$

the main concern when building the family tree, i.e., when approximating elements of $\mathscr{F}$ by elements of $\mathscr{G}_s$, lies in the approximation of elements of $\mathscr{F}$ by those of $\mathscr{F}_s$.

18.3   A Conjecture

As is indicated in the previous section, for each s, the generation $\mathscr{G}_s$ could be a finite covering of the subset $\mathscr{F}_s$ of $\mathscr{F}$ consisting of functions with strong sparsity 2^s. Once the problem is reduced to such sparsely parametrized subsets $\mathscr{F}_s$ it is enough to show that each $j \in \mathscr{G}_S$ has a sparse approximation $j_s \in \mathscr{F}_s$ such that

$$\sum_{s=1}^{S} 2^{s/2}\|j_s - j\|_2 \leq C.$$

where C is a universal constant. However, it may be the case that such a result can only be established defining sparse approximations after a change of dictionary. This is formalized in Conjecture 18.1 below.

From now on we consider a general symmetric matrix Σ_0 which will play the role of $\hat{\Sigma}$. Without loss of generality for our purposes, we can assume that Σ_0 is non-degenerate. So let Σ_0 be a positive definite $p \times p$ matrix with 1's on the diagonal. For $\beta \in \mathbb{R}^p$ we define the norm $\|X\beta\| := \sqrt{\beta^T \Sigma_0 \beta}$. We consider $\beta^T \Sigma_0 \beta$, with $\|\beta\|_1 = 1$ or with $\beta \in \mathscr{S}$ where $\mathscr{S}$ is the simplex $\mathscr{S} := \{\beta_j \geq 0 \,\forall j, \sum_{j=1}^{p} \beta_j = 1\}$.

We bring in the following conjecture.

Conjecture 18.1   There exists a universal constant C and a linear map $A : \mathbb{R}^p \to \mathbb{R}^{p^*}$ (a change of dictionary) with $\mathrm{rank}(A) = p \leq p^*$ such that for each $\beta \in \mathbb{R}^p$ with $\|\beta\|_1 = 1$ there is a sequence $\{b_s\} \subset \mathbb{R}^p$, called a *path*, such that for all s the vector Ab_s has at most 2^s non-zero elements and such that

$$\sum_{s} 2^{s/2}\|X(b_s - \beta)\| \leq C.$$

It means that we can take for $\mathscr{G}_s$ a suitable finite net in a space of dimension 2^s after choosing the 2^s basis functions of this space from a set p of possible basis functions.

This remains a conjecture: we have no proof. In Lemma 18.2 the bound still depends on the largest eigenvalue of Σ_0 and Lemma 18.4 is only able to handle β's with finite entropy. Both lemma's however uses the identity matrix $A = I$ whereas our conjecture allows for a more general matrix A which may depend on Σ_0.

18.4 Dependence on the Maximal Eigenvalue

Let β be a fixed member of the simplex $\mathcal{S}$. The following result is a minor extension of Talagrand's finding for the orthogonal case (Talagrand 2005).

Lemma 18.2 *Let $\phi_{\max}^2$ be the largest eigenvalue of Σ_0. Then there is a path $\{b_s\}$ such that each b_s has at most $2^s - 1$ non-zero coefficients, and such that*

$$\sum_s 2^{s/2}\|X(b_s - \beta)\| \le \frac{2\sqrt{2}}{\sqrt{2}-1}\phi_{\max}.$$

Proof of Lemma 18.2 The path will be formed by taking, for all s, b_s as the vector of the $2^s - 1$ largest coefficients of β, putting all other coefficients to zero. Assume without loss of generality that $\beta_1 \ge \cdots \ge \beta_p$ and that $p = 2^T - 1$ for some $T \in \mathbb{N}$. Let $I_1 := \{1\}$, $I_2 := \{2, 3\}$, $I_3 := \{4, 5, 6, 7\}$ etc., that is $I_s := \{2^{s-1}, \ldots, 2^s - 1\}$, $s = 1, 2, \ldots, T$. Take $b_s := \beta_{\cup_{k \le s} I_k}$. Since $\cup_{k \le s} I_k = \{1, \ldots, 2^s - 1\}$ the vector b_s has at most $2^s - 1$ non-zero entries.

Note that $|I_k| = 2^{k-1}$ ($k = 1, 2, \ldots$) and $|\cup_{k \le s} I_k| = 2^s - 1$ ($s = 1, 2, \ldots, T$). This implies

$$\|\beta_{I_{k+1}}\|_2^2 \le 2^k \frac{\|\beta_{I_k}\|_1^2}{2^{2(k-1)}} = \frac{4\|\beta_{I_k}\|_1^2}{2^k}.$$

It follows that

$$\|\beta - b_s\|_2 = \|\beta_{\cup_{k \ge s} I_{k+1}}\|_2 \le 2\sum_{k \ge s} 2^{-k/2}\|\beta_{I_k}\|_1.$$

Therefore

$$\|X(\beta - b_s)\| \le \phi_{\max}\|\beta - b_s\|_2 \le 2\phi_{\max}\sum_{k \ge s} 2^{-k/2}\|\beta_{I_k}\|_1.$$

But then

$$\sum_s 2^{s/2}\|X(\beta - b_s)\| \le 2\phi_{\max} \sum_s 2^{s/2} \sum_{k\ge s} 2^{-k/2}\|\beta_{I_k}\|_1$$

$$= 2\phi_{\max} \sum_k \|\beta_{I_k}\|_1 2^{-k/2} \sum_{s=1}^{k} 2^{s/2}$$

$$= 2\phi_{\max} \sum_k \|\beta_{I_k}\|_1 \sum_{s=1}^{k} 2^{-(k-s)/2}$$

$$\le \frac{2\sqrt{2}}{\sqrt{2}-1}\phi_{\max}.$$

□

18.5   Approximation Using Maurey's Lemma

We further examine for a fixed β the approximation of $X\beta$ by Xb_s where b_s has at most 2^s non-zero's. The results are however not good enough to prove the conjecture: they will lead to additional log-terms (see Problem 18.1).

The next lemma is established by random sampling from the multinomial distribution. The result is a special case of Maurey's lemma, see e.g. Carl (1985).

Lemma 18.3 *For all s there exists a vector b_s with $\|b_s\|_1 = \|\beta\|_1$ and at most 2^s non-zero coefficients such that*

$$\|X(b_s - \beta)\| \le 2^{-s/2}\|\beta\|_1.$$

Proof of Lemma 18.3 Fix some s. Suppose first that $\|\beta\|_1 = 1$. Let $m_s := 2^s$ and let $(N_1, \ldots, N_p)$ be a random vector having the multinomial distribution with parameters m_s and $(|\beta_1|, \ldots, |\beta_p|)$. Then $\mathrm{var}(N_j) = m_s|\beta_j|(1 - |\beta_j|), j = 1, \ldots, p$ and for $j \ne k$,

$$\mathrm{cov}(N_j, N_k) = -m_s|\beta_j||\beta_k|.$$

Take $\hat\beta_{j,s} := \mathrm{sign}(\beta_j)N_j/m_s, j = 1, \ldots, p$. Then $\hat\beta_s$ is an unbiased estimator of β and (writing $\Sigma_0 = (\sigma_{j,k})$)

$$\mathbb{E}\|X(\hat\beta_s - \beta)\|^2 = \sum_{j=1}^p \mathbb{E}|\hat\beta_{j,s} - \beta_j|^2 + \sum_{j\ne k}\sigma_{j,k}\mathbb{E}(\hat\beta_{j,s} - \beta_j)(\hat\beta_{k,s} - \beta_k)$$

$$= \sum_{j=1}^{p} \mathrm{var}(N_j)/m_s + \sum_{j \neq k} \sigma_{j,k} \mathrm{sign}(\beta_j) \mathrm{sign}(\beta_k) \mathrm{cov}(N_j, N_k)/m_s$$

$$= \sum_{j=1}^{p} |\beta_j|(1 - |\beta_j|)/m_s - \sum_{j \neq k} \sigma_{j,k} \beta_j \beta_k / m_s$$

$$= \|\beta\|_1/m_s - \|X\beta\|^2/m_s.$$

Hence, since $\|\beta\|_1 = 1$ and $\|X\beta\|^2 \geq 0$, it follows that

$$\mathbb{E}\|X(\hat{\beta}_s - \beta)\|^2 \leq 1/m_s.$$

Moreover, $\hat{\beta}_s$ has at most m_s non-zero elements since

$$\sum_{j=1}^{p} 1\{\hat{\beta}_{j,s} \neq 0\} = \sum_{j=1}^{p} 1\{N_j \geq 1\} \leq \sum_{j=1}^{p} N_j = m_s.$$

Furthermore, $\|\hat{\beta}_s\|_1 = \sum_{j=1}^{p} N_j/m = 1$. So there must exists a b_s satisfying $\|X(b_s - \beta)\|^2 \leq 1/m_s$, $\|b_s\|_1 = 1$ and having at most m_s elements.

If $\|\beta\|_1$ is arbitrary but non-zero, we can apply the above argument to the normalized version $\beta/\|\beta\|_1$. □

The key argument used in the above lemma is perhaps more transparent if instead of sampling from the multinomial distribution, one samples from a Poisson distribution. See Problem 18.3.

18.6   Dependence on the Entropy

Define the entropy of a vector $\beta \in \mathbb{R}^p$ as

$$\mathscr{H}(\beta) := \sum_{j=1}^{p} \log_2(1 + 1/|\beta_j|)|\beta_j|.$$

In Problem 18.4 one sees that the entropy $\mathscr{H}$ is at most $\|\cdot\|_r^r/(1-r)$ for all $0 \leq r < 1$, where $\|\beta\|_r^r := \sum_{j=1}^{p} |\beta_j|^r$ for $0 < r < 1$ and $\|\beta\|_0^0 = S_\beta$, $\beta \in \mathbb{R}^p$. In other words, if for some $0 < r < 1$ (with $r \downarrow 0$ as limiting case) the vector β is ℓ_r-weakly sparse (see also Sect. 2.10 for this concept), with $\|\cdot\|_r$-radius equal to one, then its entropy is bounded by $1/(1-r)$. However, we are interested in the case $r = 1$. Thus, the lemma below can not handle all β with ℓ_1-norm equal to one. The worst case is when all coefficients in β are equal in absolute value: $|\beta_j| = 1/p, j = 1, \ldots, p$. Then $\mathscr{H}(\beta) = \log_2 p$.

Lemma 18.4 *Fix β in the simplex $\mathscr{S}$. There exists a path $\{b_s\}$ such that each b_s has at most $2^{s+1} - 1$ non-zero elements and such that*

$$\sum_s 2^{s/2} \|X(b_s - \beta)\| \leq \mathscr{H}(\beta).$$

Proof of Lemma 18.4 Assume without loss of generality that $\beta_1 \geq \cdots \geq \beta_p$ and that $p = 2^T - 1$ for some $T \in \mathbb{N}$. As in the proof of Lemma 18.2 let $I_s := \{2^{s-1}, \ldots, 2^s - 1\}$, $s = 1, 2, \ldots, T$. Note thus that $\cup_{k \leq s} |I_k| = 2^s - 1$ for all s. By Lemma 18.3 for all s one can find a $(b_s)_{\cup_{k>s} I_k}$ having only 2^s non-zero coefficients such that

$$\|X(\beta)_{\cup_{k>s} I_k} - X(b_s)_{\cup_{k>s} I_k}\| \leq 2^{-s/2} \|(\beta)_{\cup_{k>s} I_k}\|_1.$$

We therefore have

$$\sum_s 2^{s/2} \|X(\beta)_{\cup_{k>s} I_k} - X(b_s)_{\cup_{k>s} I_k}\| \leq \sum_s \|(\beta)_{\cup_{k>s} I_k}\|_1$$

$$= \sum_s \sum_{j \geq 2^s} \beta_j.$$

Because the $\{\beta_j\}$ are in decreasing order, we know that for for all s and for $j \geq 2^s$,

$$\beta_j \leq \|(\beta)_{\cup_{k \leq s} I_k}\|_1 / (2^s - 1) \leq 1/(2^s - 1).$$

Hence

$$\sum_s \sum_{j \geq 2^s} \beta_j \leq \sum_{s=1}^S \sum_{\beta_j \leq 1/(2^s - 1)} \beta_j$$

$$= \sum_{s=1}^S \sum_{s \leq \log_2(1 + 1/\beta_j)} \beta_j \leq \sum_{j=1}^p \log_2(1 + 1/\beta_j)\beta_j = \mathscr{H}(\beta).$$

We now need to complete for all s the vector b_s with coefficients in the set $\cup_{k \leq s} I_k$. Take

$$b_s := (\beta)_{\cup_{k \leq s} I_k} + (b_s)_{\cup_{k>s} I_k}.$$

Then b_s has at most $2^s - 1 + 2^s = 2^{s+1} - 1$ non-zero elements. Moreover,

$$\|X(\beta - b_s)\| = \|X(\beta)_{\cup_{k>s} I_k} - X(b_s)_{\cup_{k>s} I_k}\|.$$

$\square$

18.7 Some Further Bounds

One may wonder whether the result of Lemma 18.3 in Sect. 18.5 can be improved. For example, instead of basing the result on the variance of an unbiased estimator, one may look for mean square error trading off bias and variance. This is done in the next two lemmas. In the first one we use Poisson sampling and in the second one we sample from the multinomial distribution. Throughout, the vector $\beta \in \mathbb{R}^p$ is fixed.

Lemma 18.5 *Fix some s and let β^* be defined as the Lasso solution*

$$\beta^* := \arg\min_{b \in \mathbb{R}^p} \left\{ \|X\beta - Xb\|^2 + 2^{-s+1}\|b\|_1 \right\}.$$

Then

$$\|X(\beta^* - \beta)\|^2 \leq \min_{J \subset \{1,\ldots,p\}} \min_{b \in \mathbb{R}^p} \left\{ \|X(b_J - b)\|^2 + 2^{-2s}|J|/\Lambda_{\min}(\Sigma_0) \right\}$$

where $\Lambda_{\min}(\Sigma_0)$ is the smallest eigenvalue of Σ_0. Moreover, there exists a vector b_s having at most $2^s\|b_s\|_1$ non-zero elements, satisfying

$$\|X(b_s - \beta)\|^2 + 2^{-s}\|b_s\|_1 \leq \|X(\beta^* - \beta)\|^2 + 2^{-s+1}\|\beta^*\|_1.$$

Note that in the above lemma, the fact that β^* is a minimizer implies that $\|X(\beta^* - \beta)\|^2 + 2^{-s+1}\|\beta^*\|_1 \leq 2^{-s+1}\|\beta\|_1$. So the solution b_s of Lemma 18.5 satisfies $\|b_s\|_1 \leq 2\|\beta\|_1$ and hence has at most $2^{s+1}\|\beta\|_1$ non-zero elements. Recall also that in context of Conjecture 18.1, one may assume without loss of generality $\Lambda_{\min}(\Sigma_0) \geq 1/2$, say.

Proof of Lemma 18.5 The first result follows from the noiseless version of Theorem 2.1. To prove the second result, we let $(N_1, \ldots, N_p)$ be independent and let N_j have the Poisson distribution with parameter $m_s|\beta_j^*|$, $j = 1, \ldots, p$ where $m_s := 2^s$. We further define $\hat{\beta}_j := \text{sign}(\beta_j^*)N_j/m_s$. Then $\mathbb{E}\hat{\beta} = \beta^*$ and

$$\sum_{j=1}^{p} \text{var}(\hat{\beta}_j) = \|\beta^*\|_1/m_s = 2^{-s}\|\beta^*\|_1.$$

Writing

$$\|X(\hat{\beta} - \beta)\|^2 = \|X(\beta^* - \beta)\|_2 + \|X(\hat{\beta} - \beta^*)\|^2 + 2(\beta^* - \beta)^T \Sigma_0(\hat{\beta} - \beta^*),$$

we see that the bias-variance decomposition

$$\mathbb{E}\|X(\hat{\beta} - \beta)\|^2 = \mathbb{E}\|X(\beta^* - \beta)\|^2 + 2^{-s}\|\beta^*\|_1$$

holds true. Moreover

$$\mathbb{E}\|\hat{\beta}\|_1 = \|\beta^*\|_1,$$

so that

$$\mathbb{E}\|X(\hat{\beta} - \beta)\|^2 + 2^{-s}\mathbb{E}\|\hat{\beta}\|_1 = \|X\beta - X\beta^*\|^2 + 2^{-s+1}\|\beta^*\|_1.$$

Note furthermore that for $S_{\hat{\beta}} = \{j : \hat{\beta}_j \neq 0\}$

$$|S_{\hat{\beta}}| \leq 2^s\|\hat{\beta}\|_1.$$

Hence there must exist a b_s with $|S_{b_s}| \leq 2^s\|b_s\|_1$ and

$$\|X(b_s - \beta)\|^2 + 2^{-s}\|b_s\|_1 \leq \|X\beta - X\beta^*\|^2 + 2^{-s+1}\|\beta^*\|_1.$$

$\square$

We may also work out a more refined scheme, based on sampling from the multinomial.

Lemma 18.6 *Fix some s and let*

$$\beta^* := \arg\min_{b \in \mathbb{R}^p}\left\{ \|X(b - \beta)\|^2 + 2^{-s}\|b\|_1^2 - 2^{-s}\|Xb\|^2 \right\}.$$

There exists a b_s with at most 2^s non-zero elements and with $\|b_s\|_1 = \|\beta^*\|_1$ such that

$$\|X(b_s - \beta)\|^2 \leq \|X(\beta^* - \beta)\|^2 + 2^{-s}\|\beta^*\|_1^2 - 2^{-s}\|X\beta^*\|^2.$$

Proof of Lemma 18.6 Let $m_s := 2^s$ and $(N_1, \ldots, N_p)$ be multinomial with parameters m_s and $(|\beta_1^*|, \ldots, |\beta_p^*|)/\|\beta^*\|_1$. Define $\hat{\beta}_j := \text{sign}(\beta_j^*)\|\beta^*\|_1 N_j/m_s$, $j = 1, \ldots, p$. Then we get by the bias-variance decomposition

$$\mathbb{E}\|X(\hat{\beta} - \beta)\|^2 = \|X(\beta^* - \beta)\|^2 + 2^{-s}\|\beta^*\|_1^2 - 2^{-s}\|X\beta^*\|^2.$$

We further have $\|\hat{\beta}\|_1 = \|\beta^*\|_1$ and $\hat{\beta}$ has at most m_s non-zero elements. $\square$

Problems

18.1 Applying Lemma 18.3, verify that it yields

$$\sum_{s=1}^{S} 2^{s/2} \|X(b_s - \beta)\|_2 = S \le 1 + \log_2 n/2.$$

(Recall that $1/(2\sqrt{n}) \le 2^S \le 1/\sqrt{n}$.)

18.2 Let $\mathscr{F} := \{f = X\beta : \|\beta\|_1 \le 1\}$ where $\mathrm{diag}(X^T X)/n = I$. Endow $\mathscr{F}$ with the metric generated by the norm $\|\cdot\|_n$.

(i) Show that the δ-covering number of $\mathscr{F}$ satisfies for some universal constant C_0

$$\log N(\delta) \le C_0 \delta^{-2}(\log p + \log(1/\delta)), \ \delta > 0.$$

Hint: use Lemma 18.3 in Sect. 18.5.
(ii) Combine this result with Problem 17.3 to obtain (with $V(f) = \epsilon^T f/\sqrt{n}, f \in \mathscr{F}$ and $1/(2\sqrt{n}) \le 2^{-S} \le 1/\sqrt{n}$)

$$\gamma_S \le C \log n \sqrt{\log p},$$

where C is a universal constant.

18.3 Consider some $\beta \in \mathscr{F}$. Let $N_1, \ldots, N_p$ be independent, N_j have the Poisson distribution with parameter $2^s \beta_j$ and let $\hat{\beta}_{j,s} := 2^{-s} N_j, j = 1, \ldots, p$.

(i) Check that $\hat{\beta}_s$ is an unbiased estimator of β.

(ii) Check that

$$\mathbb{E}\|X(\hat{\beta}_s - \beta)\|^2 = \sum_{j=1}^{p} \mathrm{var}(\hat{\beta}_{j,s}) = 2^{-s} \sum_{j=1}^{p} \beta_j = 2^{-s}.$$

(iii) Show that the estimator $\hat{\beta}_j$ has on average only 2^s non-zero elements:

$$\mathbb{E}|\{j : \hat{\beta}_{j,s} \ne 0\}| \le \sum_{j=1}^{p} \mathbb{E}N_j = 2^s \sum_{j=1}^{p} \beta_j = 2^s.$$

18.4 Consider the entropy $\mathscr{H}(\beta)$ of a vector $\beta \in \mathbb{R}^p$ as defined in Sect. 18.6. Show that for $0 < r < 1$

$$\mathscr{H}(\beta) \le \frac{1}{1-r} \sum_{j=1}^{p} |\beta_j|^r.$$

Hint: employ the inequality $\log x \le x - 1$ for all $x > 0$. Also investigate the limiting case $r \downarrow 0$.

References

F. Bach, Structured sparsity-inducing norms through submodular functions, in *Advances in Neural Information Processing Systems (NIPS)*, vol. 23 (2010)

F. Bach, R. Jenatton, J. Mairal, G. Obozinski, Optimization with sparsity-inducing penalties. Found. Trends Mach. Learn. **4**, 1–106 (2012)

A. Belloni, V. Chernozhukov, L. Wang, Square-root Lasso: pivotal recovery of sparse signals via conic programming. Biometrika **98**, 791–806 (2011)

Q. Berthet, P. Rigollet, Complexity theoretic lower bounds for sparse principal component detection, in *Conference on Learning Theory* (2013)

P. Bickel, C. Klaassen, Y. Ritov, J. Wellner, *Efficient and Adaptive Estimation for Semiparametric Models* (Johns Hopkins University, Press Baltimore, 1993)

P. Bickel, Y. Ritov, A. Tsybakov, Simultaneous analysis of Lasso and Dantzig selector. Ann. Stat. **37**, 1705–1732 (2009)

L. Birgé, P. Massart, Minimum contrast estimators on sieves: exponential bounds and rates of convergence. Bernoulli **4**, 329–375 (1998)

M. Bogdan, E. van den Berg, W. Su, E. Candes, Statistical estimation and testing via de sorted l1 norm. ArXiv:1310.1969 (2013)

C. Borell, The Brunn-Minkowski inequality in Gauss space. Invent. Math. **30**, 207–216 (1975)

S. Boucheron, G. Lugosi, P. Massart, *Concentration Inequalities: A Nonasymptotic Theory of Independence* (Oxford University Press, Oxford, 2013)

O. Bousquet, A Bennet concentration inequality and its application to suprema of empirical processes. C. R. Acad. Sci. Paris **334**, 495–550 (2002)

P. Bühlmann, S. van de Geer, *Statistics for High-Dimensional Data: Methods, Theory and Applications* (Springer, Heidelberg, 2011)

F. Bunea, J. Lederer, Y. She, The group square-root lasso: theoretical properties and fast algorithms. IEEE Trans. Inf. Theory **2**, 1313–1325 (2014)

B. Carl, Inequalities of Bernstein-Jackson-type and the degree of compactness of operators in banach spaces. Ann. Inst. Fourier **35**, 79–118 (1985)

G. Chen, M. Teboulle, Convergence analysis of a proximal-like minimization algorithm using Bregman functions. SIAM J. Optim. **3**, 538–543 (1993)

S.S. Chen, D.L. Donoho, M.A. Saunders, Atomic decomposition by basis pursuit. SIAM J. Sci. Comput. **20**, 33–61 (1998)

R. Dudley, The sizes of compact subsets of Hilbert space and continuity of Gaussian processes. J. Funct. Anal. **1**, 290–330 (1967)

L. Dümbgen, S. van de Geer, M. Veraar, J. Wellner, Nemirovski's inequalities revisited. Am. Math. Mon. **117**, 138–160 (2010)

© Springer International Publishing Switzerland 2016
S. van de Geer, *Estimation and Testing Under Sparsity*,
Lecture Notes in Mathematics 2159, DOI 10.1007/978-3-319-32774-7

J. Friedman, T. Hastie, R. Tibshirani, Sparse inverse covariance estimation with the graphical Lasso. Biostatistics **9**, 432–441 (2008)

O. Güler, On the convergence of the proximal point algorithm for convex minimization. SIAM J. Control Optim. **29**, 403–419 (1991)

T. Hastie, R. Tibshirani, M. Wainwright, *Statistical Learning with Sparsity: The Lasso and Generalizations* (CRC Press, Boca Raton, 2015)

W. Hoeffding, Probability inequalities for sums of bounded variables. J. Am. Stat. Assoc. **58**, 13–30 (1963)

J. Janková, S. van de Geer, Confidence intervals for high-dimensional inverse covariance estimation. Electron. J. Stat. **9**, 1205–1229 (2015a)

J. Janková, S. van de Geer, Honest confidence regions and optimality in high-dimensional precision matrix estimation. ArXiv:1507.02061 (2015b)

A. Javanmard, A. Montanari, Confidence intervals and hypothesis testing for high-dimensional regression. J. Mach. Learn. Res. **15**, 2869–2909 (2014)

R. Jenatton, J.-Y. Audibert, F. Bach, Structured variable selection with sparsity inducing norms. J. Mach. Learn. Res. **12**, 2777–2824 (2011)

V. Koltchinskii, *Oracle Inequalities in Empirical Risk Minimization and Sparse Recovery Problems: École d'Été de Probabilités de Saint-Flour XXXVIII-2008*, vol. 38 (Springer, Heidelberg, 2011)

V. Koltchinskii, K. Lounici, A. Tsybakov, Nuclear-norm penalization and optimal rates for noisy low-rank matrix completion. Ann. Stat. **39**, 2302–2329 (2011)

J. Lafond, Low rank matrix completion with exponential family noise, in *Conference on Learning Theory* (2015)

B. Laurent, P. Massart, Adaptive estimation of a quadratic functional by model selection. Ann. Stat. **28**, 1302–1338 (2000)

G. Lecué, S. Mendelson, Sparse recovery under weak moment assumptions. J. Eur. Math. Soc. To appear, available at arXiv preprint. arXiv:1401.2188 (2014)

M. Ledoux, Talagrand deviation inequalities for product measures. ESAIM Probab. Stat. **1**, 63–87 (1996)

M. Ledoux, *The Concentration of Measure Phenomenon*, vol. 89 (American Mathematical Society, Providence, 2005)

M. Ledoux, M. Talagrand, *Probability in Banach Spaces: Isoperimetry and Processes* (Springer, New York, 1991)

K. Lounici, Sup-norm convergence rate and sign concentration property of Lasso and Dantzig estimators. Electron. J. Stat. **2**, 90–102 (2008)

K. Lounici, M. Pontil, S. van de Geer, A. Tsybakov, Oracle inequalities and optimal inference under group sparsity. Ann. Stat. **39**, 2164–2204 (2011)

P. Massart, About the constants in Talagrand's concentration inequalities for empirical processes. Ann. Probab. **28**, 863–884 (2000)

A. Maurer, M. Pontil, Structured sparsity and generalization. J. Mach. Learn. Res. **13**, 671–690 (2012)

N. Meinshausen, P. Bühlmann, High-dimensional graphs and variable selection with the Lasso. Ann. Stat. **34**, 1436–1462 (2006)

S. Mendelson, Learning without concentration. J. ACM **62**, 21 (2015)

C. Micchelli, J. Morales, M. Pontil, A family of penalty functions for structured sparsity, in *Advances in Neural Information Processing Systems, NIPS 2010*, vol. 23 (2010)

R.I. Oliveira, Sums of random Hermitian matrices and an inequality by Rudelson. Electron. Commun. Probab. **15**, 26 (2010)

D. Pollard, *Convergence of Stochastic Processes* (Springer, New York, 1984)

P. Ravikumar, M. Wainwright, G. Raskutti, B. Yu, High-dimensional covariance estimation by minimizing ℓ_1-penalized log-determinant divergence. Electron. J. Stat. **5**, 935–980 (2011)

Z. Ren, T. Sun, C.-H. Zhang, H. Zhou, Asymptotic normality and optimalities in estimation of large Gaussian graphical models. Ann. Stat. **43**, 991–1026 (2015)

A. Rothmann, P. Bickel, E. Levina, J. Zhu, Sparse permutation invariant covariance estimation. Electron. J. Stat. **2**, 494–515 (2008)

J. Schauder, Der Fixpunktsatz in funktionalräumen. Studia **2**, 171–180 (1930)

N. Städler, P. Bühlmann, S. van de Geer, ℓ_1-penalization for mixture regression models. Test **19**, 209–256 (2010)

N. Städler, P. Bühlmann, S. van de Geer, Rejoinder ℓ_1-penalization in mixture regression models. Test **19**, 280–285 (2010)

B. Stucky, S. van de Geer, Sharp oracle inequalities for square root regularization. ArXiv:1509.04093 (2015)

T. Sun, C.-H. Zhang, Comments on: ℓ_1-penalization in mixture regression models. Test **19**, 270–275 (2010)

T. Sun, C.-H. Zhang, Scaled sparse linear regression. Biometrika **99**, 879–898 (2012)

T. Sun, C.-H. Zhang, Sparse matrix inversion with scaled lasso. J. Mach. Learn. Res. **14**, 3385–3418 (2013)

M. Talagrand, Concentration of measure and isoperimetric inequalities in product spaces. Publ. Math. IHES **81**, 73–205 (1995)

M. Talagrand, *The Generic Chaining* (Springer, Heidelberg, 2005)

R. Tibshirani, Regression analysis and selection via the Lasso. J. R. Stat. Soc. Ser. B **58**, 267–288 (1996)

J. Tropp, User-friendly tail bounds for sums of random matrices. Found. Comput. Math. **12**, 389–434 (2012)

J.A. Tropp, An introduction to matrix concentration inequalities. Found. Trends Mach. Learn. **8**, 1–230 (2015)

S. van de Geer, *Empirical Processes in M-Estimation* (Cambridge University Press, Cambridge, 2000)

S. van de Geer, Least squares estimation with complexity penalties. Math. Methods Stat. **10**, 355–374 (2001)

S. van de Geer, The deterministic Lasso, in *JSM Proceedings, 2007, 140* (American Statistical Association, Alexandria, 2007)

S. van de Geer, Weakly decomposable regularization penalties and structured sparsity. Scand. J. Stat. **41**, 72–86 (2014)

S. van de Geer, P. Bühlmann, On the conditions used to prove oracle results for the Lasso. Electron. J. Stat. **3**, 1360–1392 (2009)

S. van de Geer, J. Janková, Semi-parametric efficiency bounds and efficient estimation for high-dimensional models. ArXiv:1601.00815 (2016)

S. van de Geer, A. Muro, On higher order isotropy conditions and lower bounds for sparse quadratic forms. Electron. J. Stat. **8**, 3031–3061 (2014)

S. van de Geer, P. Bühlmann, Y. Ritov, R. Dezeure, On asymptotically optimal confidence regions and tests for high-dimensional models. Ann. Stat. **42**, 1166–1202 (2014)

R. van Handel, Chaining, interpolation and convexity. ArXiv:1508.05906 (2015)

A. van der Vaart, *Asymptotic Statistics*, vol. 3 (Cambridge University Press, Cambridge, 2000)

A.W. van der Vaart, J.A. Wellner, *Weak Convergence and Empirical Processes*. Springer Series in Statistic (Springer, New York, 1996)

G. Watson, Characterization of the subdifferential of some matrix norms. Linear Algebra Appl. **170**, 33–45 (2015)

M. Yuan, Y. Lin, Model selection and estimation in regression with grouped variables. J. R. Stat. Soc. Ser. B **68**, 49 (2006)

X. Zeng, A. Mario, The ordered weighted l1 norm: Atomic formulation, dual norm, and projections. ArXiv:1409.4271 (2014)

C.-H. Zhang, S.S. Zhang, Confidence intervals for low dimensional parameters in high dimensional linear models. J. R. Stat. Soc. Ser. B Stat. Methodol. **76**, 217–242 (2014)

Author Index

Audibert, J.-Y., 75

Bühlmann, P., 10, 11, 28, 53, 54, 61, 63, 125, 257
Bach, F., 75, 76
Belloni, A., 27
Berthet, Q., 191
Bickel, P., 24, 56, 211, 216
Birgé, L., 125
Bogdan, M., 97
Borell, C., 121, 131
Boucheron, S., 235, 239
Bousquet, O., 227, 233, 235–237
Bunea, F., 89

Candès, E., 97
Carl, B., 260
Chen, G., 10
Chen, S.S., 7
Chernozhukov, V., 27

Dümbgen, L., 196, 204, 221
Dezeure, R., 63
Donoho, D.L., 7
Dudley, R., 252

Friedman, J., 216

Güler, O., 10

Hastie, T., 2, 216
Hoeffding, W., 241

Janková, J., 57, 215, 217, 218
Javanmard, A., 63
Jenatton, R., 75

Klaassen, C., 56
Koltchinskii, V., 12, 135, 182

Lafond, J., 187
Laurent, B., 123, 126, 128
Lecué, G., 152, 230
Lederer, J., 89
Ledoux, M., 233, 235, 239
Levina, E., 211, 216
Lin, Y., 89
Lounici, K., 12, 46, 135
Lugosi, G., 235, 239

Mario, A., 97
Massart, P., 123, 125, 126, 128, 233, 235, 236, 239
Maurer, A., 75, 124
Meinshausen, N., 61
Mendelson, S., 152, 230
Michelli, C., 75, 86
Montanari, A., 63
Morales, J., 75, 86
Muro, A., 225

© Springer International Publishing Switzerland 2016
S. van de Geer, *Estimation and Testing Under Sparsity*,
Lecture Notes in Mathematics 2159, DOI 10.1007/978-3-319-32774-7

Index

ℓ_1-operator norm, 44

active set, 6
allowed set, 78, 88
allowed vector, 108
almost orthogonality, 9, 42, 50, 62
almost-differentiable, 105
anti-projection, 42, 54
approximation error, 2, 173
asymptotic linearity, 65, 218, 220

basis pursuit, 7
Bernstein's inequality, 124, 133
bias, 44
Brouwer's fixed point, 199

candidate oracle, 13, 16, 82, 110, 113, 185
chaining, 245
compatibility constant, 11, 80, 230
concentration, 235, 244, 250
confidence interval, 65, 67
contraction, 234
convex conjugate, 10, 105
covering number, 245

de-selection, 54, 204
de-sparsifying, 63, 67, 71, 93, 217
decomposable, 10
density estimation, 140, 155, 172
dual norm, 76, 89, 183
dual norm inequality, 10, 44

effective sparsity, 11, 17, 81, 108, 229
empirical process, 139, 239
empirical risk, 1, 103
entropy, 245
entropy of a vector, 261
estimation error, 33
exponential family, 140, 155, 209

family tree, 246
fixed design, 7
Frobenius norm, 182

generalized linear model, 146, 160
generic chaining, 246
Gram matrix, 8, 61, 143, 223
graphical Lasso, 210, 216
group Lasso, 89, 95, 101, 170

high-dimensional, 1, 7, 103
Hoeffding's inequality, 133, 241
Huber loss, 146

inverse, 48
irrepresentable condition, 53, 205, 214
isotropy, 224

KKT-conditions, 9, 28, 37, 41, 69, 77, 93, 101, 200, 217, 219

© Springer International Publishing Switzerland 2016
S. van de Geer, *Estimation and Testing Under Sparsity*,
Lecture Notes in Mathematics 2159, DOI 10.1007/978-3-319-32774-7

LECTURE NOTES IN MATHEMATICS   🐎 Springer

Editors in Chief: J.-M. Morel, B. Teissier;

Editorial Policy

1. Lecture Notes aim to report new developments in all areas of mathematics and their applications – quickly, informally and at a high level. Mathematical texts analysing new developments in modelling and numerical simulation are welcome.

   Manuscripts should be reasonably self-contained and rounded off. Thus they may, and often will, present not only results of the author but also related work by other people. They may be based on specialised lecture courses. Furthermore, the manuscripts should provide sufficient motivation, examples and applications. This clearly distinguishes Lecture Notes from journal articles or technical reports which normally are very concise. Articles intended for a journal but too long to be accepted by most journals, usually do not have this "lecture notes" character. For similar reasons it is unusual for doctoral theses to be accepted for the Lecture Notes series, though habilitation theses may be appropriate.

2. Besides monographs, multi-author manuscripts resulting from SUMMER SCHOOLS or similar INTENSIVE COURSES are welcome, provided their objective was held to present an active mathematical topic to an audience at the beginning or intermediate graduate level (a list of participants should be provided).

   The resulting manuscript should not be just a collection of course notes, but should require advance planning and coordination among the main lecturers. The subject matter should dictate the structure of the book. This structure should be motivated and explained in a scientific introduction, and the notation, references, index and formulation of results should be, if possible, unified by the editors. Each contribution should have an abstract and an introduction referring to the other contributions. In other words, more preparatory work must go into a multi-authored volume than simply assembling a disparate collection of papers, communicated at the event.

3. Manuscripts should be submitted either online at www.editorialmanager.com/lnm to Springer's mathematics editorial in Heidelberg, or electronically to one of the series editors. Authors should be aware that incomplete or insufficiently close-to-final manuscripts almost always result in longer refereeing times and nevertheless unclear referees' recommendations, making further refereeing of a final draft necessary. The strict minimum amount of material that will be considered should include a detailed outline describing the planned contents of each chapter, a bibliography and several sample chapters. Parallel submission of a manuscript to another publisher while under consideration for LNM is not acceptable and can lead to rejection.

4. In general, **monographs** will be sent out to at least 2 external referees for evaluation.

   A final decision to publish can be made only on the basis of the complete manuscript, however a refereeing process leading to a preliminary decision can be based on a pre-final or incomplete manuscript.

   Volume Editors of **multi-author works** are expected to arrange for the refereeing, to the usual scientific standards, of the individual contributions. If the resulting reports can be forwarded to the LNM Editorial Board, this is very helpful. If no reports are forwarded or if other questions remain unclear in respect of homogeneity etc, the series editors may wish to consult external referees for an overall evaluation of the volume.

5. Manuscripts should in general be submitted in English. Final manuscripts should contain at least 100 pages of mathematical text and should always include

   - a table of contents;
   - an informative introduction, with adequate motivation and perhaps some historical remarks: it should be accessible to a reader not intimately familiar with the topic treated;
   - a subject index: as a rule this is genuinely helpful for the reader.
   - For evaluation purposes, manuscripts should be submitted as pdf files.

6. Careful preparation of the manuscripts will help keep production time short besides ensuring satisfactory appearance of the finished book in print and online. After acceptance of the manuscript authors will be asked to prepare the final LaTeX source files (see LaTeX templates online: https://www.springer.com/gb/authors-editors/book-authors-editors/manuscriptpreparation/5636) plus the corresponding pdf- or zipped ps-file. The LaTeX source files are essential for producing the full-text online version of the book, see http://link.springer.com/bookseries/304 for the existing online volumes of LNM). The technical production of a Lecture Notes volume takes approximately 12 weeks. Additional instructions, if necessary, are available on request from lnm@springer.com.

7. Authors receive a total of 30 free copies of their volume and free access to their book on SpringerLink, but no royalties. They are entitled to a discount of 33.3 % on the price of Springer books purchased for their personal use, if ordering directly from Springer.

8. Commitment to publish is made by a *Publishing Agreement*; contributing authors of multiauthor books are requested to sign a *Consent to Publish form*. Springer-Verlag registers the copyright for each volume. Authors are free to reuse material contained in their LNM volumes in later publications: a brief written (or e-mail) request for formal permission is sufficient.

Addresses:
Professor Jean-Michel Morel, CMLA, École Normale Supérieure de Cachan, France
E-mail: moreljeanmichel@gmail.com

Professor Bernard Teissier, Equipe Géométrie et Dynamique,
Institut de Mathématiques de Jussieu – Paris Rive Gauche, Paris, France
E-mail: bernard.teissier@imj-prg.fr

Springer: Ute McCrory, Mathematics, Heidelberg, Germany,
E-mail: lnm@springer.com

Printed in the United States
By Bookmasters